LONDON MATHEMATICAL SOCIETY LECTURE NOTE SERIES

Managing Editor: Professor J.W.S. Cassels, Department of Pure Mathematics and Mathematical Statistics, University of Cambridge, 16 Mill Lane, Cambridge CB2 1SB, England

The titles below are available from booksellers, or, in case of difficulty, from Cambridge University Press.

34 Representation theory of Lie groups, M.F. ATIYAH *et al*
36 Homological group theory, C.T.C. WALL (ed)
39 Affine sets and affine groups, D.G. NORTHCOTT
46 p-adic analysis: a short course on recent work, N. KOBLITZ
50 Commutator calculus and groups of homotopy classes, H.J. BAUES
59 Applicable differential geometry, M. CRAMPIN & F.A.E. PIRANI
66 Several complex variables and complex manifolds II, M.J. FIELD
69 Representation theory, I.M. GELFAND *et al*
76 Spectral theory of linear differential operators and comparison algebras, H.O. CORDES
77 Isolated singular points on complete intersections, E.J.N. LOOIJENGA
83 Homogeneous structures on Riemannian manifolds, F. TRICERRI & L. VANHECKE
86 Topological topics, I.M. JAMES (ed)
87 Surveys in set theory, A.R.D. MATHIAS (ed)
88 FPF ring theory, C. FAITH & S. PAGE
89 An F-space sampler, N.J. KALTON, N.T. PECK & J.W. ROBERTS
90 Polytopes and symmetry, S.A. ROBERTSON
92 Representation of rings over skew fields, A.H. SCHOFIELD
93 Aspects of topology, I.M. JAMES & E.H. KRONHEIMER (eds)
94 Representations of general linear groups, G.D. JAMES
95 Low-dimensional topology 1982, R.A. FENN (ed)
96 Diophantine equations over function fields, R.C. MASON
97 Varieties of constructive mathematics, D.S. BRIDGES & F. RICHMAN
98 Localization in Noetherian rings, A.V. JATEGAONKAR
99 Methods of differential geometry in algebraic topology, M. KAROUBI & C. LERUSTE
100 Stopping time techniques for analysts and probabilists, L. EGGHE
101 Groups and geometry, ROGER C. LYNDON
103 Surveys in combinatorics 1985, I. ANDERSON (ed)
104 Elliptic structures on 3-manifolds, C.B. THOMAS
105 A local spectral theory for closed operators, I. ERDELYI & WANG SHENGWANG
106 Syzygies, E.G. EVANS & P. GRIFFITH
107 Compactification of Siegel moduli schemes, C-L. CHAI
108 Some topics in graph theory, H.P. YAP
109 Diophantine analysis, J. LOXTON & A. VAN DER POORTEN (eds)
110 An introduction to surreal numbers, H. GONSHOR
113 Lectures on the asymptotic theory of ideals, D. REES
114 Lectures on Bochner-Riesz means, K.M. DAVIS & Y-C. CHANG
115 An introduction to independence for analysts, H.G. DALES & W.H. WOODIN
116 Representations of algebras, P.J. WEBB (ed)
117 Homotopy theory, E. REES & J.D.S. JONES (eds)
118 Skew linear groups, M. SHIRVANI & B. WEHRFRITZ
119 Triangulated categories in the representation theory of finite-dimensional algebras, D. HAPPEL
121 Proceedings of *Groups - St Andrews 1985*, E. ROBERTSON & C. CAMPBELL (eds)
122 Non-classical continuum mechanics, R.J. KNOPS & A.A. LACEY (eds)
125 Commutator theory for congruence modular varieties, R. FREESE & R. MCKENZIE
126 Van der Corput's method of exponential sums, S.W. GRAHAM & G. KOLESNIK
127 New directions in dynamical systems, T.J. BEDFORD & J.W. SWIFT (eds)
128 Descriptive set theory and the structure of sets of uniqueness, A.S. KECHRIS & A. LOUVEAU
129 The subgroup structure of the finite classical groups, P.B. KLEIDMAN & M.W.LIEBECK
130 Model theory and modules, M. PREST
131 Algebraic, extremal & metric combinatorics, M-M. DEZA, P. FRANKL & I.G. ROSENBERG (eds)
132 Whitehead groups of finite groups, ROBERT OLIVER
133 Linear algebraic monoids, MOHAN S. PUTCHA
134 Number theory and dynamical systems, M. DODSON & J. VICKERS (eds)
135 Operator algebras and applications, 1, D. EVANS & M. TAKESAKI (eds)
136 Operator algebras and applications, 2, D. EVANS & M. TAKESAKI (eds)
137 Analysis at Urbana, I, E. BERKSON, T. PECK, & J. UHL (eds)
138 Analysis at Urbana, II, E. BERKSON, T. PECK, & J. UHL (eds)
139 Advances in homotopy theory, S. SALAMON, B. STEER & W. SUTHERLAND (eds)
140 Geometric aspects of Banach spaces, E.M. PEINADOR & A. RODES (eds)
141 Surveys in combinatorics 1989, J. SIEMONS (ed)

142 The geometry of jet bundles, D.J. SAUNDERS
143 The ergodic theory of discrete groups, PETER J. NICHOLLS
144 Introduction to uniform spaces, I.M. JAMES
145 Homological questions in local algebra, JAN R. STROOKER
146 Cohen-Macaulay modules over Cohen-Macaulay rings, Y. YOSHINO
147 Continuous and discrete modules, S.H. MOHAMED & B.J. MÜLLER
148 Helices and vector bundles, A.N. RUDAKOV *et al*
149 Solitons, nonlinear evolution equations and inverse scattering, M. ABLOWITZ & P. CLARKSON
150 Geometry of low-dimensional manifolds 1, S. DONALDSON & C.B. THOMAS (eds)
151 Geometry of low-dimensional manifolds 2, S. DONALDSON & C.B. THOMAS (eds)
152 Oligomorphic permutation groups, P. CAMERON
153 L-functions and arithmetic, J. COATES & M.J. TAYLOR (eds)
154 Number theory and cryptography, J. LOXTON (ed)
155 Classification theories of polarized varieties, TAKAO FUJITA
156 Twistors in mathematics and physics, T.N. BAILEY & R.J. BASTON (eds)
157 Analytic pro-*p* groups, J.D. DIXON, M.P.F. DU SAUTOY, A. MANN & D. SEGAL
158 Geometry of Banach spaces, P.F.X. MÜLLER & W. SCHACHERMAYER (eds)
159 Groups St Andrews 1989 volume 1, C.M. CAMPBELL & E.F. ROBERTSON (eds)
160 Groups St Andrews 1989 volume 2, C.M. CAMPBELL & E.F. ROBERTSON (eds)
161 Lectures on block theory, BURKHARD KÜLSHAMMER
162 Harmonic analysis and representation theory, A. FIGA-TALAMANCA & C. NEBBIA
163 Topics in varieties of group representations, S.M. VOVSI
164 Quasi-symmetric designs, M.S. SHRIKANDE & S.S. SANE
165 Groups, combinatorics & geometry, M.W. LIEBECK & J. SAXL (eds)
166 Surveys in combinatorics, 1991, A.D. KEEDWELL (ed)
167 Stochastic analysis, M.T. BARLOW & N.H. BINGHAM (eds)
168 Representations of algebras, H. TACHIKAWA & S. BRENNER (eds)
169 Boolean function complexity, M.S. PATERSON (ed)
170 Manifolds with singularities and the Adams-Novikov spectral sequence, B. BOTVINNIK
171 Squares, A.R. RAJWADE
172 Algebraic varieties, GEORGE R. KEMPF
173 Discrete groups and geometry, W.J. HARVEY & C. MACLACHLAN (eds)
174 Lectures on mechanics, J.E. MARSDEN
175 Adams memorial symposium on algebraic topology 1, N. RAY & G. WALKER (eds)
176 Adams memorial symposium on algebraic topology 2, N. RAY & G. WALKER (eds)
177 Applications of categories in computer science, M. FOURMAN, P. JOHNSTONE, & A. PITTS (eds)
178 Lower K- and L-theory, A. RANICKI
179 Complex projective geometry, G. ELLINGSRUD *et al*
180 Lectures on ergodic theory and Pesin theory on compact manifolds, M. POLLICOTT
181 Geometric group theory I, G.A. NIBLO & M.A. ROLLER (eds)
182 Geometric group theory II, G.A. NIBLO & M.A. ROLLER (eds)
183 Shintani zeta functions, A. YUKIE
184 Arithmetical functions, W. SCHWARZ & J. SPILKER
185 Representations of solvable groups, O. MANZ & T.R. WOLF
186 Complexity: knots, colourings and counting, D.J.A. WELSH
187 Surveys in combinatorics, 1993, K. WALKER (ed)
188 Local analysis for the odd order theorem, H. BENDER & G. GLAUBERMAN
189 Locally presentable and accessible categories, J. ADAMEK & J. ROSICKY
190 Polynomial invariants of finite groups, D.J. BENSON
191 Finite geometry and combinatorics, F. DE CLERCK *et al*
192 Symplectic geometry, D. SALAMON (ed)
193 Computer algebra and differential equations, E. TOURNIER (ed)
194 Independent random variables and rearrangement invariant spaces, M. BRAVERMAN
195 Arithmetic of blowup algebras, WOLMER VASCONCELOS
196 Microlocal analysis for differential operators, A. GRIGIS & J. SJÖSTRAND
197 Two-dimensional homotopy and combinatorial group theory, C. HOG-ANGELONI, W. METZLER & A.J. SIERADSKI (eds)
198 The algebraic characterization of geometric 4-manifolds, J.A. HILLMAN
199 Invariant potential theory in the unit ball of $\mathbf{C}^n$, MANFRED STOLL
200 The Grothendieck theory of dessins d'enfant, L. SCHNEPS (ed)
201 Singularities, JEAN-PAUL BRASSELET (ed)
202 The technique of pseudodifferential operators, H.O. CORDES
203 Hochschild cohomology of von Neumann algebras, A. SINCLAIR & R. SMITH
204 Combinatorial and geometric group theory, A.J. DUNCAN, N.D. GILBERT, J. HOWIE (eds)
207 Groups of Lie type and their geometries, W.M. KANTOR & L. DI MARTINO (eds)
208 Vector bundles in algebraic geometry, N.J. HITCHIN, P. NEWSTEAD & W.M. OXBURY (eds)

London Mathematical Society Lecture Note Series. 208

Vector Bundles in Algebraic Geometry

Durham 1993

Edited by

N.J. Hitchin
University of Cambridge

P.E. Newstead
University of Liverpool

W.M. Oxbury
University of Durham

CAMBRIDGE
UNIVERSITY PRESS

CAMBRIDGE UNIVERSITY PRESS
Cambridge, New York, Melbourne, Madrid, Cape Town, Singapore, São Paulo, Delhi

Cambridge University Press
The Edinburgh Building, Cambridge CB2 8RU, UK

Published in the United States of America by Cambridge University Press, New York

www.cambridge.org
Information on this title: www.cambridge.org/9780521498784

First published 1995

A catalogue record for this publication is available from the British Library

ISBN 978-0-521-49878-4 paperback

Transferred to digital printing 2009

Contents

List of Participants

I.V. Artamkin (Moscow), V. Balaji (SPIC, Madras), R.N. Barlow (Durham), S. Bauer (Oxford), A. Beauville (Paris-Sud), U.N. Bhosle (TIFR, Bombay), F.J. Bloore (Liverpool), F. Bottacin (Paris-Sud), S.B. Bradlow (Illinois), L. Brambila-Paz (UAM-Iztapalapa), R. Braun (Bayreuth), R. Brussee (Bayreuth), U. Bruzzo (Genova), R. Cannings (Imperial, London), P. Cook (Liverpool), W. Crawley-Boevey (Oxford), S. Crotty (Newcastle), A. Dancer (Cambridge), G. Daskalopoulos (Princeton), W. Decker (Saarland), S.K. Donaldson (Oxford), J.-M. Drézet (Paris VII), R. Earl (Oxford), O. García-Prada (Paris-Sud), P. Gothen (Warwick), I.M. Grzegorczyk (Oregon), G. Hein (Berlin), D. Hernandez-Ruiperez (Salamanca), N.J. Hitchin (Warwick), J. Hoggart (Liverpool), M.P. Holland (Sheffield), G. Horrocks (Newcastle), K. Hulek (Hannover), L. Jeffrey (Cambridge), M.M. Kapranov (Northwestern), A.D. King (Liverpool), Y. Laszlo (Paris-Sud), R. Ledgard (Manchester), M. Lehn (Zurich), J. Le Potier (Paris VII), A. Maciocia (Edinburgh), A. Matuschke (Berlin), G. Megyesi (Cambridge), J.R. Merriman (Kent), R. Miro-Roig (Zaragoza), S. Mukai (Nagoya), M.S. Narasimhan (ICTP, Trieste), B. Nasatyr (Cambridge), P.E. Newstead (Liverpool), G. Ottaviani (Firenze), W.M. Oxbury (Durham), V. Pidstrigach (Moscow), J. Poritz (IAS, Princeton), E. Previato (Boston), Z. Ran (California), E.G. Rees (Edinburgh), M. Reid (Warwick), N. Rhagavendra (SPIC, Madras), G.K. Sankaran (Cambridge), A. Schofield (Bristol), N.I. Shepherd-Barron (Cambridge), A. Sinha (NERIST, Itanagar), I. Sols Lucia (Madrid), C. Sorger (Paris VII), A. Szenes (MIT), M. Teixidor i Bigas (Tufts), G. Trautmann (Kaiserslautern), A.N. Tyurin (Steklov Institute, Moscow), K. Ueno (Kyoto), H. Verrill (Cambridge), C. Walter (Nice), R.A. Wentworth (Harvard), P.M.H. Wilson (Cambridge), P. Zelewski (Newcastle), K. Zuo (Kaiserslautern).

Introduction

There are few areas of pure mathematics which have profited more from the influx of new ideas from other disciplines than the subject of this book. The study of vector bundles over algebraic varieties has been stimulated by successive waves of migrant concepts over the last few years, largely coming from mathematical physics. It nevertheless also retains its roots in old questions concerning subvarieties of projective space, and this is a continuing source of challenging problems. The 1993 Durham Symposium *Vector Bundles in Algebraic Geometry*, sponsored by the LMS and SERC, had as its aim the goal of bringing together the leading researchers in the field to explore further these interactions: to see how old problems would yield to new techniques, and to present new opportunities for the already highly developed subject of algebraic geometry.

The present book is not, however, simply the Proceedings of that Symposium. Its purpose is certainly to reflect what was said and done there, but we hope that it also presents to the mathematical world an overview of the key areas of research involving vector bundles. Some of the principal speakers have been encouraged to expand their talks to give surveys of their respective areas. The reader can thus find here not only reports of recent progress, but also a perspective on where the new ideas have come from, what they are doing at the moment, and what they might be capable of in the future.

The incursions from mathematical physics and differential geometry have taken several forms, and all are represented here. Probably the first one began twelve years ago with Donaldson's proof of the theorem of Narasimhan and Seshadri using gauge-theoretic methods and ideas from symplectic geometry. The legacy of that approach can be seen in the article of Bradlow, Daskalopoulos, Garcia-Prada and Wentworth, where the concept of moment map gives on the one hand a secure analytical foundation to the construction of moduli spaces of bundles with extra structure, and also its link with the algebraic geometric viewpoint of Geometric Invariant Theory. One of the remarkable features of the symposium was the way in which various forms of 'augmented structure' were appearing spontaneously in the content of many talks, as a response to different types of problems. This particular paper gives a statement of the state of the art in this general area. One of these structures, arising in a purely algebro-geometric setting, is what Le Potier calls a 'coherent system'. The discussion of these systems, including some applications, forms part of his contribution to this book. In particular, he uses coherent systems together with the concept of determinant bundle (which itself has already arisen in several contexts) to determine the Picard groups of some moduli spaces.

The input of gauge theory into the subject is not confined to symplectic geometry, and Donaldson's polynomial invariants, yielding information on the differentiable structure of algebraic surfaces, form an extremely active area of research. It relies of course on the study of moduli spaces of stable holomorphic bundles, but by embedding the subject in the more general one of instantons on four-manifolds, it lends it an extra degree of flexibility. The 'expected dimensions' acquire a reality of interpretation which can be put to use to prove very deep results. There are two contributions here in this field. The first is by Donaldson himself on Floer homology and the second by Tyurin on the impact of Donaldson polynomials in algebraic geometry. Donaldson's paper includes some concrete calculations for elliptic surfaces and a preview of the 'quantum cohomology' of the moduli space of bundles on a curve, surely a topic for rapid development in the next few years, while Tyurin advances the 'Jacobian' of an algebraic surface (the Gieseker closure of the component of the moduli space of bundles which contains the cotangent bundle) as an essential tool for the study of surface geometry. It is clear by now for various reasons that moduli spaces of vector bundles have an exceedingly rich internal geometry which rivals the classical theory of the Jacobian of a curve. Tyurin's terminology is quite apposite. In a similar vein, the paper of Balaji and Vishwanath discusses the analogues of the Picard bundles over moduli spaces for curves.

In recent years a second influx of ideas from physics has come from conformal field theory and in particular the challenge of understanding the Verlinde formulae which describe the dimension of the space of sections of an ample line bundle over the moduli space of bundles on a curve. There are many aspects to this problem, and by now many proofs using different methods. Here, we restrict ourselves to two discussions of the topic. In one, Ueno describes for algebraic geometers the construction of 'conformal blocks' from the physicists' description of conformal field theory. In the other contribution, Szenes discusses the structure of the formulae themselves, and in particular in the case of groups other than SL_2. The whole subject is very broad, and a complete picture requires the establishment of rigorous links between different languages. It happened that, in the course of the symposium, key results providing some of these were announced by Narasimhan, Beauville and Laszlo but these will appear elsewhere.

In any mathematical discipline, it is unwise to concentrate unduly on the general at the expense of the particular. Algebraic geometry in particular is a mature subject which is blessed with a wealth of beautiful examples, and techniques which make even non-generic objects subject to close analysis. Within the realm of holomorphic vector bundles, the Horrocks-Mumford bundle is highly distinguished; the only known indecomposable rank 2 bundle on $\mathbf{P}^4$, with 15,000 symmetries and closely related to modular surfaces, it stands out like the icosahedron as an object of endless fascination. Hulek gives here a survey of what is known about this bundle and its relationship with other areas of mathematics since its discovery in 1972. A section of the Horrocks-Mumford bundle gives an abelian surface in $\mathbf{P}^4$ which is not a complete intersection. This general problem in $\mathbf{P}^4$ and $\mathbf{P}^5$ is considered by Decker and Popescu in their attack on the classification of codimenson 2 subvarieties. Although a subject of some considerable antiquity, there are now new results emanating from computer algebra working on Beilinson's spectral sequence. The basic information about the moduli space of stable bundles, its dimenson and smoothness, is given by the sheaf cohomology groups $H^i(X, Ad(E))$ for $0 \leq i \leq 2$. A bundle for which the groups vanish for all i is called *exceptional* and is the subject of Drézet's paper. Exceptional bundles are building blocks for semi-stable sheaves on projective spaces, and the author uses them to give information about the moduli spaces for projective spaces of arbitrary dimension.

The Durham symposium was a successful forum for discussing all these new approaches to the subject of vector bundles in algebraic geometry. We hope that this book will make available to a wider mathematical public the essential ideas in this rapidly developing area.

Acknowledgements

Our thanks are due to all of the participants of the Durham Symposium for contributing to a lively and enjoyable meeting. This was made possible by the support of the London Mathematical Society and the Science and Engineering Research Council; and with the help of the Department of Mathematical Sciences and Grey College, Durham. In particular we would like to thank John Bolton, Iain MacPhee and Tony Scholl; and Mary Bell and Sue Nesbitt for their tireless efficiency with the organisation. Finally, we would like to thank Peter Craig for his assistance with some of the TeX files.

On the Deformation Theory of Moduli Spaces of Vector Bundles

V. Balaji and P.A. Vishwanath

§0 Introduction

This article is of the nature of a discussion of various results in the deformation theory of moduli spaces of vector bundles and the techniques involved in their proofs. We have concentrated mostly on the conceptual points of proofs and largely ignored all technical and computational details with needed references to enable the interested reader to fill in the gaps.

The layout of the article is as follows: in section 1, we discuss results on the deformations of the moduli spaces of rank 2 vector bundles. The topic of section 2 concerns a study of intermediate Jacobians of these objects with a view to getting Torelli type theorems for moduli spaces and finally, in section 3 we discuss the deformations of Picard bundle on the moduli space of vector bundles.

§1 Deformations of the moduli space

Let C be a smooth projective curve of genus $g \geq 3$ defined over the complex number field $\mathbb{C}$. Let ξ be a line bundle on C. Let

$$M_\xi := M_C(2, \xi)$$

denote the moduli space of semi-stable vector bundles of rank 2 and determinant isomorphic to ξ on C. In this section we consider M_ξ for $\xi \cong \mathcal{O}_C$ and $\xi \cong \mathcal{O}_C(x_0)$, $x_0 \in C$. These we abbreviate by M_0 and M_1 respectively. Both M_0 and M_1 are projective varieties which are unirational and normal of dimension $3g-3$. Further, M_1 is a smooth variety. The theorem of Narasimhan and Ramanan computes the cohomology of the tangent bundle of M_1. More precisely,

Theorem 1.1. (cf [NR])[1] *If T_{M_1} denotes the tangent bundle of M_1, we have*

$$h^i(M_1, T_{M_1}) := \dim H^i(M_1, T_{M_1}) = \begin{cases} 3g-3 & \text{if } i = 1 \\ 0 & \text{otherwise} \end{cases}$$

Remark 1.2.

(a) Note that the group $H^0(M_1, T_{M_1})$, of infinitesimal automorphisms of M_1, is trivial implies that the group of automorphisms, $\mathrm{Aut}(M_1)$, of M_1 is discrete. But as M_1 is known to be canonically polarised (i.e. the canonical class is negatively ample), $\mathrm{Aut}(M_1)$ is an algebraic group and hence is finite.

(b) The number of moduli, $h^1(M_1, T_{M_1})$, of M_1 (cf. [KS]) is equal to the number of moduli of C. This intuitively means that any small variation of M_1 is again a moduli space of vector bundles on a variation of C.

(c) Again, because the canonical class of M_1 is negatively ample, Nakano's vanishing theorem (cf. [KS]) gives the vanishing of $h^i(M_1, T_{M_1})$ for $i \geq 2$. So it suffices, to prove the Theorem 1.1, to compute h^0 and h^1.

The proof of Theorem 1.1 uses correspondence techniques.[2] That is, let V be the universal bundle on $M_1 \times C$. Let ad V denote the bundle of traceless endomorphisms of V. Let p and q denote the projections from $M_1 \times C$ onto M_1 and C respectively. We then have two Leray spectral sequences

$$\begin{array}{ccc} H^{i+j}(M_1 \times C, \mathrm{ad}\, V) & \Longleftarrow & H^i(M_1, R^j p_*(\mathrm{ad}\, V)) \\ \Uparrow & & \\ H^i(C, R^j q_*(\mathrm{ad}\, V)) & & \end{array}$$

[1] Of course, the theorem is true more generally in all rank n and degree d such that $(n, d) = 1$ and curves of all genera $g > 1$.

[2] We should add here that there is a different proof of Theorem 1.1, due to N.Hitchin, available now (cf. [H]).

On the one hand, deformation theory of stable bundles on a curve yields

$$R^j p_*(\text{ad } V) \cong \begin{cases} 0 & \text{if } i \neq 1 \\ T_{M_1} & \text{if } i = 1. \end{cases}$$

When this is fed into the Leray spectral sequence arising from the projection p, we get

$$H^i(M_1, T_{M_1}) \cong H^i(M_1 \times C, \text{ad } V). \tag{1.1}$$

On the other hand, if we could somehow study the deformation theory of the family $\{V_x\}_{x \in C}$ (where $V_x := V|_{M_1 \times \{x\}}$) of bundles on M_1 parametrised by C, we could hope to connect $H^*(M_1 \times C, \text{ad } V)$ with $H^*(C, T_C)$ and try to compute $h^i(M_1, T_{M_1})$. This is provided by

Proposition 1.3. (cf. [NR], [S1]) *Let* $\text{ad}_x V := \text{ad}(V_x)$, $x \in C$. *Then we have:*

(*a*) $\dim H^i(M_1, \text{ad}_x V) = \begin{cases} 1 & \text{if } i = 1 \\ 0 & \text{if } i = 0, 2 \end{cases}$

(*b*) *The Kodaira-Spencer map* (cf. [KS], [NR]) *for the family* $\{V_x\}_{x \in C}$

$$\rho_x : T_{C,x} \longrightarrow H^1(M_1, \text{ad}_x V)$$

is an isomorphism for all $x \in X$.

Proof of Theorem 1.1: Note that Proposition 1.3 has as a consequence:

$$R^j q_*(\text{ad } V) \simeq \begin{cases} 0 & \text{if } i \neq 1 \\ T_C & \text{if } i = 1. \end{cases}$$

This, together with the Leray spectral sequence arising from the projection q, gives

$$H^i(C, T_C) \cong H^i(M_1 \times C, \text{ad } V). \tag{1.2}$$

Now put (1.1) and (1.2) together to complete the proof of Theorem 1.1.

The proof of Proposition 1.3 uses a construction which has come to be familiar as "Hecke Correspondence" (cf. [NR], [MS], [B]). Briefly, if $P := \mathbf{P}(V_x)$, it can be shown, using "elementary transformations" that P parametrises a family $\{W_t\}_{t \in P}$ of rank 2, trivial determinant semistable vector bundles on C. Consider

$$\begin{array}{ccc} & P & \\ {\scriptstyle\psi} \swarrow & & \searrow {\scriptstyle\phi} \\ M_1 & & M_0 \end{array}$$

where ψ is the canonical projection and ϕ is the characteristic map given by the family $\{W_t\}_{t\in P}$. It is clear that ϕ is a P^1-bundle when restricted to the open set, U, of stable bundles in M_0. We then have

(a) The codimension of $P - \phi^{-1}(U)$ in P is at least $g-1$ where g is the genus of the curve.

(b) Let T_ψ and T_ϕ denote the relative tangent sheaves of the maps ψ and ϕ respectively. We have, on $\phi^{-1}(U)$, $T_\psi \simeq T_\phi^*$.

These facts, together with the Leray spectral sequence and a repeated application of Hartog type theorem for cohomology (see [S1] for details), give

$$H^i(M_1, \mathrm{ad}_x V) \cong H^{i-1}(M_0, \mathcal{O}) \quad 0 \le i \le 2.$$

But M_0 is known to be unirational and complete. This proves (a). For a proof of (b) see [NR] or [S1].

Let N be the desingularisation of M_0 constructed in [S2]. Let PV_4 denote the category of parabolic semi-stable bundles (V, Δ), where V is of rank 4 and $\det V \simeq \mathcal{O}_C$, and Δ is a parabolic structure at the marked point $x_0 \in C$ with small weights (α_1, α_2). Then N parametrises isomorphism classes of vector bundles in P_4 such that $\mathrm{End}\, V$, the endomorphism algebra, is a specialisation of the (2×2)-matrix algebra $\mathcal{M}_2$.

Theorem 1.4. (cf. [BV1]) *If T_N denotes the tangent bundle of N, we have*

$$h^i(N, T_N) = \begin{cases} 3g-3 & \text{if } i = 1 \\ 0 & \text{if } i = 0, 2 \end{cases}$$

Before proceeding with an idea of proof of Theorem 1.4, we collect some facts about the variety N we need for the proof (cf. [BS], [BV1]).

(a) The variety N represents a natural moduli functor and hence there exists a universal family of rank 4 bundles on $C \times N$.

(b) There also exists a family of quadratic forms $\{Q_t\}_{t\in N}$ on a fixed 3-dimensional vector space parametrised by N. This stratifies N into subvarieties $\{N_i\}_{i=1}^3$ defined by

$$N_i = \{t \in N \mid \mathrm{rank}\, Q_t \le 3 - i\}.$$

Thus $N \supset N_1 \supset N_2 \supset N_3$. If we denote $N - N_2$ by Z and $N_1 - N_2$ by Y, then Y is a smooth divisor in Z with $Z - Y \simeq U$ (U is the open set of stable bundles in M_0). Further, Y is a $\mathbf{P}^{g-2} \times \mathbf{P}^{g-2}$ - bundle on $K - K_0$; where K is the Kummer variety of $\dim g$ associated to the Jacobian of C, K_0 is its singular locus. The normal bundle, n_Y, of Y in Z, is isomorphic to $\mathcal{O}(-1,-1)$ when restricted to the fibres of Y over $K - K_0$.

(c) Finally, the codimension of N_2 in N is 3.

We can now outline very briefly the key steps involved in proving Theorem 1.4.

Step 1. By a switching trick in the Hecke correspondence, we can compute the dimension of $H^i(U, T_U)$. In fact

$$h^i(U, T_U) = \begin{cases} 0 & i = 0, 2 \\ 3g - 3 & i = 1. \end{cases}$$

Step 2. By the use of a Hartog type argument, using part (c) above, we see that it is enough to compute $h^i(Z, T_Z)$.

Step 3. $H^i(Z, T_Z) \cong H^i(U, T_U) \quad i = 0, 1, 2.$

Here we need the following cohomological result (cf. [G]).

$$\varinjlim_n H^i(Z, \mathcal{O}_Z(nY) \otimes T_Z) \cong H^i(U, T_U)$$

where $Y \subset Z$ is the divisor and $U = Z - Y$.

Consider now the following exact sequence:

$$A_k : 0 \longrightarrow \mathcal{O}_Z((k-1)Y) \longrightarrow \mathcal{O}_Z(kY) \longrightarrow n_Y^k \longrightarrow 0$$

for all $k > 0$. Here n_Y is the normal bundle of Y in Z.

Using the explicit description of n_Y in (b) above, we prove that

$$H^i(Z, T_Z \otimes n_Y^k) = 0, \quad i = 0, 1, 2 \quad \forall k > 0.$$

Then, by step 3 and the long exact sequence of A_k, we get, for all $k > 0$

$$H^i(Z, T_Z \otimes \mathcal{O}(kY)) \simeq H^i(Z, T_Z \otimes \mathcal{O}((k-1)Y))$$

$$\Longrightarrow H^i(Z, T_Z) \simeq H^i(U, T_U).$$

§2 Intermediate Jacobians, Torelli-type theorems

This section, as the title indicates, deals with Torelli type theorems for the moduli spaces $M := M_1$ and N, the natural desingularisation of M_0. We sometimes write M_C and N_C to stress the dependence of these spaces on the curve C.

The Weil-Griffiths intermediate Jacobian associated to the third cohomology group of a unirational variety V is an interesting invariant (cf. [Gr]) and is especially suited to study Torelli type theorems for the moduli spaces of vector bundles. The key concept involved here is the so called 'Weil map'. We recall in brief the definition of the Weil map.

Let V be a smooth projective unirational variety and let T be a parameter space which we assume to be a smooth projective variety. Let A be an algebraic cycle on $V \times T$ of codimension 2. Then we have its fundamental class $\alpha \in H^4(V \times T, \mathbf{Z})$. Assume $H^2(V, \mathbf{Z})_{tor} = 0$. Then the (1,3)-Kunneth component

$$\alpha_{1,3} \in H^1(V, \mathbf{Z}) \otimes H^3(T, \mathbf{Z})$$

gives a homomorphism

$$\alpha_{1,3} : H_1(T, \mathbf{Z}) \longrightarrow H^3(V, \mathbf{Z});$$

from which we get a map of real tori

$$\phi_A : \frac{H_1(T, \mathbf{R})}{H_1(T, \mathbf{Z})} \longrightarrow \frac{H^3(V, \mathbf{R})}{H^3(V, \mathbf{Z})}. \tag{2.3}$$

The real vector spaces $H^3(V, \mathbf{R})$ and $H_1(T, \mathbf{R})$ are given complex structures through the C-operator in Hodge theory. Now the fact that the form α is of Hodge type $(2,2)$, since it comes from an algebraic cycle, implies that ϕ_A is actually a holomorphic map between the complex tori $Alb(T) := \frac{H_1(T,\mathbf{R})}{H_1(T,\mathbf{Z})}$, the Albanese of T and $J^2(V) := \frac{H^3(V,\mathbf{R})}{H^3(V,\mathbf{Z})}$, the 2nd intermediate Jacobian of V. This is termed the Weil map and we again denote the map by ϕ_A. One of the very important properties of this map is its functorial behaviour with respect to maps between parameter spaces (cf. [L]). We will have occasion to return to this point later in section 3. To turn these compex tori into Abelian varieties, let L be an ample line bundle on V and if w is the Kälher form on V associated with L, define a pairing

$$H^3(V,\mathbf{C}) \times H^3(V,\mathbf{C}) \longrightarrow \mathbf{C}$$

$$(\alpha,\beta) \longrightarrow \int_V w^{n-3} \wedge \alpha \wedge \beta \qquad (n = \dim .V).$$

It can be shown that this pairing satisfies Hodge-Riemann conditions turning $J^2(V)$ into a polarised Abelian variety. Note that we have tacitly assumed that all classes in $H^3(V,\mathbf{R})$ are primitive. This is true because of our assumption that V is unirational. This assumption is satisfied for the examples we shall consider. We also let Ψ_L to denote the ample line bundle on $J^2(V)$ defined by the above pairing and often refer to Ψ_L as the polarisation on $J^2(V)$ induced from L.

In the context of the moduli spaces of vector bundles, if W is a universal bundle on $C \times M$, its second chern class $c_2(W) \in H^2(C \times M,\mathbf{Z})$ gives the Weil map

$$\phi_W : Alb(C) \to J^2(M).$$

We remark that the unirationality of M ensures that this map is independent of the choice of the universal bundle on $C \times M$. Similar considerations apply to the variety N: if E denotes *the* rank 4 bundle on $C \times N$, we get the Weil map

$$\phi_E : Alb(C) \to J^2(N).$$

We then have

Theorem 2.1. (cf. [MN],[NR],[B]) *The Weil maps*

(a) $\phi_W : Alb(C) \to J^2(M)$ *is an isomorphism of Abelian varieties .*

(b) $\phi_E : Alb(C) \to J^2(N)$ *is an isogeny of degree* 2^g.

The proof of the theorem is rather technical and we concentrate only on its consequences. One natural question in the context of the Torelli type theorem for moduli spaces is whether the polarisation on $J^2(M)$ (resp. $J^2(N)$) induced from M (resp. N) is independent of the choice of an ample line bundle on M (resp. N). Before we discuss this we make a definition.

Definition 2.2. *Let A be an Abelian variety and L_1 and L_2 be two line bundles on A. We say that L_1 is equivalent to L_2 (written $L_1 \equiv L_2$) if a* *power of* L_1 *is algebraically equivalent to a power of* L_2. *Further, if* A *(resp.* A'*) is given the polarisation* L *(resp.* L'*), we say that a map* $f : A \to A'$ *of Abelian varieties is polarisation preserving if* $f^*L' \equiv L$ *on* A.

In the case of M, since its anticanonical class is known to be ample, there is a canonical choice of a polarisation on $J^2(M)$; namely, the one induced from the dual of the canonical bundle. But, unfortunately, this fact is not known for the variety N ($Pic(N) \simeq \mathbf{Z} \oplus \mathbf{Z}$). So, let C_0 be a smooth curve for which the Neron-Severi group of its Jacobian, $NS(J(C_0))$, is isomorphic to $\mathbf{Z}$. It is well known that such curves exist. If $\Psi_1(C_0)$ and $\Psi_2(C_0)$ are two polarisations on $J^2(N)$ induced from two ample line bundles on N_{C_0}, we have, from part (b) of Theorem 2.1,

$$\Psi_1(C_0) \equiv \Psi_2(C_0).$$

Now, if C is any smooth curve and L_1 and L_2 are ample line bundles on N_C, connect C to C_0 in a holomorphic one parameter family $\{C_t\}$. It can then be shown, from the nature of construction of N, that L_1 and L_2 can be spread out to the whole of the family. Then, $\phi_E^*\Psi_2(C_t)$ and $\phi_E^*\Psi_1(C_t)$ are sections, over the family, of the local system formed from the cohomology groups $\{H^2(N_{C_t}, \mathbf{Z})\}$. And at curve C_0, some powers of these sections agree—thereby implying that the same powers of $\phi_E^*\Psi_1(C)$ and $\phi_E^*\Psi_2(C)$ agree on $J^2(N_C)$. That is, the equivalence class of the induced polarisation on $J^2(N)$ is independent of the choice of an ample line bundle on N. In what follows we always identify $Alb(C)$ with the Jacobian, $J(C)$, of C and give the natural polarisation $\Theta(C)$, afforded by the "theta divisor", on $Alb(C)$. Note that the above argument also proves that the map ϕ_E is polarisation preserving. A similar statement holds for the map ϕ_W.

Once these preliminary details regarding polarisations are fixed, a Torelli type theorem for the space N (along similar lines for the space M) can be proved as follows: suppose that the moduli spaces N_{C_1} and N_{C_2} are isomorphic via an isomorphism f and let E_1 (resp. E_2) be the universal rank 4 bundle on $C \times N_1$ (resp. $C \times N_2$). If $\overline{f}$ is the induced isomorphism between $J^2(N_1)$ and $J^2(N_2)$, using the explicit nature of the isogenies ϕ_{E_i}, one can show the existence of a commutative diagram

$$\begin{array}{ccc} J^2(N_{C_1}) & \xrightarrow{\overline{f}} & J^2(N_{C_2}) \\ \phi_{E_1}\uparrow & & \uparrow\phi_{E_2} \\ Alb(C_1) & \xrightarrow{u} & Alb(C_2) \end{array}$$

where u is again an isomorphism. And because ϕ_{E_i} and $\overline{f}$ are polarisation preserving, it easily follows, from the commutativity of the diagram, that $u^*\Theta(C_2) \equiv \Theta(C_1)$. But then, by the classical Torelli theorem, we find that the curves C_1 and C_2 are isomorphic.

§3 Deformations of the Picard bundle

In this section we let M denote the moduli space of stable bundles on the curve C of rank 2 and determinant isomorphic to a fixed line bundle ξ of degree d. We consider only the case where d is a fixed odd integer greater than $4g-3$. We also let W denote a universal family on $M \times C$. Let p and q stand for the canonical projections from $M \times C$ to M and C respectively. Then the direct image sheaf $\mathcal{U} := p_*(W)$ is a locally free sheaf and is referred to as a Picard bundle on M. Note that this depends on the choice of the universal bundle on $M \times C$. So we work with a fixed choice of a universal family. This construction also comes with an obvious family of deformations of the Picard bundle, $\{\mathcal{W}(j)\}_{j \in J}$, parameterised by the Jacobian , $J := J(C)$, of C. Namely, for $j \in J(C)$, set

$$\mathcal{W}(j) := p_*(W \times q^* L_j)$$

(here L_j is the line bundle corresponding to $j \in J$. The main theorem of this section is

Theorem 3.1. (cf. [BV2]) *For a smooth curve C, without automorphisms, of genus g, $g > 2$, we have*

(a) $\dim H^i(M, \mathrm{ad}\mathcal{W}) = \begin{cases} g & if\, i = 1 \\ 0 & if\, i = 0, 2 \end{cases}$

(b) The g-dimensional family $\{\mathcal{W}(j)\}_{j \in J}$, defined above, is injective.

Remark 3.2. A detailed study of Picard bundles, $\mathcal{P}_d$, on $J^{(d)}$, the component of $Pic(C)$ parametrising degee d ($d > 2g-1$) line bundles, has been studied extensively. See [M], [K1], [K2] for questions regarding the topology, deformations and cohomology of these objects.

Before proceeding with the idea of proof, we spend some time on the results of Thaddeus (cf. [T]) which realises the projective bundle of the

Picard bundle, $\mathbf{P}(\mathcal{W})$, as an end product of a series of blow-ups and blow-downs starting with a projective space. More precisely, let P_i $(i \in \mathbf{Z}; 0 \leq i \leq \frac{d-1}{2})$ denotes the moduli space of pairs (V, s); where V is a point in M and $s \in \mathbf{P}(H^0(C, V))$. The pair (V, s) is assumed to be stable with respect to a weight $\alpha \in \{max(0, \frac{d}{2} - i - 1), \frac{d}{2} - i\}$. These moduli spaces exist as smooth projective varieties and carry universal families. We summarise the results that we need in this work as follows.

(a) For $i = 0$, the space P_0 can be identified with the projective space $\mathbf{P}(H^0(K_C \otimes \xi)^*)$; where K_C is the canonical bundle of the curve C. Also, the space P_1 can be obtained from P_0 by blowing up the space P_0 along the curve C embedded in P_0 via the complete linear series $|K_C \otimes \xi|$. This blow up map we denote by ψ_1.

(b) For $i > 1$, one can pass from P_{i-1} to P_i by a blow up and a blow down. More precisely, we have a diagram

$$\begin{array}{ccc} \tilde{P}_i & \xrightarrow{\phi_i} & P_i \\ \downarrow{\scriptstyle \psi_i} & & \\ P_{i-1} & & \end{array}$$

Here ψ_i (resp. ϕ_i) is a blow up of P_{i-1} (resp. P_i) with smooth centre B_i (resp. A_i) with the same exceptional divisor E_i. The centres of the blow up, B_i and A_i, admit explicit descriptions as projective bundles over the i-th symmetric power, S^iC, of C.

(c) Finally, the variety $P := P_i$ $(i = \frac{d-1}{2})$ can be identified with $\mathbf{P}(\mathcal{W})$.

In order to compute the numbers $h^i(M, \mathrm{ad}(\mathcal{W}))$; $0 \leq i \leq 2$, we remark that it suffices to compute $h^i(P, T_P)$; where T_P denotes the tangent bundle of P. These two numbers can be easily related by an application of Theorem 1.1. But, since the codimension of A_i in P_{i-1} (resp. B_i in P_i) is at least five for $i > 2$, using a Hartog type argument it suffices to compute $h^i(\tilde{P}_2, T_{\tilde{P}_2})$. Now, by an argument very similar in spirit to step 3 in the proof of Theorem 1.4, we reduce to computing $h^i(\tilde{P}_2 - E_2, T_{\tilde{P}_2})$—which is equal to $h^i(P_1, T_{P_1})$ as the map ψ_1 is an isomorphism outside of E_2 and the codimension of B_1 in P_1 is large (cf. [T]). This number can in fact be computed for all values of i

and we get [3]

$$h^i(P_1, T_{P_1}) \simeq \begin{cases} 4g-3 & \text{if } i=1 \\ 0 & \text{if } i \neq 1. \end{cases}$$

This completes the proof of (a).

To prove (b), it is enough to show that the codimension two cycles $\{C_2(\mathcal{W}(j))\}_{j\in J}$ on M given by the 2nd Chern class of the family $\{\mathcal{W}(j)\}_{j\in J}$ of deformations of the Picard bundle are all distinct in the Chow ring of the moduli space M. Let $\overline{\mathcal{W}}$ be the bundle on $J \times M$ defining the family $\{\mathcal{W}(j)\}$. Consider the Weil map

$$\phi := \phi_{\overline{\mathcal{W}}} : Alb(J) \to J^2(M)$$

given by the codimension two algebraic cycle $C_2(\overline{\mathcal{W}})$ on $J \times M$.

On the other hand, the modified family $\{C_2(\mathcal{W}(j)) - C_2(\mathcal{W}(0))\}$ (here 0 denotes the identity element of J) of cycles algebraically equivalent to zero on M and parametrised by J, gives an Abel-Jacobi map (cf. [W], [Gr])

$$w : J \to J^2(M).$$

Similarly the cycle $\Delta - \{0\} \times J$ (here Δ is the diagonal cycle) on $J \times J$ gives an Abel-Jacobi map $w' : J \to Alb(J)$. It is easily seen that w' is an isomorphism. The functoriality of the Abel-Jacobi map gives $\phi \circ w' = w$ (cf. [L]).

A fundamental property of the map w, closely related to the functorial property, is that it is holomorphic. So if two cycles are rationally equivalent, they occur in a family of cycles parametrised by the projective line $\mathbf{P}^1$. The holomorphicity of w implies that these two cycles have the same image by w. So, to prove our claim, it is enough to show that the map w, or, what is the same, ϕ is injective. For this we need to compute the $(1,3)$ Kunneth component of $c_2(\overline{\mathcal{W}})$ in $H^4(J \times M, \mathbf{Z})$. One may argue, because of the functorial nature of Chern classes, that it suffices to compute it on the space $J \times P$. Further, we may reduce to computing the $(1,3)$ component of $c_2(\overline{\mathcal{W}_1})$ in $H^4(J \times P_1, \mathbf{Z})$—because the codimensions of centres of blow-up are large enough to ensure this. Here $\overline{\mathcal{W}_1}$ is the bundle on $J \times P_1$ defined by the requirement that if W_1 is the universal bundle on $C \times P_1$ and p and

[3]Note that for $i = 1$, this means that the space of first-order deformations of P_1 coincides with the space of embedded first-order deformations of C in P_0 modulo projective transformations.

q are the projections of $C \times P_1$ onto P_1 and C respectively, then

$$\overline{\mathcal{W}_1}|_{\{j\}\times P_1} \cong p_*(W_1 \otimes q^* L_j)$$

(L_j is the line bundle corresponding to the point j on J). Finally, one can even reduce to making this computation on $J \times E_1$. Recall that E_1 is the exceptional divisor in P_1. But the universal family on $C \times P_1$ when restricted to $C \times E_1$ can be described explicitly (cf. [T]). Now one applies the Grothendieck-Riemann-Roch theorem to write down $c_2(\overline{\mathcal{W}_1}|_{J\times E_1})$ and isolates the $(1,3)$ component to show that it is an unimodular element (cf. [BV2] for details).

Remark 3.3.

(a) Note that the proof of Theorem 3.1(b) also shows that the 2nd intermediate Jacobian of pair spaces, $J^2(P_i)$ are all isomorphic as polarised abelian varieties to the Jacobian, $J(C)$ of the curve C. This can be proved analogous to the proof for N in section 2 once we note that $Pic(P_i) \cong \mathbf{Z} \oplus \mathbf{Z}$. So Torelli-type theorem holds for these spaces too. One may also recover Theorem 2.1(a) from the proof of Theorem 3.1. Also, the proof of part (a) computes the number of moduli and the group of infinitesimal automorphisms of the pair spaces P_i. Note that the number of moduli turns out to be $4g-3$. This suggests that the moduli for P_i should be the universal Picard variety of the curve C.

(b) The algebraic cycles $C_2(\mathcal{W}(j))$, $j \in J$, are closely related to the twisted Brill-Noether cycles $W^0_{2,2g-3}(j)$ defined on the moduli space M. In fact, one may regard the isomorphism of $J(C)$ with $J^2(M)$ as being given by $j \to W^0_{2,2g-3}(j)$, just as in the case of the Jacobian of a curve—where the auto-duality is given by the translates of the theta divisor.

References

[B] V. Balaji, Intermediate Jacobian of some moduli spaces of vector bundles on curves, Amer. J. Math., Vol. 112, 1990, pp. 611-630.

[BS] V. Balaji and C.S. Seshadri, Cohomology of a moduli space of vector bundles, The Grothendieck Festschrift, Vol. 1, Birkhauser, 1990, 87-119.

[BV1] V. Balaji and P.A. Vishwanath, On the deformations of certain moduli spaces of vector bundles, Amer. J. Math., Vol. 115, no. 2, April 1993, pp 279-303.

[BV2] V. Balaji and P.A. Vishwanath, Deformations of Picard sheaves and moduli of pairs, to appear in Duke Math. Journal.

[G] A. Grothendieck, Eléments de Géométrie Algébrique, Springer-Verlag, 1972.

[Gr] P. Griffiths, Periods of integrals on algebraic manifolds I, Amer. J. math., 90, 1968,568-626.

[H] N.J. Hitchin, Stable bundles and integrable systems, Duke Math. J., 54, 1987, 91-114.

[K1] G. Kempf, Toward inversion of abelian integrals-I, Ann. of Math., Vol. 110, 1979, pp. 243-273.

[K2] G. Kempf, Inversion of Abelian integrals, Bull. of Amer. Math. Society, vol. 6, 1987, 25-32.

[KS] Kodaira K., and Spencer D.C., On deformations of complex analytic structures I, II, Ann. Math., II. Ser. 67, 328-466.

[L] D. Lieberman, Intermediate Jacobians, Proceedings of 5th Nordic Summer-School, Oslo, 1970.

[M] A. Mattuck, Symmetric products and Jacobians, Amer. J. Math., 83, 1961, 189-206.

[MN] D. Mumford and P.E. Newstead, Periods of moduli spaces of vector bundles on curves, Amer. J. Math., 90,1968, pp 1201-1208.

[MS] V. Mehta and C.S. Seshadri, Moduli of vector bundles on curves with parabolic structures, Math. Ann., 248, 1980, 205-239.

[NR] M.S. Narasimhan and S. Ramanan, Deformations of the moduli space of vector bundles over an algebraic curve, Ann. of Math., Vol. 101, 1975, pp. 391-417.

[S1] C.S.Seshadri, Fibrés vectoriels sur les courbes algébriques, Astérisque 96.

[S2] C.S. Seshadri, Desingularisation of moduli varieties of vector bundles on curves, Proceedings of the kyoto conference on Alg. Geom., 1977, 155-184.

[T] M. Thaddeus, Stable pairs, linear systems and the Verlinde formula, to appear in the Invent. Math.

[W] A. Weil, On Picard varieties, American Journal of Mathematics, Vol. 74 (1952) pp. 865-893.

School of Mathematics
SPIC Science Foundation
92 G.N. Chetty Road, T. Nagar
Madras-600 017, INDIA.

e-mail: balaji@ssf.ernet.in, vish@ssf.ernet.in

STABLE AUGMENTED BUNDLES OVER RIEMANN SURFACES

STEVEN BRADLOW
Department of Mathematics
University of Illinois
Urbana, IL 61801

GEORGIOS D. DASKALOPOULOS
Department of Mathematics
Princeton University
Princeton, NJ 08544

OSCAR GARCÍA-PRADA
Département de Mathématiques
Université de Paris-Sud
91405 Orsay Cedex, France

RICHARD WENTWORTH
Department of Mathematics
Harvard University
Cambridge,MA 02138

Introduction

At the symposium in Durham, the proceedings of which are reflected in this volume, there was a significant number of talks on what can generally be called "augmented holomorphic bundles". What we mean by this term is a holomorphic object which consists of one or more holomorphic bundles together with certain extra holomorphic data, typically in the form of prescribed holomorphic sections. We were ourselves responsible for discussions of so-called holomorphic pairs (i.e. a single bundle with one prescribed section), holomorphic k-pairs (i.e. a single bundle with k prescribed sections), and holomorphic triples (i.e. two bundles plus a holomorphic map between them). There were also discussions of Higgs bundles (i.e. bundles together with a section of a specific associated bundle), and of objects consisting of a bundle plus a k-dimensional linear subspace of its space of holomorphic sections (called "coherent systems" by Le Potier, and "Brill-Noether pairs" by King and Newstead).

While each variant has special features, there are important aspects common to all these types of augmented bundles. Perhaps the most significant is the fact that all admit definitions of stability which extend the usual notion of stability for a holomorphic bundle, and which allow the construction of moduli spaces. Furthermore, except for the case of Higgs bundles, the definitions each involve a real parameter. By varying the parameter, this leads to a chain of birationally equivalent moduli spaces. An equally significant shared characteristic is a Hitchin-Kobayashi correspondence between the definitions of stability and the conditions for existence of solutions to certain coupled Hermitian-Einstein equations. This leads to descriptions of the moduli spaces as Marsden-Weinstein reductions, and in particular gives rise to Kähler structures.

It was immediately clear that it would be useful to collect together these different examples in a single survey, and that the inter-relations between them should be elucidated. In part, that is the purpose of this paper. We have however gone somewhat beyond a review of previously published work, and a number of the results we will describe are new, i.e. were not known (or at least were not clear) at the time of the Durham symposium. In particular, we will describe a set of equations which determine special metrics for coherent systems, and discuss the associated Hitchin-Kobayashi correspondence. We will also describe the symplectic interpretation of these equations and use this to show how the moduli spaces of stable coherent systems relate to the moduli spaces of stable k-pairs.

We should point out that in the interests of brevity and clarity, any proofs we give are somewhat sketchy - detailed proofs will, where necessary, be given in a separate publication. We should also mention that our survey is not exhaustive. We have concentrated mostly on the gauge theoretic, or analytic, point of view, and furthermore have not covered all known examples of augmented bundles.

Important examples not discussed include parabolic bundles, framed bundles (or bundles with level structures), and the pairs of Huybrechts and Lehn. Apart from the constraints of space and time, our main reason for omitting these is that we have limited our attention to bundles over smooth closed Riemann surfaces. The cases we have excluded involve base spaces that are singular or non-compact or of dimension greater than one. We have however included at the end (in Section 6) a few short comments on these and other omitted examples.

In Section 1 we give the basic definitions and describe some relationships between the various examples of augmented bundles. In Section 2 we review their analytic descriptions, the equations for special metrics, and the associated symplectic moment maps. We use this analytic machinery in Section 4, where the construction of moduli spaces is discussed. Before doing so, however, some important features of the parameters involved in the definitions of stability are described in Section 3. The results of Sections 3 and 4 are combined in Section 5, where we give a description of what we call the "master space construction".

Acknowledgements. The authors would like to thank the organizers of the Durham symposium for the opportunity to participate in what was an extremely interesting, informative and enjoyable occasion. They would also like to thank Alastair King for many helpful conversations and ideas.

1. Stability for Augmented Bundles

Let X be a closed Riemann surface of genus g. We will denote a complex vector bundle over X by E. If E is given a holomorphic structure, we denote both the holomorphic bundle and the associated coherent analytic sheaf by $\mathcal{E}$. If the degree and rank of $\mathcal{E}$ are d and R respectively, then the slope of $\mathcal{E}$ is given by $\mu(\mathcal{E}) = d/R$. Similarly for any subbundle $\mathcal{E}' \subset \mathcal{E}$, we define the slope by $\mu(\mathcal{E}') = \deg(\mathcal{E}')/\mathrm{rank}(\mathcal{E}')$.

In this section we give the definitions of stability for the augmented bundles we will be considering in this paper. As will be seen, when the definition involves a parameter, there is sometimes more than one way to formulate the stability condition. The choices depend on exactly how one defines the parameter, and for this there seem to be two main strategies. The first is dictated by the correspondence between stability and equations (see Section 2), which suggests that the parameters in the definitions of stability should correspond to the parameters in the equations. The parameters which we have denoted by the letter τ are all of this sort. This is not always the most convenient strategy since in some cases it leads to stability conditions that, at least superficially, bear no resemblance to the usual slope stability condition for pure holomorphic bundles. For the purpose of comparison with the ordinary Mumford-Takemoto stability, it is thus sometimes more convenient to define the parameter in a different way. The various forms of the definitions are of course equivalent. In Table 1 we have listed the stability conditions in the form which is most convenient for comparison with equations, but in the discussion below we have tried to give a more comprehensive survey.

§*1.1 k-Pairs*

Perhaps the simplest augmentation one can add to a holomorphic bundle $\mathcal{E}$ is a single holomorphic section $\phi \in \mathcal{E}$. We will reserve the term **holomorphic pair** for such a pair, which we will denote by $(\mathcal{E}, \phi)$. As will be seen, a single section can be considered as a special case of various more general types of augmentation. One such generalization is from a single section to more than one section. We begin with this type of augmented bundle, i.e.

Definition 1.1. *A holomorphic* **k-pair** $(\mathcal{E}; \phi_1, \phi_2, \ldots, \phi_k)$ *consists of a holomorphic bundle,* $\mathcal{E} \longrightarrow X$*, together with* k *sections* $\{\phi_1, \phi_2, \ldots, \phi_k\}$ *in*

$H^0(X,\mathcal{E})$.

Two k-pairs $(\mathcal{E};\phi_1,\phi_2,\dots,\phi_k)$ *and* $(\mathcal{E}';\phi'_1,\phi'_2,\dots,\phi'_k)$ *are* **isomorphic** *if there is an isomorphism* $I:\mathcal{E}\longrightarrow\mathcal{E}'$ *such that* $I(\phi_i)=\phi'_i$ *for* $1\le i\le k$.

Definition 1.2a. *Given* $\tau\in\mathbb{R}$, *we say the k- pair* $(\mathcal{E};\phi_1,\phi_2,\dots,\phi_k)$ *is* **τ-stable** *if the following two conditions apply to the subbundles of* $\mathcal{E}$:

(1) $\mu(\mathcal{E}')<\tau$ *for all holomorphic subbundles* $\mathcal{E}'\subseteq\mathcal{E}$, *and*

(2) $\mu(\mathcal{E}/\mathcal{E}_\phi)>\tau$ *for any proper subbundle* $\mathcal{E}_\phi\subseteq\mathcal{E}$ *such that* $\phi_i\in\mathcal{E}_\phi$ *for all* $1\le i\le k$.

If both strong inequalities are replaced by weak ones, then we say the k-pair is τ-semistable.

This can be reformulated as follows if we set $\sigma=\tau-\mu(\mathcal{E})$.

Definition 1.2b. *We say the k-pair* $(\mathcal{E};\phi_1,\phi_2,\dots,\phi_k)$ *is* **σ-stable** *if*

(1) $\mu(\mathcal{E}')<\mu(\mathcal{E})+\sigma$ *for all holomorphic subbundles* $\mathcal{E}'\subseteq\mathcal{E}$, *and*

(2) $\mu(\mathcal{E}_\phi)<\mu(\mathcal{E})-\frac{(R-R')}{R'}\sigma$ *for any proper subbundle* $\mathcal{E}_\phi\subseteq\mathcal{E}$ *such that* $\phi_i\in\mathcal{E}_\phi$ *for all* $1\le i\le k$.

In the special case when $k=1$ and $R=2$ this reduces to the stability condition given in [T].

§*1.2 Coherent Systems*

If we view a section $\phi\in H^0(X,E)$ as determining a one dimensional subspace, then the natural generalization of the holomorphic pair comes from replacing the section by a k-dimensional linear subspace of $H^0(X,\mathcal{E})$. Following Le Potier, we shall call these coherent systems, but the definition of stability we give is due to [RV].

Definition 1.3. *A dimension k* **coherent system** $(\mathcal{E},V)$ *consists of a holomorphic bundle,* $\mathcal{E}\longrightarrow X$, *together with a k-dimensional linear subspace* $V\subset H^0(X,\mathcal{E})$.

Two coherent systems $(\mathcal{E},V)$ *and* $(\mathcal{E}',V')$ *are* **isomorphic** *if there is an isomorphism* $I:\mathcal{E}\longrightarrow\mathcal{E}'$ *such that* $I(V)=V'$.

We define the **subobjects** *of* $(\mathcal{E},V)$ *to be subbundles* $\mathcal{E}'\subset\mathcal{E}$ *together with subspaces* $V'\subseteq V\cap H^0(X,\mathcal{E}')$.

For a given $\alpha \in \mathbb{R}$, we define the **α-degree** *of $(\mathcal{E}', V')$ to be*

$$\deg_\alpha(\mathcal{E}', V') = \deg(\mathcal{E}') + \alpha \dim(V') \ . \tag{1.1}$$

The **α-slope** *of $(\mathcal{E}', V')$ is then*

$$\mu_\alpha(\mathcal{E}', V') = \frac{\deg_\alpha(\mathcal{E}', V')}{\operatorname{rank}(\mathcal{E}')} \ . \tag{1.2}$$

Definition 1.4. *We say the coherent system $(\mathcal{E}, V)$ is* **α-stable** *if*

$$\mu_\alpha(\mathcal{E}', V') < \mu_\alpha(\mathcal{E}, V)$$

for all subsystems $(\mathcal{E}', V')$, though it is clearly enough to check this condition on subsystems where $V' = V \cap H^0(X, \mathcal{E}')$. If the strong inequality is replaced by a weak one, then we say the coherent system is α-semistable.

Remark. The definitions of a k-pair and a k-dimensional coherent system are clearly closely related, in a way similar to that in which k-*frames* in $\mathbb{C}^n$ and k-*planes* in $\mathbb{C}^n$ are related. Indeed, given a k-pair $(\mathcal{E}; \phi_1, \phi_2, \ldots, \phi_k)$ in which the sections are linearly independent in $H^0(X, \mathcal{E})$, we get a k-dimensional coherent system with $V = \operatorname{span}\{\phi_1, \ldots, \phi_k\}$, and conversely, given a coherent system $(\mathcal{E}, V)$, we get a k-pair by taking a basis for V. A natural question to ask is how the notions of stability compare for k-pairs and coherent systems which are related in this way. One finds

Proposition 1.5. *Let $\{\phi_1, \ldots, \phi_k\}$ be k linearly independent sections in $H^0(X, \mathcal{E})$, and let $V = \operatorname{span}\{\phi_1, \ldots, \phi_k\}$. If $(\mathcal{E}, V)$ is an α- stable coherent system, then $(\mathcal{E}; \phi_1, \phi_2, \ldots, \phi_k)$ is a τ-stable k- pair, where α and τ are related by*

$$\tau = \mu_\alpha(\mathcal{E}, V) = \mu(E) + \alpha \frac{k}{\operatorname{rank}(E)} \ . \tag{1.3}$$

Proof. This can be shown directly from the definitions. It emerges much more transparently, however, as an immediate consequence of Propositions 1.13 and 1.14 (in Section 1.5). □

The converse is not true, that is the τ-stability of the k-pair does not imply the α-stability of the coherent system. Counter-examples can be constructed, for instance, as follows.

Example 1.6. Let $\mathcal{L}$ be a line bundle, and let $\mathcal{E}_0$ be a semistable rank $(R-1)$ bundle, such that $2g-1 < \deg(\mathcal{L}) < \mu(\mathcal{E}_0)$. Then both bundles have vanishing first sheaf cohomology, and are generated by global sections. Take any extension

$$0 \longrightarrow \mathcal{E}_0 \longrightarrow \mathcal{E} \longrightarrow \mathcal{L} \longrightarrow 0$$

and sections $\phi_1, \phi_2, \ldots, \phi_{k-1}, \phi_k$ such that

(1) $\phi_1, \phi_2, \ldots, \phi_{k-1}$ generate $\mathcal{E}_0$, and
(2) ϕ_k is not contained in $H^0(\mathcal{E}_0)$.

Notice that $0 \subset \mathcal{E}_0 \subset \mathcal{E}$ is the Harder-Narasimhan filtration for $\mathcal{E}$, so that $\mu(\mathcal{E}') \leq \mu(\mathcal{E}_0)$ for any subbundle $\mathcal{E}' \subset \mathcal{E}$. In addition, there are clearly no proper subbundles which contain all k sections $\phi_1, \phi_2, \ldots, \phi_k$. It follows that for any $\tau > \mu(\mathcal{E}_0)$, the k-pair $(\mathcal{E}, \phi_1, \phi_2, \ldots, \phi_k)$ is τ-stable .

Now consider the coherent system with $V = \text{span}\{\phi_1, \phi_2, \ldots, \phi_k\}$. If we set $V_0 = \text{span}\{\phi_1, \phi_2, \ldots, \phi_{k-1}\}$, then $(\mathcal{E}_0, V_0)$ is a proper subsystem of $(\mathcal{E}, V)$. It is easily checked that $\mu_\alpha(\mathcal{E}_0, V_0) > \mu_\alpha(\mathcal{E}, V)$ for any $\alpha > 0$. Hence for any $\alpha > 0$, the coherent system $(\mathcal{E}, V)$ is NOT α-stable.

In section 4 we will examine how the moduli spaces of stable k-pairs and coherent systems are related.

§1.3 Holomorphic Triples

Using the dimensional reduction description of holomorphic pairs given in [GP], one arrives at a quite different generalization to those given by k-pairs or coherent systems. The basic idea is that the data contained in the pair $(\mathcal{E}, \phi)$ over X is equivalent to the data contained in certain holomorphic extensions over $X \times \mathbb{P}^1$. The extensions in question are of the form

$$0 \longrightarrow p^*\mathcal{E} \longrightarrow \mathcal{F} \longrightarrow q^*\mathcal{O}(2) \longrightarrow 0\ ,$$

where p, q denote the projections from $X \times \mathbb{P}^1$ to X and $\mathbb{P}^1$ respectively, and $\mathcal{O}(2)$ is the degree 2 line bundle on $\mathbb{P}^1$. The point is that such extensions are classified by $H^1(X \times \mathbb{P}^1, p^*\mathcal{E} \otimes q^*\mathcal{O}(-2))$, which, by the Kunneth formula, is isomorphic to $H^0(X, \mathcal{E})$. This construction can be generalized to include extensions of the form

$$0 \longrightarrow p^*\mathcal{E}_1 \longrightarrow \mathcal{F} \longrightarrow p^*\mathcal{E}_2 \otimes q^*\mathcal{O}(2) \longrightarrow 0\ ,$$

where $\mathcal{E}_1$ and $\mathcal{E}_2$ are both bundles over X. Again by the Künneth formula, such extensions are classified by $H^0(X, \mathcal{E}_1 \otimes \mathcal{E}_2^*)$. Such extensions thus correspond to the triples $(\mathcal{E}_1, \mathcal{E}_2, \Phi)$ over X, where

Definition 1.7. *A* **holomorphic triple** $(\mathcal{E}_1, \mathcal{E}_2, \Phi)$ *consists of two holomorphic bundles,* $\mathcal{E}_i \longrightarrow X$*, together with a holomorphic homomorphism* $\Phi : \mathcal{E}_2 \longrightarrow \mathcal{E}_1$.

Two holomorphic triples $(\mathcal{E}_1, \mathcal{E}_2, \Phi)$ *and* $(\mathcal{E}_1', \mathcal{E}_2', \Phi')$ *are* **isomorphic** *if there are isomorphisms* $I_1 : \mathcal{E}_1 \longrightarrow \mathcal{E}_1'$ *and* $I_2 : \mathcal{E}_2 \longrightarrow \mathcal{E}_2'$ *such that* $\Phi' \circ I_2 = I_1 \circ \Phi$*, i.e. such that the following diagram commutes*

$$\begin{array}{ccc} \mathcal{E}_2 & \xrightarrow{I_2} & \mathcal{E}_2' \\ {\scriptstyle\Phi}\downarrow & & \downarrow{\scriptstyle\Phi'} \\ \mathcal{E}_1 & \xrightarrow[I_1]{} & \mathcal{E}_1' \end{array}$$

We define the **subobjects** *of* $(\mathcal{E}_1, \mathcal{E}_2, \Phi)$ *to be triples* $T' = (\mathcal{E}_1', \mathcal{E}_2', \Phi')$ *where* $\mathcal{E}_1'(\mathcal{E}_2')$ *is a subbundle of* $\mathcal{E}_1(\mathcal{E}_2)$ *and* Φ' *is the restriction of* Φ *to* $\mathcal{E}_2'$.

Given $\tau \in \mathbb{R}$, we assign a real valued quantity to each subtriple by

$$\theta_\tau(T') = (\mu(\mathcal{E}_1' \oplus \mathcal{E}_2') - \tau) - \frac{R_2'}{R_2}\frac{(R_1 + R_2)}{(R_1' + R_2')}(\mu(\mathcal{E}_1 \oplus \mathcal{E}_2) - \tau)\,, \tag{1.5}$$

where R_1, R_2, R_1', R_2' are the ranks of $\mathcal{E}_1, \mathcal{E}_2, \mathcal{E}_1', \mathcal{E}_2'$ respectively, and μ denotes the ordinary slope.

Definition 1.8a. *We say* $T = (\mathcal{E}_1, \mathcal{E}_2, \Phi)$ *is* τ**-stable** *if*

$$\theta_\tau(T') < 0$$

for all proper subtriples $T' = (\mathcal{E}_1', \mathcal{E}_2', \Phi')$ *of* T*. If the strong inequality is replaced by a weak one, then we say the triple is* τ*-semistable.*

For comparison with the usual slope conditions, a more convenient reformulation of this condition is as follows. Given a parameter α, we define the α**-degree** and α**-slope** of a subtriple by

$$\deg_\alpha(\mathcal{E}_1', \mathcal{E}_2', \Phi') = \deg(\mathcal{E}_1' \oplus \mathcal{E}_2') + \alpha R_2'\,, \tag{1.6}$$

$$\mu_\alpha(\mathcal{E}_1', \mathcal{E}_2', \Phi') = \frac{\deg_\alpha(\mathcal{E}_1', \mathcal{E}_2', \Phi')}{R_1' + R_2'}\,. \tag{1.7}$$

Definition 1.8b. *A triple $T = (\mathcal{E}_1, \mathcal{E}_2, \Phi)$ is said to be* **α-stable** *(resp.* **semistable***) if for all proper subtriples $T' \subset T$ we have*

$$\mu_\alpha(T') < \mu_\alpha(T) \text{ (resp. } \mu_\alpha(T') \leq \mu_\alpha(T)) \ .$$

Lemma 1.9. *(cf. [BGP])These two notions of stability for a triple are equivalent when α and τ are related by*

$$\tau = \mu_\alpha(T) = \mu(\mathcal{E}_1 \oplus \mathcal{E}_2) + \frac{R_2}{R_1 + R_2}\alpha \ . \tag{1.8}$$

§1.4 Higgs Bundles

The final type of augmented bundle which we will consider can be thought of as a kind of holomorphic pair in which the notion of section has been generalized. In this generalization we allow the ϕ in $(\mathcal{E}, \phi)$ to be a holomorphic section, not necessarily of $\mathcal{E}$, but of an associated bundle or more generally any bundle naturally related to $\mathcal{E}$. If the associated bundle is $\mathrm{End}(\mathcal{E}) \otimes T^{*\prime}{}_X$, where $T^{*\prime}{}_X$ is the holomorphic cotangent bundle of X, then the resulting object is a called a Higgs bundle.

Definition 1.10. *A* **Higgs Bundle** *$(\mathcal{E}, \Theta)$ consists of a holomorphic bundle, $\mathcal{E} \longrightarrow X$, together with a Higgs field $\Theta \in H^0(X, \mathrm{End}(\mathcal{E}) \otimes K)$, where K is the canonical bundle on X.*

Two Higgs bundles $(\mathcal{E}, \Theta)$ and $(\mathcal{E}', \Theta')$ are **isomorphic** *if there is an isomorphism $I : \mathcal{E} \longrightarrow \mathcal{E}'$ such that $\Theta' \circ I = I \circ \Theta$, i.e. such that the following diagram commutes*

$$\begin{array}{ccc} \mathcal{E} & \xrightarrow{I} & \mathcal{E}' \\ {\scriptstyle\Theta}\downarrow & & \downarrow{\scriptstyle\Theta'} \\ \mathcal{E} \otimes K & \xrightarrow[I]{} & \mathcal{E}' \otimes K \end{array}$$

Definition 1.11. *A Higgs Bundle $(\mathcal{E}, \Theta)$ is defined to be* **stable** *if*

$$\mu(\mathcal{E}') < \mu(\mathcal{E})$$

for all Θ-invariant subbundles, i.e. for all subbundles $\mathcal{E}' \subset \mathcal{E}$ such that $\Theta(\mathcal{E}') \subset \mathcal{E}' \otimes K$. If the strong inequality is replaced by a weak one, then we say the Higgs bundles is semistable.

§1.5 *Specialized triples*

It is interesting to note that the augmented bundles discussed above can all be realized as "specialized triples". The specializations required to do this are partly in the choice of the underlying bundles of the triple. This is not however sufficient to yield the stability criteria we want as special cases of the stability for triples. The further specialization that is required can be understood either as a restriction on the notion of isomorphism, or as a restriction on the complex gauge group (i.e. the group of complex automorphisms) of the triple.

(a) k-Pairs. The k-pair $(\mathcal{E}; \phi_1, \phi_2, \ldots, \phi_k)$ can be described as a triple by using the fact that a section $\phi \in H^0(X, \mathcal{E})$ is equivalent to a map $\phi : \mathcal{O} \longrightarrow \mathcal{E}$, where $\mathcal{O} \longrightarrow X$ is the structure sheaf. Since we have k sections, we should consider holomorphic triples in which the bundle $\mathcal{E}_2$ is the trivial rank k complex bundle with a fixed trivial holomorphic structure, i.e. $E_2 = \mathcal{O}^k$. Since a map $\Phi : \mathcal{O}^k \longrightarrow \mathcal{E}$ is equivalent to a set of k sections $\{\phi_1, \ldots, \phi_k\}$, we get:

Proposition 1.12. *There is a bijective correspondence between k-pairs of the form $(\mathcal{E}; \phi_1, \phi_2, \ldots, \phi_k)$ and holomorphic triples of the form $(\mathcal{E}, \mathcal{O}^k, \Phi)$*

Notice that this bijection does not give a 1-1 correspondence between the *isomorphism classes* of the k-pairs and the *isomorphism classes* of the corresponding triples. The reason is that an automorphism of $(\mathcal{E}, \mathcal{O}^k, \Phi)$ can in general change the holomorphic structure (though not the isomorphism class!) of both $\mathcal{E}$ and $\mathcal{O}^k$, but it is only those which leave $\mathcal{O}^k$ unchanged that correspond to automorphisms of the k-pair. This restriction on the allowed automorphisms leads to a restriction on the notion of stability for the triples $(\mathcal{E}, \mathcal{O}^k, \Phi)$. It is stability in this restricted sense that corresponds to the notion of stability for a k- pair. More precisely:

Proposition 1.13. *Let $(\mathcal{E}, \phi_1, \ldots, \phi_k)$ be a holomorphic k-pair, and let $(\mathcal{E}, \mathcal{O}^k, \Phi)$ be the corresponding holomorphic triple. Then the following are* *equivalent:*

(1) *the k-pair* $(E, \phi_1, \ldots, \phi_k)$ *is* τ*- stable,*

(2) *the triple* $(\mathcal{E}, \mathcal{O}^k, \Phi)$ *is* τ*-stable in the restricted sense that the stability condition is satisfied, but only by subtriples of the form* $(\mathcal{E}', \mathcal{O}', \Phi)$ *where* $\mathcal{E}'$ *can be any subbundle of* $\mathcal{E}$*, and* $\mathcal{O}'$ *is 0 or* $\mathcal{O}^k$*.*

Proof. Under the correspondence between $(\mathcal{E}, \phi_1, \ldots, \phi_k)$ and the triple $(\mathcal{E}, \mathcal{O}^k, \Phi)$, subbundles $\mathcal{E}' \subset \mathcal{E}$ can be identified with subtriples of the form $(\mathcal{E}', 0, \Phi)$. Furthermore, subbundles which contain all k sections can be identified precisely with the subtriples of the form $(\mathcal{E}', \mathcal{O}^k, \Phi)$. The result then follows by checking that the condition $\theta_\tau(\mathcal{E}', 0, \Phi) < 0$ is equivalent to the condition $\mu(\mathcal{E}') < \tau$, and that the condition $\theta_\tau(\mathcal{E}', \mathcal{O}^k, \Phi) < 0$ is equivalent to the condition $\mu(\mathcal{E}/\mathcal{E}') > \tau$. □

(b) Coherent Systems. By setting $V = \mathrm{span}\{\phi_1, \ldots, \phi_k\}$, we get a map from the triples $(\mathcal{E}, \mathcal{O}^k, \Phi)$ to coherent systems. A subsystem $(\mathcal{E}', V')$ corresponds to a subtriple $(\mathcal{E}', \mathcal{O}', \Phi)$ where $\mathcal{O}'$ is a trivial subbundle of $\mathcal{O}^k$ and $V' = \Phi(H^0(X, \mathcal{O}'))$. By checking the stability condition on such subtriples, we get an analog of Proposition 1.13 , namely

Proposition 1.14. *Let* $(\mathcal{E}, V)$ *be a k-dimensional coherent system, and let* $(\mathcal{E}, \mathcal{O}^k, \Phi)$ *be a holomorphic triple such that* $V = \Phi(H^0(X, \mathcal{O}^k) = \mathrm{span}\{\phi_1, \ldots, \phi_k\}$*. Let* τ *and* α *be such that*

$$\tau = \mu(E) + \alpha \frac{\dim(V)}{\mathrm{rank}(E)} .$$

Then the following are equivalent:

(1) *the coherent system* $(\mathcal{E}, V)$ *is* α*-stable,*

(2) *the triple* $(\mathcal{E}, \mathcal{O}^k, \Phi)$ *is* τ*-stable in the restricted sense that the stability condition is satisfied, but only by subtriples of the form* $(\mathcal{E}', \mathcal{O}', \Phi)$ *where* $\mathcal{E}'$ *can be any subbundle of* $\mathcal{E}$*, and* $\mathcal{O}'$ *is any trivial subbundle of* $\mathcal{O}^k$*.*

(c) Fixed $\mathcal{E}_1$ triples. In both of the previous examples (i.e. k-pairs and coherent systems), the associated triples $(\mathcal{E}_1, \mathcal{E}_2, \Phi)$ had the bundle $\mathcal{E}_2$ fixed. One can equally well consider specialized triples in which $\mathcal{E}_1$ is fixed, and in fact there is a duality between these and the fixed $\mathcal{E}_2$ triples. With $\mathcal{E}_1$ fixed, one gets a special case of the objects studied by Huybrechts and Lehn. In

[HL] they consider pairs $(\mathcal{E},\Phi)$ consisting of a coherent sheaf over a smooth projective variety plus a homomorphism $\Phi:\mathcal{E}\longrightarrow\mathcal{E}_0$ to a fixed sheaf $\mathcal{E}_0$ over the same variety. They give a definition of stability with respect to a *polynomial*, δ. In the special case that the sheaves are holomorphic bundles over a smooth curve, the polynomial has degree zero and the definition of stability reduces to the conditions

(1) $\mu(\mathcal{E}') < \mu(\mathcal{E}) - \frac{1}{R}\delta$ for all subbundles in the kernel of $\Phi:\mathcal{E}\longrightarrow\mathcal{E}_0$,
(2) $\mu(\mathcal{E}') < \mu(\mathcal{E}) + \frac{R-R'}{RR'}\delta$ for all proper subbundles,

where R (resp. R') is the rank of $\mathcal{E}$ (resp. $\mathcal{E}'$).

The two cases in this definition correspond to subtriples $(\mathcal{E}_0',\mathcal{E}',\Phi)\subseteq(\mathcal{E}_0,\mathcal{E},\Phi)$ in which $\mathcal{E}_0'$ is respectively trivial or $\mathcal{E}_0$. One thus finds that

Proposition 1.15. *Let $\mathcal{E}_0\longrightarrow X$ be a fixed bundle over a smooth algebraic curve. Let $\mathcal{E}\longrightarrow X$ be a bundle over X, and let $(\mathcal{E},\Phi)$ be a pair in the sense of Huybrechts and Lehn, with $\Phi:\mathcal{E}\longrightarrow\mathcal{E}_0$. Let $(\mathcal{E}_0,\mathcal{E},\Phi)$ be the corresponding triple. Let τ and δ be real parameters such that*

$$\tau=\mu(\mathcal{E}_0)-\frac{1}{\mathrm{rank}(\mathcal{E}_0)}\delta\ .$$

Then the following are equivalent:

(1) *pair $(\mathcal{E},\Phi)$ is δ-stable in the sense of [HL],*
(2) *the triple $(\mathcal{E},\mathcal{E}_0,\Phi)$ is τ-stable in the restricted sense that the stability condition for triples is satisfied, but only by subtriples of the form $(\mathcal{E}_0',\mathcal{E}',\Phi)$ where $\mathcal{E}'$ can be any subbundle of $\mathcal{E}$, but $\mathcal{E}_0'$ is either trivial or $\mathcal{E}_0$.*

(d) Higgs Bundles. To describe Higgs bundles as specialized triples, we must consider triples $(\mathcal{E}_1,\mathcal{E}_2,\Phi)$ in which the two bundles are related by $\mathcal{E}_1=\mathcal{E}_2\otimes K$, where K is the canonical bundle of the Riemann Surface X. It follows immediately that

Proposition 1.16. *There is a bijective correspondence between Higgs bundles $(\mathcal{E},\Theta)$ and holomorphic triples $(\mathcal{E}\otimes K,\mathcal{E},\Theta)$.*

The correspondence between the stability criteria for Higgs bundles and such triples has one unusual aspect, having to do with the role of parameters in the definitions. Recall that the stability criterion for a triple involves a parameter, τ, while that for a Higgs bundle does not. For the triple

$(\mathcal{E}\otimes K,\mathcal{E},\Theta)$, τ-stability is defined in terms of the quantity θ_τ evaluated on subtriples. Notice however, that for subtriples of the form $(\mathcal{E}'\otimes K,\mathcal{E}',\Theta)$, we get

$$\theta_\tau(\mathcal{E}'\otimes K,\mathcal{E}') = \mu((\mathcal{E}'\otimes K)\oplus\mathcal{E}') - \mu((\mathcal{E}\otimes K)\oplus\mathcal{E}) \ ,$$

with the cancellation of τ being caused by the fact that the subbundles in the subtriple have the same rank. Furthermore, subtriples of this kind are precisely the ones which correspond to sub-Higgs bundles. We thus get

Proposition 1.17. *Let $(\mathcal{E},\Theta)$ be a Higgs bundle, and let $(\mathcal{E}\otimes K,\mathcal{E},\Theta)$ be the corresponding holomorphic triple. Then the following are equivalent:*

(1) *the Higgs bundle $(\mathcal{E},\Theta)$ is stable,*
(2) *the triple $(\mathcal{E}\otimes K,\mathcal{E},\Theta)$ is stable in the restricted sense that (for any τ) the τ-stability condition is satisfied only by subtriples of the form $(\mathcal{E}'\otimes K,\mathcal{E}')$.*

2. Analytic Aspects

By virtue of the correspondence between holomorphic structures and $\bar\partial$-operators on a fixed smooth bundle $E \longrightarrow X$, the augmented bundles in Section 1 can each be given an analytic description . Since X is a Riemann surface, the set of holomorphic structures corresponds to the set of all such $\bar\partial$-operators. This set, which we denote by $\mathcal{C}$, is an infinite dimensional complex affine space, modelled on the space of $(0,1)$-forms with values in E, i.e. on $\Omega^{0,1}(X,E)$. We will use the notation $\bar\partial_E$ for elements of $\mathcal{C}$.

The sets of all holomorphic augmented bundles can likewise be given descriptions as infinite dimensional complex spaces. In each case these are subspaces of an ambient "configuration space", with the subspaces cut out by a holomorphicity condition. For example, the space of holomorphic pairs on E is the subspace $\mathcal{H}\subset\mathcal{C}\times\Omega^0(X,E)$ determined by the condition $\bar\partial_E(\phi) = 0$. The details for the other types of augmented bundle are collected together in Table 2.

The complex gauge groups in Table 2 are, by definition, the groups of complex bundle automorphisms covering the identity map on the base. Notice that in each case there is an action of the the complex gauge group on the space of all holomorphic augmented bundles. Via this action, two augmented bundles are isomorphic if and only if they are represented by points on the same complex gauge orbit.

§2.1 *Equations for Special Metrics*

On a complex vector bundle $E \longrightarrow X$ with $\overline{\partial}$-operator $\overline{\partial}_E$, one can define the Hermitian-Einstein equations. These equations relate $\overline{\partial}_E$ and a hermitian metric H on E, and take the form

$$(2.1) \qquad \sqrt{-1}\Lambda F_{\overline{\partial}_E,H} = \mu \mathbf{I}.$$

Here $F_{\overline{\partial}_E,H} \in \Omega^2(\mathrm{End}(E))$ is the curvature of the unique connection determined by $\overline{\partial}_E$ and H, $\Lambda F_{\overline{\partial}_E,H} \in \Omega^0(X, \mathrm{End}(E))$ is a contraction of $F_{\overline{\partial}_E,H}$ and the Kähler form ω of the metric on X, μ is a constant equal to the slope $\mu(E)$, and $\mathbf{I}$ is the identity section of $\mathrm{End}(E)$.

There are analogous equations for special metrics on each of the augmented bundles discussed in Section 1. These now involve the extra holomorphic data as well as the $\overline{\partial}_E$-operators. Table 1 contains a complete listing. Except for the orthonormal τ-vortex equations on coherent systems, these equations have all previously been studied (in the references given in Table 1). The equations for coherent systems can be motivated by moment map considerations, and we will say more about this in the next section.

Notice that the Hermitian-Einstein equations can be defined on complex bundles over Kähler manifolds of any dimension. Together with the integrability condition $(\overline{\partial}_E)^2 = 0$, they give the absolute minima of the Yang-Mills functional. The other equations in Table 1 can similarly be related to minimization criteria for gauge theoretic functionals. The functionals are now of Yang-Mills-Higgs type, i.e. have terms depending on the extra data on the bundle. On a given type of augmented bundle, the equations for the minima of the functional can be identified with the appropriate equations from Table 1 together with the integrability condition on $\overline{\partial}_E$, plus the holomorphicity condition on the data which defines the augmented bundle (cf., for example, [B], [GP]).

§2.2 *Symplectic Structures and Moment Maps*

If we fix a Hermitian metric on $E \longrightarrow X$, then the spaces $\Omega^{p,q}(X, E)$ and $\Omega^{p,q}(X, \mathrm{End}(E))$ acquire fiber-wise, and hence L^2, Hermitian inner products. This gives rise to Kähler and hence symplectic structures on the configuration spaces for the augmented bundles. This is immediate in the case of k- pairs, Higgs bundles, and also triples (if metrics are fixed on both the underlying smooth bundles in a triple). The case of coherent systems is only slightly less direct, and will be explained below.

In each case, given the fixed metrics on the bundles, one can define a reduction from the full complex gauge groups to the (real) unitary gauge groups. We will denote these by $\mathfrak{G}_{\mathbb{C}}$ and $\mathfrak{G}$ respectively. The $\mathfrak{G}_{\mathbb{C}}$- actions on the configuration spaces restrict to give *symplectic* actions of the unitary gauge groups on the underlying symplectic manifolds.

Table 3 lists the moment maps for these actions, where we have identified the Lie algebra of $\mathfrak{G}$ and its dual via the Ad-invariant inner product. Notice that in all cases the moment maps are essentially the left hand sides of the equations for special metrics in Table 1. These equations involve a Hermitian bundle metric and the holomorphic data which specify the augmented bundle. When the holomorphic data is fixed, the equations are interpreted as defining special metrics. Alternatively, they become moment map conditions when the metric is fixed and the holomorphic data are the arguments in the equation. Consider, for the sake of illustration, the case of holomorphic pairs. This duality between the moment map condition and the τ-vortex equation leads to

Proposition 2.1. *Let $\Psi_1 : \mathcal{H} \longrightarrow \mathfrak{g}$ be the moment map given in Table 3, and let $\mathcal{V}_\tau$ be the set of all $(\overline{\partial}_E, \phi) \in \mathcal{H}$ for which there exists a solution (i.e. a metric) to the τ-vortex equations. Then*

(1) *$\mathcal{V}_\tau$ is the $\mathfrak{G}_{\mathbb{C}}$-saturation of level set $\Psi_1^{-1}(\tau)$, i.e. it is the union of the $\mathfrak{G}_{\mathbb{C}}$ orbits through $\Psi_1^{-1}(\tau)$, and*

(2) *there is a bijective correspondence*

$$\mathcal{V}_\tau/\mathfrak{G}_{\mathbb{C}} \leftrightarrow \Psi_1^{-1}(\tau)/\mathfrak{G}.$$

Similar correspondences apply for the other moment maps in Table 3. These will be exploited in Section 4 to give complementary descriptions of moduli spaces.

The symplectic structure and moment map for coherent systems arise as follows. As indicated in Table 2, the configuration space for k-dimensional coherent systems on a given complex bundle $E \longrightarrow X$ is constructed by considering holomorphic k-frames modulo the action of GL(k). The set of all holomorphic k-frames is the subspace of $\mathcal{H}^k$ given by

$$\mathcal{ST}_k = \{(\overline{\partial}_E; \phi_1, \phi_2, \ldots, \phi_k) \in \mathcal{H}^k \mid \text{the sections are linearly independent}\}$$

The group $GL(k)$ acts on the sections, and the configuration space for the coherent systems is thus $\mathcal{H}^{CS} = \mathcal{ST}_k/GL(k)$.

If we fix a metric on the bundle, then $\mathcal{C} \times (\Omega^0(X, E))^k$, and thus $\mathcal{H}^k$, acquires a symplectic structure. We can also then define the unitary gauge group for E, and restrict the above $GL(k)$ action on $\mathcal{H}^k$ to an action of the unitary group $U(k)$. Furthermore, a calculation shows that:

Proposition 2.2. *The $U(k)$ action on $\mathcal{H}^k$ is symplectic, and has a moment map given by*

$$\Psi_U(\overline{\partial}_E, \phi_1, \phi_2, \ldots, \phi_k) = -i < \phi_i, \phi_j > , \tag{2.2}$$

that is, $\Psi_U(\overline{\partial}_E, \phi_1, \phi_2, \ldots, \phi_k)$ is the skew-Hermitian matrix whose (i, j) entry is $-i < \phi_i, \phi_j >$.

For any real $\alpha > 0$ there is clearly a bijective correspondence between the complex quotient $\mathcal{ST}_k/GL(k, \mathbb{C})$ and the symplectic quotient $\Psi_U^{-1}(-i\alpha\mathbf{I})/U(k)$. This induces a symplectic structure on

$$\mathcal{H}^{CS} = \mathcal{ST}_k/GL(k, \mathbb{C}) = \Psi_U^{-1}(-i\alpha\mathbf{I})/U(k) . \tag{2.3}$$

If we denote the corresponding symplectic form by ω_α, then it follows from the symplectic quotient constructions that

$$\omega_\alpha = \alpha^2 \omega_1 . \tag{2.4}$$

We can now consider the action of the unitary gauge group $\mathfrak{G}$ on the symplectic manifold $(\mathcal{H}^{CS}, \omega_\alpha)$. From the above identification, this is the same as considering the action on the quotient $\Psi_U^{-1}(-i\alpha\mathbf{I})/U(k)$. It follows that

Proposition 2.3. *The moment map $\Psi_{CS} : \mathcal{H}^{CS} \longrightarrow \mathbf{g}^*$ with respect to the Kähler form ω_α is given by*

$$\Psi_{CS}(\overline{\partial}_E, V) = \Psi_k(\overline{\partial}_E, \phi_1, \ldots, \phi_k) , \tag{2.5}$$

where $\{\phi_1, \ldots, \phi_k\}$ *is any basis for* V *such that* $\Psi_U(\overline{\partial}_E, \phi_1, \ldots, \phi_k) = -i\alpha\mathbf{I}$, *i.e. such that* $< \phi_i, \phi_j >= \alpha\mathbf{I}$.

We can now see how the orthonormal τ-vortex equations arise. In order to have the right relation between the moment map and the equations for special metrics, we require that the equations *characterize* the $\mathfrak{G}_{\mathbb{C}}$-saturation of the level sets $\Psi_{CS}^{-1}(-i\tau\mathbf{I}) \in \mathcal{H}^{CS}$. But the set $\mathcal{V}_\tau^{CS} = \mathfrak{G}_{\mathbb{C}} \circ \Psi_{CS}^{-1}(-i\tau\mathbf{I})$ consists of all coherent systems $(\mathcal{E}, V)$ for which the following condition is satisfied:

Condition 2.4a. *There is a basis* $\{\phi_1, \phi_2, \ldots, \phi_k\}$ *for* V *and a complex gauge transformation* $g \in \mathfrak{G}_{\mathbb{C}}$ *such that*

$$\Psi_k(g \circ (\overline{\partial}_E, \phi_1, \phi_2, \ldots, \phi_k)) = -i\tau\mathbf{I} ,$$

where

$$\Psi_U(\overline{\partial}_E, \phi_1, \ldots, \phi_k) = -i\alpha\mathbf{I} .$$

By using $g \in \mathfrak{G}_{\mathbb{C}}$ to transform the metric K on E, we see that this is equivalent to the condition that

Condition 2.4b. *There is a basis* $\{\phi_1, \phi_2, \ldots, \phi_k\}$ *for* V *and a metric* H *on* E *such that*

(2.6a) $$< \phi_i, \phi_j >_H = \alpha\mathbf{I} ,$$

and

(2.6b) $$i\Lambda F_{\overline{\partial}_E, H} + \Sigma_{i=1}^k \phi_i \otimes \phi_i^* = \tau\mathbf{I} .$$

The parameters τ and α in equations (2.6) are not independent. By integrating the trace of the k-τ-vortex equation, and using the Chern-Weil formula for the degree of the bundle, one gets the third condition

(2.6c) $$\deg(E) + k\alpha = \mathrm{rank}(E)\tau$$

Equations (2.6) are the orthonormal τ-vortex equations.

Remark. If we combine Proposition 2.2 with the calculation for the $\mathfrak{G}$-action on $\mathcal{H}^k$, we can obtain the moment map for the action of the product $\mathfrak{G} \times U(k)$ on $\mathcal{H}^k$. The result is the map

$$\Psi : \mathcal{H}^k \longrightarrow \mathfrak{g}^* \oplus \mathfrak{u}(k)^*$$

given by

(2.7) $$\Psi = (\Psi_k, \Psi_U) .$$

We have the following equivalence.

Proposition 2.5. *Let $q : \Psi_U^{-1}(-i\alpha\mathbf{I}) \longrightarrow \mathcal{H}^{CS}$ denote the quotient (2.3). Then*

$$q^{-1}(\Psi_{CS}^{-1}(-i\tau\mathbf{I})) = \Psi^{-1}(-i\tau\mathbf{I}, -i\alpha\mathbf{I}) \,, \tag{2.8}$$

We will return to this in Section 4, where we will use it to relate the moduli spaces of k-dimensional coherent systems and that of k-pairs.

§2.3 Hitchin-Kobayashi Correspondences

The relation between the Hermitian-Einstein equation and ordinary slope stability of a holomorphic bundle is given by the following result, known as the Hitchin- Kobayashi correspondence.

Theorem 2.6 [UY,D2,NS,Ko,L]. *An irreducible holomorphic bundle admits a Hermitian-Einstein metric if and only if the bundle is stable.*

For all of our examples of augmented bundles we have both stability criteria and special equations, with obvious formal similarities to the case of plain bundles. We can thus expect correspondences similar to the one above for each of the augmented bundle types in Table 1. Indeed, in all cases except for that of coherent systems, precise statements and proofs of such Hitchin-Kobayashi type correspondences can be found in the references listed in Table 1. One has, for example, the result that:

Theorem 2.7 [B]. *For a generic permitted value of τ, the following are equivalent*

(A) *the holomorphic pair $(\overline{\partial}_E, \phi)$ admits a Hermitian metric satisfying the τ-Vortex -equation,*

(B) *the holomorphic pair $(\overline{\partial}_E, \phi)$ is τ-stable.*

The implication (B)$\Rightarrow$ (A) holds for all τ.

The non-generic values of τ are those for which there can be "reducible or split" solutions to the τ-Vortex equation. This will be explained further in Section 3. Notice that an immediate corollary of this theorem is

Corollary 2.8. *For generic τ, one can identify $\mathcal{V}_\tau = \mathcal{H}_\tau^s$, where $\mathcal{H}_\tau^s$ denotes the set of τ-stable holomorphic pairs in $\mathcal{H}$.*

In the case of coherent systems, one direction in the equivalence is readily established, namely:

Proposition 2.9. *Let $(\mathcal{E}, V)$ be a k-dimensional coherent system, and let τ be generic (in the sense described in section 3). Suppose that $(\mathcal{E}, V)$ admits a solution to the orthonormal τ-vortex equations. Then $(\mathcal{E}, V)$ is α-stable in the sense of Definition 1.4, with α and τ related by (2.6c).*

Sketch of Proof. Let $\{\phi_1, \phi_2, \ldots, \phi_k\}$ and H be the basis for V and the metric on E giving the solution to equations (2.6). Consider a subsystem $(\mathcal{E}', V')$ with $V' = V \cap H^0(X, \mathcal{E}')$ and $\dim(V') = k'$. By making a unitary change of basis if necessary, we can assume that $\{\phi_1, \ldots, \phi_{k'}\}$ is a basis for V'. Using the C^∞ splitting $\mathcal{E} = \mathcal{E}' \oplus \mathcal{E}/\mathcal{E}'$, we can extract from equation (2.6b) its projection onto $\mathrm{Hom}(\mathcal{E}', \mathcal{E}')$. This gives the equation

$$(2.9) \qquad i\Lambda F'_{\overline{\partial}_E, H} + B + \Sigma_{i=1}^{k'} \phi_i \otimes \phi_i^* + \Phi' = \tau \mathbf{I}' ,$$

where the term B comes from the second fundamental form for the inclusion $\mathcal{E}' \hookrightarrow \mathcal{E}$, and Φ' denotes the projection of $\Sigma_{i=k'+1}^{k} \phi_i \otimes \phi_i^*$ onto $Hom(\mathcal{E}', \mathcal{E}')$. Taking $\int_X Tr$ of this equation, and using the positivity of the terms B and Φ', yields the condition

$$(2.10) \qquad \deg(\mathcal{E}') + \Sigma_{i=1}^{k'} ||\phi_i||^2 \leq \tau \mathrm{rank}(\mathcal{E}') ,$$

with equality only in the case of a decomposable coherent system. Now by equations (2.6a) and (2.6c) we see that $\Sigma_{i=1}^{k'} ||\phi_i||^2 = \alpha k'$, and $\tau = \mu_\alpha(\mathcal{E}, V)$. Equation (2.10) thus implies that the subsystem $(\mathcal{E}', V')$ satisfies the α-stability condition. □

Suppose conversely that we are given an α-stable k- dimensional coherent system $(\mathcal{E}, V)$. To prove the converse to Proposition 2.9, we need to show that we can choose a basis $\{\phi_1, \phi_2, \ldots, \phi_k\}$ of V and a metric H on E such that the orthonormal τ-vortex equations are satisfied. One strategy for doing this is to convert the problem into one about metrics on a holomorphic triple. As discussed in Section 1.5, we can do this by viewing a basis $\{\phi_1, \ldots, \phi_k\}$ as the image of a map $\Phi : \mathcal{O}^k \longrightarrow \mathcal{E}$. The triples associated to $(\mathcal{E}, V)$ are thus of the form $(\mathcal{E}, \mathcal{O}^k, \Phi)$, with $\Phi(\mathcal{O}^k) = V$. We then use

Proposition 2.10. *Let $(\mathcal{E}, V)$ be a k-dimensional coherent system, and let $(\mathcal{E}, \mathcal{O}^k, \Phi)$ be a corresponding triple with $\Phi(\mathcal{O}^k) = V$. Then the following*

are equivalent:

(1) *One can choose a basis* $\{\phi_1, \phi_2, \ldots, \phi_k\}$ *of* V *and a metric* H *on* E *such that the orthonormal* τ*-vortex equations are satisfied,*

(2) *One can find metrics* H *and* h *on* E *and* $X \times \mathbb{C}^k$ *respectively such that the fiber metric* h *is independent of the fiber, and the following two equations are satisfied*

$$i\Lambda F_{\overline{\partial}_E, H} + \Phi\Phi^* = \tau \mathbf{I} \tag{2.11a}$$

$$\int_X (i\Lambda F_{\overline{\partial}_0, h} - \Phi^*\Phi) = \alpha \mathbf{I}_k, \tag{2.11b}$$

Here $\overline{\partial}_E$*(resp.*$\overline{\partial}_0$*) is the* $\overline{\partial}$*-operator giving the holomorphic structure on* $\mathcal{E}$*(resp.* $\mathcal{O}^k$*). Also, in (2.11b) we have identified* $\Omega^0(X, \mathrm{End}(\mathcal{O}^k)) \simeq C^\infty(X, GL(k))$ *and* $\mathbf{I}_k$ *is the identity element in* $GL(k)$*.*

Equations (2.11) are clearly related to the coupled vortex equations for triples of the form $(\mathcal{E}, \mathcal{O}^k, \Phi)$, and these latter equations can in turn be related by dimensional reduction to the Hermitian-Einstein equations on a holomorphic extension over $X \times \mathbb{P}^1$. The basic idea is then to adapt the proof of Theorem (2.6), specifically the methods of Simpson in [Si1], so that it applies not to the full Hermitian-Einstein equations, but to the modified version which corresponds to equations (2.11). We leave a complete treatment of this result to a future publication.

Appendix: Relation to Specialized Triples

It is interesting to observe that the moment maps for k-pairs, coherent systems and Higgs bundles can each be related to the moment map for triples. For example:

k-Pairs.

Let $\mathcal{H}^T(E, \mathbb{C}^k)$ be the space of triples $(\mathcal{E}_1, \mathcal{E}_2, \Phi)$ in which the underlying smooth bundles of $\mathcal{E}_1$ and $\mathcal{E}_2$ are E and the trivial rank k bundle respectively. This has a subspace $\mathcal{H}_0^T(E, \mathbb{C}^k)$ consisting of triples of the form $(\mathcal{E}, \mathcal{O}^k, \Phi)$, i.e. in which $\mathcal{E}_2 = \mathcal{O}^k$. Recall from Section 1.5 that this subspace consists of the triples which correspond to k-pairs on E. Now fix a metric k_0 on the trivial rank k bundle such that the standard frame is an orthonormal frame, and also fix a background metric K on E. Use these metrics to induce the symplectic structure on $\mathcal{H}^T(E, \mathbb{C}^k)$, and to define the unitary gauge groups $\mathfrak{G}$ for E and $\mathfrak{G}_0 = C^\infty(X, U(k))$ for the trivial bundle. The moment map for the action of $\mathfrak{G} \times \mathfrak{G}_0$ on $\mathcal{H}^T(E, \mathbb{C}^k)$ is as in Table 3, i.e. is $\Psi_T = (\Psi_1, \Psi_2)$ with $\Psi_1(\mathcal{E}, \mathcal{O}^k, \Phi) = \Lambda F_{\bar{\partial}_E, K} - i\Phi\Phi^*$ and $\Psi_2(\mathcal{E}, \mathcal{O}^k, \Phi) = \Lambda F_{\bar{\partial}_0, k_0} + i\Phi^*\Phi$. Now consider the subgroup of $\mathfrak{G} \times \mathfrak{G}_0$ given by $\mathfrak{G}_k = \mathfrak{G} \times \mathbf{I}$, where $\mathbf{I}$ is the identity element in $\mathfrak{G}_0$. The subspace $\mathcal{H}_0^T(E, \mathbb{C}^k)$ is a $\mathfrak{G}_k$-invariant set. Furthermore the moment map for the action of this subgroup is obtained from Ψ_T by projecting onto the Lie subalgebra of $\mathfrak{G}^k$, and this projection is simply projection of (Ψ_1, Ψ_2) onto its first factor. We can restrict this map to the subspace $\mathcal{H}_0^T(E, \mathbb{C}^k)$ and thus show

Proposition 2.11. *Let $\Psi_T^{(k)}$ denote the moment map for $\mathfrak{G}_k$ on $\mathcal{H}^T(E, \mathbb{C}^k)$, and let Ψ_k be the moment map for $\mathfrak{G}$ on $\mathcal{H}^k$. Let $(\mathcal{E}, \mathcal{O}^k, \Phi) \in \mathcal{H}_0^T(E, \mathbb{C}^k)$ be the triple corresponding to the k-pair $(\mathcal{E}, \phi_1, \dots, \phi_k) \in \mathcal{H}^k$. Then*

$$\Psi_T^{(k)}(\mathcal{E}, \mathcal{O}^k, \Phi) = \Psi_k(\mathcal{E}, \phi_1, \dots, \phi_k) .$$

Proof. Since the standard frame for $E_2 = X \times \mathbb{C}^k$ is orthonormal with respect to the metric k_0, we compute that $\Phi\Phi^* = \Sigma_{i=1}^k \phi_i \otimes \phi_i^*$.

Coherent Systems.

For the case of coherent systems, we need to consider the action of the subgroup $\mathfrak{G}_{CS} = \mathfrak{G} \times U(k)$ on the same space as for k-pairs, i.e. on $\mathcal{H}^T(E, \mathbb{C}^k)$. Here $U(k) \subset \mathfrak{G}_0$ is the subgroup of globally constant gauge transformations. The projection from $\mathfrak{g} \oplus \mathfrak{g}_0$ onto the Lie subalgebra of $\mathfrak{G}_{CS}$ is given by $(u, v) \mapsto (u, \int_X v)$, so the moment map for $\mathfrak{G}_{CS}$ has $\int_X(\Psi_2)$ as its second component. When evaluated on a triple of the form $(\mathcal{E}, \mathcal{O}^k, \Phi)$ we get

$$\Psi_T^{(CS)}(\mathcal{E}, \mathcal{O}^k, \Phi) = (\Lambda F_{\bar{\partial}_E, K} - i\Sigma_{i=1}^k \phi_i \otimes \phi_i^* \ , \ i < \phi_i, \phi_j >) \ , \tag{2.13}$$

and hence

Proposition 2.12. *Let $\Psi_T^{(CS)}$ denote the moment map for $\mathfrak{G}_{CS}$ on $\mathcal{H}^T(E, \mathbb{C}^k)$, and let Ψ be the moment map in equation (2.7) for the $\mathfrak{G} \times U(k)$-action on*

$\mathcal{H}^k$. Let $(\mathcal{E}, \mathcal{O}^k, \Phi) \in \mathcal{H}^T(E, \mathbb{C}^k)$ be the triple corresponding to the k-Pair $(\mathcal{E}, \phi_1, \ldots, \phi_k) \in \mathcal{H}^k$. Then

$$\Psi_T^{(CS)}(\mathcal{E}, \mathcal{O}^k, \Phi) = \Psi(\mathcal{E}, \phi_1, \ldots, \phi_k) \, .$$

Suppose further that $\mathcal{H}^{CS}$ is given the symplectic structure coming from $\Psi_U^{-1}(-i\alpha\mathbf{I})/U(k)$, and that Ψ_{CS} and q are as in Propositions 2.3 and 2.5. Let $(\mathcal{E}, V)$ be a coherent system in $\mathcal{H}^{CS}$, and let $(\mathcal{E}, \mathcal{O}^k, \Phi)$ be the triple corresponding to a k-pair in $q^{-1}(\mathcal{E}, V)$. Then

$$\Psi_{CS}(\mathcal{E}, V) = \Psi_T^{(CS)}(\mathcal{E}, \mathcal{O}^k, \Phi) \, .$$

Higgs Bundles.

To understand Higgs bundles from this point of view, we must consider the space of triples in which $\mathcal{E}_1 = \mathcal{E} \otimes K$ and $\mathcal{E}_2 = \mathcal{E}$. Fix a metric on the underlying smooth bundle E, and let k be the metric on K such that $\Lambda F_k = constant$. Let $\mathfrak{G}_1$ and $\mathfrak{G}$ be the unitary gauge groups of $E \otimes K$ and E respectively. The relevant subgroup of $\mathfrak{G}_1 \times \mathfrak{G}$ is the copy of $\mathfrak{G}$ "diagonally" embedded by $g \mapsto (g \otimes \mathbf{I}, g)$. At the Lie algebra level, the projection map will take (u, v) to $\frac{u+v}{2}$, and it can be shown that the moment map for this subgroup, evaluated on a triple $(\mathcal{E} \otimes K, \mathcal{E}, \Theta)$, corresponds to the moment map for $\mathfrak{G}$, evaluated on the Higgs bundle $(\mathcal{E}, \Theta)$.

3. The Stability Parameters

The parameter involved in the definition of stability for augmented bundles can be interpreted in several ways. These different interpretations lead, as we shall see in Section 4, to different constructions of the corresponding moduli spaces. On the one hand the parameter appears naturally in the equations corresponding to the various augmented bundles. In contrast with the parameter appearing in the Hermitian-Einstein equation or Hitchin self-duality equations (see Section 2.1), this parameter is not fixed by the topology, and one can actually solve the equations for different values of the parameter. The parameter can also be interpreted from the point of view of Geometric Invariant Theory: When trying to construct an algebraic moduli space parametrizing equivalence classes of augmented bundles, due to the additional structure, one has some freedom in the choice of a linearization necessary to perform the GIT quotient. This freedom —which does not exist in the case of the moduli space of semistable bundles—is precisely the choice of a one-parameter family of linearizations. A third interpretation, at least for pairs and triples, is related to the correspondence between these objects and certain $SU(2)$-equivariant vector bundles on $X \times P^1$ (see Sections 1.3 and 4.3). The parameter can be encoded in the choice of a Kähler metric on $X \times P^1$, in such a way that the τ-stability of the augmented bundle is equivalent to the slope stability of the equivariant bundle with respect to the corresponding Kähler polarization.

§3.1 *Upper and Lower Bounds*

Although the parameter that appears in the definition of stability for the augmented bundles can be in principle any real number, it turns out that the stability condition forces it to be bounded from below and in most cases also bounded from above. In other words, in most cases no stable objects exist for values of the parameter outside of a certain interval. In this section we show how to obtain these bounds. They are always expressible in terms of the numerical invariants of the augmented bundle (ranks, degrees, etc.).

Proposition 3.1 *Let $(\mathcal{E};\phi_1,\dots,\phi_k)$ be a τ-semistable k-pair of rank R and degree d, then*

(1) $\tau \geq \mu(\mathcal{E}) = \frac{d}{R}$.

(2) *If $k < R$, then* $\tau \leq \frac{d}{R-k}$.

(3) *If $k \geq R$, then in general τ is unbounded from above. Moreover, if $\tau > d$, then all τ-semistable k-pairs are τ-stable, and a k-pair is τ-stable if and only if the sections generically generate the fiber of $\mathcal{E}$.*

If $(\mathcal{E};\phi_1,\dots,\phi_k)$ is in fact τ-stable, then the inequalities in (1) *and* (2) *are strict.*

Proof. The lower bound follows from applying (1) in Definition 1.2a to the subobject $(\mathcal{E},0)$. If $k < R$, the subsheaf $\mathcal{E}'$ generated by the sections has of course rank less than or equal to k and if $k < R$, then $\mathcal{E}'$ is a proper subsheaf of $\mathcal{E}$ and we can apply condition (2) in Definition 1.2a to $(\mathcal{E}', \phi_1, ..., \phi_k)$ to obtain the upper bound for τ. For the proof of (4) we refer to [BeDW]. □

Corollary 3.2 *For $k < R$, if $(\mathcal{E}; \phi_1, ..., \phi_k)$ is τ-semistable (τ-stable) for some value of τ, then $d \geq 0$ ($d > 0$).*

Similarly one can prove

Proposition 3.3 *Let $(\mathcal{E}, V)$ be an α-semistable coherent system of rank R, degree d and dimension k, then*

(1) $\alpha \geq 0$.

(2) *If $k < R$, then* $\alpha \leq \frac{d}{R-k}$.

If $(\mathcal{E}, V)$ is in fact α-stable, then the above inequalities are strict.

Corollary 3.4 *For $k < R$, if $(\mathcal{E}, V)$ is α-semistable (α-stable) for some value of α, then $d \geq 0$ ($d > 0$).*

Proposition 3.5 *Let $(\mathcal{E}_1, \mathcal{E}_2, \Phi)$ be a τ-semistable triple, then*

(1) $\tau \geq \mu(\mathcal{E}_1)$.

(2) *If $R_1 \neq R_2$, then* $\tau \leq \mu(\mathcal{E}_1) + \frac{R_2}{|R_1 - R_2|}(\mu(\mathcal{E}_1) - \mu(\mathcal{E}_2))$.

Equivalently, if the triple is α-semistable, then

(1') $\alpha \geq 0$.

(2') *If $R_1 \neq R_2$, then* $\alpha \leq (1 + \frac{R_1 + R_2}{|R_1 - R_2|})(\mu(\mathcal{E}_1) - \mu(\mathcal{E}_2))$.

Proof. (1) follows from applying the stability condition to the subtriple $T' = (\mathcal{E}_1, 0, 0)$, and (2) from applying it to the subtriples

$$T_1 = (0, \operatorname{Ker}\Phi, \Phi) \quad \text{and} \quad T_2 = (\operatorname{Im}\Phi, \mathcal{E}_2, \Phi).$$

(See [BGP] for details). □

Combining the lower and upper bounds on τ (or α) we can deduce the following.

Corollary 3.6 *If* $\operatorname{rank}(\mathcal{E}_1)$ *and* $\operatorname{rank}(\mathcal{E}_2)$ *are unequal, then a triple $(\mathcal{E}_1, \mathcal{E}_2, \Phi)$ cannot be stable unless $\mu(\mathcal{E}_2) < \mu(\mathcal{E}_1)$.*

Furthermore,

Proposition 3.7 *Let $(\mathcal{E}_1, \mathcal{E}_2, \Phi)$ be τ-stable and suppose that $R_1 = R_2$. If Φ is not an isomorphism, then $d_1 > d_2$. In particular, in any τ-stable triple $(\mathcal{E}_1, \mathcal{E}_2, \Phi)$, the bundle map Φ is an isomorphism if and only if $R_1 = R_2$ and $d_1 = d_2$.*

Proof. The fact that $d_1 > d_2$ follows from the inequality

$$(R_1 - R_2)\tau < d_1 - d_2 ,$$

which applies if Φ is not an isomorphism (cf. [BGP]). In particular, if Φ is not an isomorphism then $d_1 \neq d_2$. Conversely, if Φ is an isomorphism, then clearly $R_1 = R_2$ and $d_1 = d_2$. □

When Φ is an isomorphism the range for τ can fail to be bounded. One has for example the following result (cf. [BGP]).

Proposition 3.8 *Suppose $\mathcal{E}_1 \cong \mathcal{E}_2$. Then for any $\tau > \mu(\mathcal{E}_1)$, the triple $(\mathcal{E}_1, \mathcal{E}_2, \Phi)$ is τ-stable if and only if Φ is an isomorphism and $\mathcal{E}_1$ is stable.*

§3.2 Critical values

In principle the parameter involved in the definition of stability is a continuously varying parameter. However, the stability properties of a given augmented bundle do not likewise vary continuously, but can change only at certain rational values of the parameter, that we shall call *critical values.* This is due to the fact that, except for the parameter itself, all numerical quantities in the definition of stability are rational numbers with bounded denominators. In the case of k-pairs this has the additional consequence that for a generic (i.e. non-critical) value of the parameter there is no distinction between stability and semistability. This is in contrast to the case of pure bundles, where the notions of stability and semistability coincide only when the rank and degree of the bundle are coprime. We shall show in the next propositions that for k-coherent systems with $k > 1$ as well as for triples both the value of the parameter and the greatest common divisor of the rank and degree are relevant.

Proposition 3.9 *The critical values of τ for k-pairs of rank R and degree d are the rational numbers whose denominator is less than or equal to R. Moreover if τ is not a critical value then all τ-semistable k-pairs are τ-stable.*

Proof. It is clear that only for critical values of τ can we have a subbundle $\mathcal{E}' \subseteq \mathcal{E}$ satisfying $\mu(\mathcal{E}') = \tau$ or $\mu(\mathcal{E}/\mathcal{E}') = \tau$. □

Proposition 3.10 *Let $(\mathcal{E}, V)$ be a k-coherent system of rank R and degree d, and let $(\mathcal{E}', V')$ be a subsystem such that $\mu_\alpha(\mathcal{E}', V') = \mu_\alpha(\mathcal{E}, V)$. Then either*

$$Rk' = R'k \quad \textit{and} \quad \mu(\mathcal{E}') = \mu(\mathcal{E}),$$

or

$$\alpha = \frac{R'd - Rd'}{Rk' - R'k}.$$

In particular, if R and d are coprime, or R and k are coprime, and α is not a rational number with denominator of magnitude less than or equal to Rk, then all α-semistable coherent systems are α-stable.

Proposition 3.11 *Let $T = (\mathcal{E}_1, \mathcal{E}_2, \Phi)$ be a τ-semistable triple, and let $T' = (\mathcal{E}'_1, \mathcal{E}'_2, \Phi')$ be a subtriple such that $\theta_\tau(T') = 0$. Then either*

$$R_1 R'_2 = R_2 R'_1 \quad and \quad \mu(\mathcal{E}'_1 \oplus \mathcal{E}'_2) = \mu(\mathcal{E}_1 \oplus \mathcal{E}_2),$$

or

$$\frac{R_2(R'_1 + R'_2)\mu(\mathcal{E}'_1 \oplus \mathcal{E}'_2) - R'_2(R_1 + R_2)\mu(\mathcal{E}_1 \oplus \mathcal{E}_2)}{R_2 R'_1 - R_1 R'_2} = \tau.$$

In particular, if $R_1 + R_2$ and $d_1 + d_2$ are coprime, and τ is not a rational number with denominator of magnitude less than $R_1 R_2$, then all τ-semistable triples are τ-stable.

The number of critical values inside the range is clearly finite if this is bounded above. Although for k-pairs and k-coherent systems, in the case where $k > R$, and for triples when $R_1 = R_2$, there is no upper bound for the parameter, the following Propositions show that the number of critical values is finite also in this case.

Proposition 3.12 *Suppose that $k > R$. Then there are no critical values for the τ-stability of a k-Pair for τ in the range (d, ∞).*

Proof. This follows from (3) of Proposition 3.1, which shows that τ-stability and τ-semistability coincide for τ in the range (d, ∞). □

Proposition 3.13 *Suppose that $k > R$. Then there is some finite α_0 such that α-semistability and α-stability coincide for a coherent system $(\mathcal{E}, V)$ whenever $\alpha \geq \alpha_0$ (and $(R, d) = 1$ or $(R, k) = 1$). In particular, when $k > R$ there are no critical values for α in the range (α_0, ∞).*

Proof. Suppose that $(\mathcal{E}, V)$ is α-semistable and that $\alpha > \frac{d(R-1)}{k}$.(With $\tau = \mu_\alpha(\mathcal{E}, V)$ this corresponds to $\tau > d$). If $(\mathcal{E}, V)$ is not stable then there is a subsystem $(\mathcal{E}', V')$ with $\mu_\alpha(\mathcal{E}', V') = \mu_\alpha(\mathcal{E}, V)$. It follows that either $Rd' - R'd = 0$ and $Rk' - R'k = 0$, or that we must have

$$\alpha = \frac{R'd - Rd'}{Rk' - R'k} .$$

However, by considering any k-pair obtained from $(\mathcal{E}, V)$, we can conclude (from the τ-stability of the k-pair) that $\mathcal{E}$ must be generically generated by global sections in V. In particular, V cannot lie completely in $H^0(X, \mathcal{E}')$ and the quotient $\mathcal{E}/\mathcal{E}'$ must be generically generated by the projections of sections in V. Thus (as in the previous Proposition), $d' \leq d$. It follows that

$$|\alpha| \leq d \frac{R + R'}{|Rk' - R'k|} \leq 2Rd .$$

Thus the result applies for any $\alpha_0 > 2Rd$. □

Remark. In fact, since coherent systems arise as U(k)-symplectic reduction of k-pairs, we expect that there should be no further critical points after $\alpha = \frac{d(R-1)}{k}$.

Similarly, for triples $(\mathcal{E}_1, \mathcal{E}_2, \Phi)$ in which $R_1 = R_2$ and $d_1 = d_2$, the range for τ is a semi-infinite interval, but, at least when the rank and degree are coprime, we have

Proposition 3.14 *Let $(\mathcal{E}_1, \mathcal{E}_2, \Phi)$ be a triple with $R_1 = R_2 = R$, $d_1 = d_2 = d$, and $(R, d) = 1$. If $\tau > \frac{d}{R}$, then τ-stability is equivalent to τ-semistability. In particular, there are no critical values of τ in the range $(d/R, \infty)$.*

Proof. Suppose that $\tau > d/R$. We first show that if Φ is not an isomorphism, then $(\mathcal{E}_1, \mathcal{E}_2, \Phi)$ cannot be τ-semistable. Indeed if Φ is not an isomorphism, then the kernel of Φ, K, is a non-trivial subbundle of $\mathcal{E}_2$ and the image I has rank strictly less than R. By applying the τ-semistability condition to the proper subtriples $(0, K, \Phi)$ and $(I, \mathcal{E}_2, \Phi)$, and using the fact that

$$\mu(K) \geq \frac{d_2 - \deg(I)}{R_2 - \operatorname{rank}(I)} = \mu(\mathcal{E}_1/I) ,$$

we thus get

$$\frac{d}{R} = \mu(\mathcal{E}_1) \leq \tau \leq \mu(\mathcal{E}_1/I) \leq \mu(K) \leq \tau' \leq \mu(\mathcal{E}_2) = \frac{d}{R} ,$$

where here τ' is related to τ by $R(\tau + \tau') = 2d$. Suppose then that $(\mathcal{E}_1, \mathcal{E}_2, \Phi)$ is τ-semistable with $R_1 = R_2 = R$ and Φ an isomorphism. Since Φ is an isomorphism, we can find subtriples $(\mathcal{E}_1', \mathcal{E}_2', \Phi')$ with $\mathcal{E}_1' \simeq \mathcal{E}_2' \simeq \mathcal{E}'$ for any subbundle $\mathcal{E}' \subset \mathcal{E}$. For these we get $\theta_\tau(\mathcal{E}_1', \mathcal{E}_2', \Phi') = (\mu(\mathcal{E}') - \mu(\mathcal{E}))$. The τ-semistability of the triple thus implies the semistability (and hence stability) of $\mathcal{E}$. The proposition now follows from Proposition 3.8. □

Summarising, we see that the set of critical values divides the range for the parameter into a finite number of subintervals, such that for values of the parameter in the interior of any of these subintervals stability and semistability coincide—provided that some extra coprimality condition is satisfied in the case of coherent systems and triples. We have then a finite number of essentially different stability conditions, and hence a finite number of moduli spaces.

§3.3 Extremal Values

One of the main questions at this point is what is the relation between the stability of the augmented bundle and the stability of the bundle itself (or bundles). It turns out that they are closely related precisely when the parameter is "small", that is when it lies in the interval between the lower bound and the next critical value. This fact will allow us, when studying in Section 4 the moduli spaces of augmented bundles for different values of the parameter, to define a map from the moduli space of augmented bundles for small value of the parameter to the moduli space of semistable bundles. These maps are the higher rank generalizations of the classical Abel-Jacobi map from the space of effective divisors—the moduli space of pairs when $\mathcal{E}$ is a line bundle—to the Jacobian of the Riemann surface.

First we consider the case in which the parameter is equal to the lower bound. It follows immediately from the definition of τ-semistability that

Proposition 3.15 *Let $(\mathcal{E};\phi_1,\dots,\phi_k)$ be a k-pair of rank R and degree d, and let $\tau = d/R$. Then $(\mathcal{E};\phi_1,\dots,\phi_k)$ is τ-semistable if and only if $\mathcal{E}$ is semistable. Moreover, for such a value of τ there are no τ-stablek-pairs.*

One has similar results for coherent systems and triples. For the "small" range of the parameter one has the following.

Proposition 3.16 [BD1],[BDW] *Let $(\mathcal{E};\phi_1,\dots,\phi_k)$ be a non-degenerate k-pair. Let τ_1 be the first critical value after $\mu(\mathcal{E})$. Then for $\mu(\mathcal{E}) < \tau < \tau_1$*

(1) *If $(\mathcal{E};\phi_1,\dots,\phi_k)$ is a τ-stable k-pair, then $\mathcal{E}$ is a semistable bundle.*
(2) *Conversely, if $\mathcal{E}$ is stable, then $(\mathcal{E};\phi_1,\dots,\phi_k)$ will be a τ-stable k-pair for any choice of $\phi_i \in H^0(\mathcal{E})$.*

Proposition 3.17 [KN,RV] *Let $(\mathcal{E},V)$ be a non-degenerate k-coherent system. Let α_1 be the first critical value after $\mu(\mathcal{E})$. Then for $0 < \alpha < \alpha_1$*

(1) *If $(\mathcal{E},V)$ is a α-stable k-coherent system, then $\mathcal{E}$ is a semistable bundle.*
(2) *Conversely, if $\mathcal{E}$ is stable, then $(\mathcal{E},V)$ will be a α-stable k-coherent system for any choice of k-linear subspace $V \subset H^0(\mathcal{E})$.*

Proposition 3.18 [BGP] *Let $(\mathcal{E}_1,\mathcal{E}_2,\Phi)$ be a non-degenerate holomorphic triple. Let τ_1 be the first crititical value after $\mu(\mathcal{E}_1)$. Then for $\mu(\mathcal{E}_1) < \tau < \tau_1$*

(1) *If $(\mathcal{E}_1,\mathcal{E}_2,\Phi)$ is a τ-stable triple, then both $\mathcal{E}_1$ and $\mathcal{E}_2$ are semistable bundles*
(2) *Conversely, if $\mathcal{E}_1$ and $\mathcal{E}_2$ are stable bundles, then $(\mathcal{E}_1,\mathcal{E}_2,\Phi)$ will be a τ-stab triple for any choice of $\Phi \in H^0(\mathrm{Hom}(\mathcal{E}_2,\mathcal{E}_1))$.*

The analysis of the situation when the parameter is equal to the upper bound—when this exists—or lies in the "large" range (the open interval between the upper bound and the immediate smaller critical value) is a little bit more involved. We consider here only the case of pairs (cf. [BDW], [T]). Similar results should hold for the other augmented bundles.

Proposition 3.19 *Let $(\mathcal{E},\phi)$ be a holomorphic pair of rank R and degree d, and let $\tau = \frac{d}{R-1}$. Then $(\mathcal{E},\phi)$ is τ-semistable if and only if $\mathcal{E}$ splits as $\mathcal{E} = \mathcal{O} \oplus \mathcal{E}_s$, where $\mathcal{E}_s$ is a semistable bundle of degree d and rank $R-1$, and ϕ is a (constant) section of $\mathcal{O}$.*

Proposition 3.20 *Let $(\mathcal{E}, \phi)$ be a holomorphic pair of rank R and degree d, and let $\tau \in (\tau_N, \frac{d}{R-1})$ (where τ_N is the biggest critical value before $\frac{d}{R-1}$). Then $(\mathcal{E}, \phi)$ is τ-stable if and only if $\mathcal{E}$ is a non trivial extension of the form*

$$0 \longrightarrow \mathcal{O} \longrightarrow \mathcal{E} \longrightarrow \mathcal{E}_s \longrightarrow 0,$$

where $\mathcal{E}_s$ is a semistable bundle of rank $R-1$ and degree d.

When there is no upper bound, we have seen above that there are no more critical values after a certain finite value and the stability condition "stabilizes". The description of what happens in this situation, for example in the case of k-pairs, is given by (3) in Proposition 3.1.

4. Moduli spaces

Having given the definitions of stability, we can now consider the construction of the corresponding moduli spaces for the various augmented bundles discussed in the previous sections. There are many possible approaches, but in keeping with the rest of this paper we shall concentrate mainly on those that come from differential and symplectic geometry. In particular we shall not describe Geometric Invariant Theory constructions. Such algebraic constructions apply for rational values of the parameters in the definitions of stability, and yield projective (or quasi projective) varieties. These can be found in [Be], [T] (for pairs), [HL] for (k-pairs), and [KN], [LeP1,2],[RV] (for coherent systems).

We will describe three different moduli space constructions. The first method is quite general (it generalizes, for example, to augmented bundles over arbitrary Kähler manifolds and works for all values of the parameters), and yields moduli spaces which admit the structure of complex analytic spaces with compatible Kähler structures away from the singularities. The method is by now standard and uses the realization of these spaces on the one hand as complex quotients and on the other hand as symplectic quotients (Marsden-Weinstein reductions). Of course, one must show that the two structures are compatible. We will illustrate these techniques in the case of the moduli spaces of k-pairs, but there is no reason why they cannot be applied in other cases.

The second construction we will describe is more specific, and gives an interesting relation between the moduli spaces for k-pairs and those for k-dimensional coherent systems.

Finally, we will describe a construction which is based on the technique of dimensional reduction. In particular we will outline how this can be used to construct the moduli spaces of stable triples as fixed point sets within larger moduli spaces of stable bundles.

While the role of the parameters is slightly different in each of the methods discussed - reflecting the various interpretations that can be given to these (cf. the Introduction to section 3)- the end result is the same. One obtains families of distinct moduli spaces. In the next Section we will discuss the nature of the dependence of these spaces on the parameters.

§4.1 *k-pairs*

Most of what follows is contained in the references [BD1], [BDW], and [BeDW] to which we refer for more details. For the sake of simplicity we assume first that $k=1$. Keeping the notation from Section 2, we start with the construction for the complex quotient of the space of holomorphic pairs, $\mathcal{H}$, on a bundle E of degree d and rank R. To do this we have to construct slices for the action of the complex gauge group $\mathfrak{G}_{\mathbb{C}}$. This can be done at all points of $\mathcal{H}$ which have trivial isotropy group, and this leads to the notion of **simple pairs** generalizing the notion of simple bundles. More precisely, consider the elliptic complex

$$(C_\phi^{\overline{\partial}_E}) \qquad 0 \longrightarrow \Omega^0(\mathrm{End}(E)) \xrightarrow{d_1} \Omega^{0,1}(\mathrm{End}(E)) \oplus \Omega^0(E) \xrightarrow{d_2} \Omega^{0,1}(E) \longrightarrow 0$$

where

$$d_1(u) = (-\overline{\partial}_E u, u\phi)$$
$$d_2(\alpha, \eta) = \overline{\partial}_E(\eta) + \alpha\phi\ .$$

This complex was first introduced in [BD1]. It can easily be checked that in all cases $H^2(C_\phi^{\overline{\partial}_E}) = 0$ (cf. [BDW], Corollary 2.7). We say that the pair $(\overline{\partial}_E, \phi)$ is **simple** if it also satisfies $H^0(C_\phi^{\overline{\partial}_E}) = 0$. Then the standard deformation theory implies

Theorem 4.1. (see [BD1], Corollary 2.9) *Let $\mathcal{H}^\sigma$ denote the subspace consisting of simple pairs. Then $\mathcal{H}^\sigma/\mathfrak{G}_{\mathbb{C}}$ is a complex manifold (possibly non-Hausdorff) of complex dimension $d + (R^2 - R)(g-1)$. Moreover:*

$$T_{[\overline{\partial}_E,\phi]}(\mathcal{H}^\sigma/\mathfrak{G}_{\mathbb{C}}) \simeq H^1(C_\phi^{\overline{\partial}_E})\ .$$

In order to put Kähler structures on our moduli spaces we have to realize them as symplectic quotients. As indicated in Section 2.2, the standard Kähler form on $\mathcal{H} \subset \mathcal{C} \times \Omega^0(E)$ is preserved by the action of the real gauge group $\mathfrak{G} \subset \mathfrak{G}_{\mathbb{C}}$, and has a $\mathfrak{G}$-equivariant moment map given (see Table 3) by

$$(4.1) \qquad \Psi_1(\overline{\partial}_E, \phi) = \Lambda F_{\overline{\partial}_E,H} - i\phi \otimes \phi^*$$

For non-critical values of τ, let

$$\mathcal{B}_\tau(d,R) = \Psi_1^{-1}(-i\tau\mathbf{I})/\mathfrak{G}\ ,$$

denote the symplectic quotient. It follows from Uhlenbeck's weak compactness theorem that $\mathcal{B}_\tau(d,R)$ is compact and Hausdorff (cf. [BD1], §5).

For generic τ, we have $\Psi_1^{-1}(-i\tau\mathbf{I})/\mathfrak{G} = \mathcal{V}_\tau/\mathfrak{G}_{\mathbb{C}}$ (cf. section 2), and thus these two quotient constructions give complementary descriptions of the same object. Furthermore, by Theorem 2.7, $\mathcal{V}_\tau/\mathfrak{G}_{\mathbb{C}}$ is homeomorphic to $\mathcal{H}^s_\tau/\mathfrak{G}_{\mathbb{C}}$, where $\mathcal{H}^s_\tau$ denotes the subspace of τ-stable pairs. It is not difficult to show that τ-stable implies simple (cf.[BD1]), and it follows that

Theorem 4.2. (see [BD1], Theorem 5.5) *With respect to the complex and symplectic structures defined above, the moduli space $\mathcal{B}_\tau(d,R)$ of τ-stable pairs on E is, for non-critical values of τ, a compact, Hausdorff, Kähler manifold of dimension $d+(R^2-R)(g-1)$.*

Furthermore, by applying Siu's criterion for Moishezon manifolds (see [Siu]), we obtain

Theorem 4.3. (see [BDW], Theorem 6.3) *For all non-critical values of τ, $\mathcal{B}_\tau$ is a non-singular projective algebraic variety.*

If $k>1$, then it is no longer automatic that $H^2(C_\phi^{\bar\partial_E})=0$. For this we need to make the following

Assumption 4.4. *Assume that $d>R(2g-2)$. The parameter τ is called* **admissible** *if*

$$\frac{d}{R}<\tau<\frac{d-(2g-2)}{R-1}\ .$$

Under this assumption, it can be shown (see [BeDW], Lemma 3.8) that τ-stability of the k-pair $(\mathcal{E};\phi_1,\phi_2,\ldots,\phi_k)$ implies $H^1(E)=0$, from which $H^2(C_\phi^{\bar\partial_E})=0$ follows. Then Theorems 4.1 - 4.3 above generalize without difficulty. We thus have

Theorem 4.5. (see [BeDW], Theorem 3.20) *For all generic and admissible values of τ, the moduli space $\mathcal{B}_\tau^k(d,R)$ of τ-stable k-pairs on E is a non-singular projective variety of dimension $kd-R(k-R)(g-1)$.*

As discussed in Section 3, an interesting natural question is the dependence of the holomorphic, Kähler, and topological structure of the spaces $\mathcal{B}_\tau(d,R)$ and $\mathcal{B}_\tau^k(d,R)$ on the parameter τ. We will have more to say about this in Section 5, however for the moment notice that Proposition 3.16 indicates the existence of a map for τ near $\mu(E)=d/R$

$$\mathcal{B}_\tau^k(d,R)\longrightarrow\mathcal{M}(d,R)\ , \tag{4.2}$$

where $\mathcal{M}(d, R)$ denotes the moduli space of semistable bundles of rank R and degree d, given by projection onto the holomorphic structure $\overline{\partial}_E$. The map (4.2) indeed exists as a morphism of algebraic varieties, and in the case where d is coprime to R, $\mathcal{B}_\tau^k(d, R)$ is the projectivization of the sum of k copies of the push forward of the universal bundle on $X \times \mathcal{M}(d, R)$ by projection onto the second factor, and (4.2) is the bundle projection. (see [BD1, Theorem 6.4], [BeDW, Propositions 3.25 and 3.26], and also the discussion below). For the case $k = 1$, one might regard the map (4.2) as a higher rank version of the Abel-Jacobi map.

Similarly, if d is coprime to $R - 1$, then the space $\mathcal{B}_\tau(d, R)$ for τ near $d/R - 1$ may be identified with the projectivization of a vector bundle over $\mathcal{M}(d, R-1)$. In this case, the bundle $\mathcal{E}$ corresponds to a non-trivial extension of a stable bundle $\mathcal{F}$ of rank $R - 1$ by the trivial line bundle (see [BDW], Corollary 6.5).

When there is no upper bound – for example, when $k > R$ and $\tau > d > R(2g - 2)$ (see Proposition 3.1 (3)) – it is shown in [BeDW] that the moduli space of τ-stable k-pairs has the structure of a projective variety which compactifies the moduli space of holomorphic degree d maps from X to the Grassmannian of complex R-planes in $\mathbb{C}^k$. This result is not obtained from the gauge theoretic discussion presented here, however, due to the singularities (see Assumption 4.4), but rather from a construction of Grothendieck.

A second important property of the spaces $\mathcal{B}_\tau(d, R)$ and $\mathcal{B}_\tau^k(d, R)$ is that they are fine moduli spaces. What is needed to prove this is a construction of universal objects. Again we will outline the construction for $k = 1$, the generalization to k-pairs being straightforward. Let

$$pr_1 : X \times \mathcal{C} \times \Omega^0(E) \longrightarrow X$$

denote the projection onto the first factor. Let $\tilde{U}_\tau = pr_1^*(E)$ be the pull-back bundle with its tautological holomorphic structure (cf. [AB]), and let $\tilde{\Phi}$ denote the tautological section, which is evidently holomorphic (see [BeDW , Section 3.2]). By restricting to $X \times \mathcal{H}_\tau^\sigma$, it follows that $\mathfrak{G}_\mathbb{C}$ acts freely on both the base and on $\tilde{U}_\tau$ and that $\tilde{\Phi}$ is equivariant with respect to this action. It follows that $\tilde{U}_\tau$ descends to a vector bundle $U_\tau(d, R)$ on $X \times \mathcal{B}_\tau(d, R)$, and $\tilde{\Phi}$ descends to a holomorphic section Φ of U_τ. These form a universal pair in the following sense:

Theorem 4.6. (see [BeDW], Proposition 3.30) *For non-critical values of τ, there is a pair $(U_\tau(d,R),\Phi)$ on $X \times \mathcal{B}_\tau(d,R)$ satisfying the property that*

$$(U_\tau(d,R),\Phi)\big|_{X\times(\bar\partial_E,\phi)} \simeq (\mathcal{E},\phi)\,.$$

A similar statement holds for k-pairs.

§4.2 Coherent Sytems

Definition 4.7. *Given a fixed bundle $E \longrightarrow X$, and parameter α, let $\mathcal{B}^{CS}_\alpha(k)$ denote the space of isomorphism classes of α-stable k-dimensional coherent systems on E.*

In this section we will exhibit the symplectic structure on $\mathcal{B}^{CS}_\alpha(k)$, and show how the relationship (described in Section 2.2) between coherent systems and k-pairs can be used to describe $\mathcal{B}^{CS}_\alpha(k)$ as a complex analytic space with a Kähler structure.

Recall that with moment maps $\Psi_k : \mathcal{H}^k \longrightarrow \mathfrak{g}^*$ and $\Psi_{CS} : \mathcal{H}^{CS} \longrightarrow \mathfrak{g}^*$ as in Section 2.2, we have $\mathcal{B}^k_\tau = \Psi_k^{-1}(-i\tau\mathbf{I})/\mathfrak{G}$ and $\mathcal{B}^{CS}_\alpha(k) = \Psi_{CS}^{-1}(-i\tau\mathbf{I})/\mathfrak{G}$, where $\tau = \mu(E) + \alpha\frac{k}{\mathrm{rank}(E)}$. Recall also that we have a moment map

$$\Psi : \mathcal{H}^k \longrightarrow \mathfrak{g}^* \oplus \mathfrak{u}(k)^*$$

given by $\Psi = (\Psi_k, \Psi_U)$ (cf. equation (2.7)) for the action of the product $\mathfrak{G} \times U(k)$. It follows from the definitions that

Proposition 4.8. *Let α and τ be related by (2.6c). Then*

$$\Psi_{CS}^{-1}(-i\tau\mathbf{I})/\mathfrak{G} = \Psi^{-1}(-i\tau\mathbf{I}, -i\alpha\mathbf{I})/\mathfrak{G} \times U(k)\,.$$

The reduction by the product $\mathfrak{G} \times U(k)$ can be done in stages in two different ways. To obtain the description in (2.3) we first reduce by the $U(k)$ action to get $\mathcal{H}^{CS} = \Psi_U^{-1}(\lambda)/U(k)$, and then take a symplectic quotient by the action of the gauge group $\mathfrak{G}$. On the other hand, starting with the action of the gauge group, the first reduction produces $\mathcal{B}^k_\tau$ as the quotient $\Psi_k^{-1}(-i\tau\mathbf{I})/\mathfrak{G}$. The residual U(k) action descends to this quotient, where it acts symplectically with respect to the induced symplectic form. Let

$$\overline{\Psi}_U : \mathcal{B}^k_\tau \longrightarrow \mathfrak{u}^*(k)$$

denote the moment map for this action on $\mathcal{B}^k_\tau$. This is evaluated at a point $[\mathcal{E};\phi_1,\phi_2,\ldots,\phi_k]$ in $\Psi_k^{-1}(-i\tau\mathbf{I})/\mathfrak{G}$ by

$$\overline{\Psi}_U([\mathcal{E};\phi_1,\phi_2,\ldots,\phi_k]) = \Psi_U(\mathcal{E};\phi_1,\phi_2,\ldots,\phi_k)\,,$$

where $\Psi_k(\mathcal{E};\phi_1,\phi_2,\ldots,\phi_k) = -i\tau\mathbf{I}$. We thus find

Corollary 4.9. *Fix $\tau \in \mathbb{R}$ and take α such that the relation (2.6c) is satisfied. Then*

$$\mathcal{B}_\alpha^{CS}(k) = \overline{\Psi}_U^{-1}(-i\alpha\mathbf{I})/U(k)\ .$$

That is, $\mathcal{B}_\alpha^{CS}(k)$ is a symplectic reduction of $\mathcal{B}_\tau^k$ by the action of the unitary group.

By standard methods (see for example [Ki]) it follows that the reduced space $\overline{\Psi}_U^{-1}(-i\alpha\mathbf{I})/U(k)$ has the structure of a complex analytic space with a symplectic structure away from the singularities, and hence

Theorem 4.10. *For admissible α (i.e. corresponding to admissible τ) the moduli space $\mathcal{B}_\alpha^{CS}(k)$ of α-stable coherent systems on X is a complex analytic space with a Kähler structure away from the singularities.*

§4.3 Moduli space of stable triples

In this section we shall sketch a construction for the moduli spaces of triples based on the fact that a triple can be regarded as a "dimensional reduction" of a certain $SU(2)$-equivariant vector bundle over $X \times \mathbb{P}^1$. More precisely, as shown in Section 1.3, one can associate to a triple $(\mathcal{E}_1, \mathcal{E}_2, \Phi)$ a holomorphic bundle over $X \times \mathbb{P}^1$ of the form

$$0 \longrightarrow p^*\mathcal{E}_1 \longrightarrow \mathcal{F} \longrightarrow p^*\mathcal{E}_2 \otimes q^*\mathcal{O}(2) \longrightarrow 0\ . \tag{4.3}$$

Consider the action of $SU(2)$ on $X \times \mathbb{P}^1$, given by the trivial action on X and the standard one on $\mathbb{P}^1$. This action can be lifted to the trivial action on $p^*\mathcal{E}_1$ and $p^*\mathcal{E}_2$ and the standard one on $q^*\mathcal{O}(2)$. On the other hand $SU(2)$ acts trivially on the extension class of (4.3)—since this is essentially Φ—and hence we can lift the action of $SU(2)$ to a holomorphic action on the whole $\mathcal{F}$, which becomes in this way an $SU(2)$-equivariant vector bundle.

This construction allows one to interpret the stability of $(\mathcal{E}_1, \mathcal{E}_2, \Phi)$ in terms of the Mumford–Takemoto slope-stability of $\mathcal{F}$. Notice that since the base manifold of $\mathcal{F}$ is of complex dimension 2, in order to talk about the stability of $\mathcal{F}$ one needs to fix a Kähler metric on $X \times \mathbb{P}^1$. The parameter τ (or equivalently α) is encoded in this choice of metric.

Let us choose a metric on X with Kähler form ω_X, and volume normalized to one. The metric we shall consider on $X \times \mathbb{P}^1$ will be the product of the metric on X with a coefficient $\alpha/2$, with $\alpha > 0$, and the Fubini–Study metric on $\mathbb{P}^1$ with volume also normalized to one. The Kähler form corresponding to this metric is

$$\omega_\alpha = \frac{\alpha}{2} p^* \omega_X \oplus q^* \omega_{\mathbb{P}^1} \ .$$

The precise relation between the stability of $(\mathcal{E}_1, \mathcal{E}_2, \Phi)$ and that of $\mathcal{F}$ is given by the following theorem.

Theorem 4.11. [B-GP, Theorem 4.1] *Let $(\mathcal{E}_1, \mathcal{E}_2, \Phi)$ be a holomorphic triple over a compact Riemann surface X. Let $\mathcal{F}$ be the holomorphic bundle over $X \times \mathbb{P}^1$ associated to $(\mathcal{E}_1, \mathcal{E}_2, \Phi)$, and let*

$$\alpha(\tau) = \frac{(R_1 + R_2)\tau - (\deg \mathcal{E}_1 + \deg \mathcal{E}_2)}{R_2} \ . \tag{4.4}$$

Suppose that $\mathcal{E}_1$ and $\mathcal{E}_2$ are not isomorphic. Then $(\mathcal{E}_1, \mathcal{E}_2, \Phi)$ is τ-stable (equivalently α-stable) if and only if $\mathcal{F}$ is stable with respect to ω_α. In the case that $\mathcal{E}_1 \cong \mathcal{E}_2 \cong \mathcal{E}$, the triple $(\mathcal{E}, \mathcal{E}, \Phi)$ is τ-stable (equivalently α-stable) if and only if $\mathcal{F}$ decomposes as a direct sum

$$\mathcal{F} = p^*\mathcal{E} \otimes q^*\mathcal{O}(1) \oplus p^*\mathcal{E} \otimes q^*\mathcal{O}(1) \ ,$$

and $p^*\mathcal{E} \otimes q^*\mathcal{O}(1)$ is stable with respect to ω_α.

Let F be the smooth underlying bundle of $\mathcal{F}$ in (4.3), and let $\mathcal{M}_\alpha$ be the moduli space of stable holomorphic structures on F with respect to ω_α. Let us assume first that either $R_1 \neq R_2$ or $d_1 \neq d_2$. As a result of Theorem 4.11, there is a map

$$\mathcal{B}_\tau^T \longrightarrow \mathcal{M}_\alpha \ , \tag{4.5}$$

where α is related to τ by (4.4). The action of $SU(2)$ on $X \times \mathbb{P}^1$ defined above induces an action on $\mathcal{M}_\alpha$ and since the bundle $\mathcal{F}$ associated to $(\mathcal{E}_1, \mathcal{E}_2, \Phi)$ is $SU(2)$-equivariant the image of the above map is contained in $\mathcal{M}_\alpha^{SU(2)}$—the set of fixed points of $\mathcal{M}_\alpha$ under the $SU(2)$ action. As proved in [GP, Proposition 5.3] the set $\mathcal{M}_\alpha^{SU(2)}$ can be described as a disjoint union of a finite number of sets

$$\mathcal{M}_\alpha^{SU(2)} = \bigcup_{i \in I} \mathcal{M}_\alpha^i,$$

where the indexing set I ranges over the set of equivalence classes of different lifts of the action of $SU(2)$ on the smooth bundle underlying $\mathcal{F}$ in (4.3). In fact $\mathcal{B}_\tau^T$ can be identified via the map (4.5) with a subset $\mathcal{M}_\alpha^{i_0}$ of the fixed-point set, where i_0 is the lift defined at the beginning of the section.

If $R_1 = R_2 = R$ and $d_1 = d_2 = d$ by Propositions 3.7 and 3.8 we can identify $\mathcal{B}_\tau^T(d,R;d,R)$ with the moduli space of stable bundles of rank R and degree d on X.

This construction enables us to apply standard facts about the more familiar moduli spaces of stable bundles $\mathcal{M}_\alpha$ (cf. [G],[M],[Ko]), and more particularly of the fixed-point sets $\mathcal{M}_\alpha^i$ (see [GP, Theorem 5.6] for details), as well as the much studied moduli spaces of stable bundles over a Riemann surface, to obtain the following result.

Theorem 4.12. [BGP, Theorem 6.1] *Let X be a compact Riemann surface of genus g and let us fix ranks R_1 and R_2 and degrees d_1 and d_2. The moduli space of τ-stable triples $\mathcal{B}_\tau^T(d_1,R_1;d_2,R_2)$ is a complex analytic space with a natural Kähler structure outside of the singularities. Its dimension at a smooth point is*

$$1 + R_2 d_1 - R_1 d_2 + (R_1^2 + R_2^2 - R_1 R_2)(g-1)\ .$$

The moduli space $\mathcal{B}_\tau^T(d_1,R_1;d_2,R_2)$ is non-empty if and only if τ is inside the interval

$$(\mu(E_1), \mu_{MAX})$$

where

$$\mu_{MAX} = \mu(E_1) + \frac{R_2}{|R_1 - R_2|}(\mu(E_1) - \mu(E_2))$$

if $R_1 \neq R_2$, and $\mu_{MAX} = \infty$ if $R_1 = R_2$. Moreover $\mathcal{B}_\tau^T(d_1,R_1;d_2,R_2)$ is in general a quasi-projective variety. It is in fact projective if $R_1 + R_2$ and $d_1 + d_2$ are coprime and τ is generic.

Remark. From the construction of the moduli space of triples one can in fact obtain the moduli space of stable pairs. To do this we choose $E_2 = L$ to be the trivial line bundle and $E_1 = E$ of rank R and degree d. The moduli space $\mathcal{B}_\tau^T(d,R;0,1)$ is almost the moduli space of stable pairs $\mathcal{B}_\tau^P(d,R)$. Recall that a triple $(\mathcal{E},\mathcal{L},\Phi)$ is τ-stable if and only if the pair $(\mathcal{E}\otimes\mathcal{L}^*,\Phi)$ is τ-stable. The Picard group of X acts on $\mathcal{B}_\tau^T$ by the rule

$$\mathcal{U}(\mathcal{E},\mathcal{L},\Phi) = (\mathcal{E}\otimes\mathcal{U}^*, \mathcal{L}\otimes\mathcal{U}^*, \Phi),$$

for $\mathcal{U} \in \mathrm{Pic}^0(X)$, and it is clear that $\mathcal{B}^P_\tau(d,R)$ can be identified with $\mathcal{B}^T_\tau(d,R;0,1)/\mathrm{Pic}^0(X)$.

The action of $\mathrm{Pic}^0(X)$ is free and proper and we can recover Theorem 4.3 as a corollary of Theorem 4.12. The smoothness of $\mathcal{B}^P_\tau(d,R)$ follows from the following proposition.

Proposition 4.13. [BGP Proposition 6.3] *Let $(\mathcal{E}_1,\mathcal{E}_2,\Phi)$ be a τ-stable holomorphic triple such that Φ is injective, then $[(\mathcal{E}_1,\mathcal{E}_2,\Phi)]$ is a smooth point of the moduli space $\mathcal{B}^T_\tau$.*

§4.4 The topology of the moduli spaces

To end this section, we will make some remarks concerning the topology of the moduli spaces described above. Dating back to the fundamental work of Atiyah and Bott on Yang-Mills equations over Riemann surfaces [AB], it has been recognized that the topology of certain moduli spaces is closely related to properties of certain associated functionals (see also [D], [Ki]). For example, it is shown in the references cited that the Betti numbers of the moduli of vector bundles can be calculated from the Morse theory of the Yang-Mills functional

$$\|\sqrt{-1}\Lambda F_{\overline{\partial}_E,H} - \mu I\|^2_{L^2}$$

on the space $\mathcal{C}$ of $\overline{\partial}$-operators on E. In all the examples of augmented bundles described in this paper one can replace the Yang-Mills functional by the norm square of the corresponding moment map. For example, for pairs we take

$$\|\sqrt{-1}\Lambda F_{\overline{\partial}_E,H} + \phi\otimes\phi^* - \tau I\|^2_{L^2}$$

on the space $\mathcal{H}$ of holomorphic pairs. It follows from [AB] and [D] that the Yang-Mills functional is a perfect Morse function, and by the results of [Ki] it is quite plausible that the same is true in the other cases as well. However, unlike the Yang-Mills functional, in the other examples the topology of the domain of definition of the functional may not be known. This excludes using an inductive formula for the cohomology as in [AB]. This problem arises, for example, with Higgs bundles, where to our knowledge the cohomology for the non-coprime case has yet to be calculated (cf. [H], [Go]).

In the coprime case, our moduli spaces admit a locally trivial fibration (for minimal τ) over the moduli of vector bundles with compact fiber (see

(4.2)), and one can therefore make use of the Serre spectral sequence. In the non-coprime case, while the map (4.2) still exists, some fibers are very complicated and not much can be said in general. However, one can compute certain low dimensional homotopy and homology groups by use of transversality theory as described in [DU]. For example, for pairs we have the following

Theorem 4.14. (cf. [BD2], Theorem 3.13) *Assume that $i \leq 2(R-1)(g-1)-2$ and that τ is in the minimal range. Then for rank R, τ-stable pairs on a Riemann surface of genus g, $\pi_i(\mathcal{B}_\tau) \simeq \pi_{i-1}(\mathfrak{G})$. In particular, if $(R,g) \neq (2,2)$, then $\pi_1(\mathcal{B}_\tau) \simeq H_1(X,\mathbb{Z}) \simeq \mathbb{Z}^{2g}$, and $\pi_2(\mathcal{B}_\tau) \simeq \mathbb{Z} \oplus \mathbb{Z}$.*

One should be able to proceed for non-minimal τ by using the birational transformations described in the next section.

We expect that transversality results of this type can be generalized in all the examples of augmented bundles to compute the low dimensional homotopy and cohomology groups of their moduli spaces, though this remains to be carried out. We also mention that the case of parabolic bundles (see Section 6) was treated with these techniques in [DW] and [P].

5. Master spaces

In this section we briefly outline the argument given in [BDW] which provides a framework for understanding the dependence of the moduli spaces $\mathcal{B}_\tau$ on the parameter τ. In part, the motivation comes from the work of Thaddeus [T] in which he gives an explicit description for how the spaces $\mathcal{B}_\tau$ change as τ passes through a critical value. His characterization, which is in terms of specific modifications along smooth subvarieties is a refinement of the results of Bertram [Be] on a resolution of the rational map from the space of extensions (the "large τ" moduli space) to the moduli of bundles. The method gives a useful way of computing quantities on one space $\mathcal{B}_\tau$ by moving them over to another $\mathcal{B}_{\tau'}$, where the calculation is possibly simpler, and keeping track of the changes. This technique was used in [T] to give a proof of a special case of the Verlinde formula and in [BeDW] to compute certain intersection numbers for the space of holomorphic maps from Riemann surfaces to Grassmannians.

§*5.1 Reduction by a subgroup*

For simplicity, we shall mostly deal with case of stable pairs, although the construction we describe is more general, and versions for k-pairs, coherent systems, and triples are possible. The basic idea is that one would like to realize the parameter τ as the Morse function for a circle action on some larger Kähler manifold (dubbed a "masterspace"), which in some sense parametrizes all stable pairs, and not just the pairs which are τ-stable for some particular choice of τ.

To this end, one notices that there is an obvious choice of circle action; namely, multiplication of the section

$$e^{i\theta}(\mathcal{E},\phi) = (\mathcal{E}, e^{i\theta}\phi) \,. \tag{5.1}$$

This is similar to the circle action on the moduli space of Higgs bundles [H]. Note, however, that the action (5.1) is trivial on the moduli space $\mathcal{B}_\tau$, since it can be realized by an element of the gauge group $\mathfrak{G}$, and on the infinite dimensional space $\mathcal{H}$ of holomorphic pairs the action is free. In order to get something interesting we need to separate the action (5.1) from the rest of the gauge group.

The trick now is to find a closed subgroup $\mathfrak{G}_0 \subset \mathfrak{G}$ whose quotient is S^1 acting as in (5.1). The master space $\hat{\mathcal{B}}$ will then be taken to be the reduction of $\mathcal{H}$ by $\mathfrak{G}_0$. Since $\hat{\mathcal{B}}$ is to parametrize stable pairs for any choice of τ, it is clear that the moment map for the $\mathfrak{G}_0$ action on $\mathcal{H}$ should be

$$\Psi_0 = \pi^\perp \left(\Lambda F_{\bar\partial_E,H} - i\phi\otimes\phi^*\right) , \tag{5.2}$$

where $\pi^\perp$ denotes the orthogonal projection in Lie $\mathfrak{G}$ to the space perpendicular to the constant multiples of the identity (see Table 2). Thus, looking at the complex gauge groups, we should find $\mathfrak{G}^0_{\mathbb{C}} \subset \mathfrak{G}_{\mathbb{C}}$ such that

$$\mathrm{Lie}(\mathfrak{G}^0_{\mathbb{C}}) = \left\{ u \in \Omega^0(\mathrm{End}(E)) : \int_X \mathrm{tr}\ u = 0 \right\} .$$

This is obtained as follows. Let $\mathfrak{G}_{\mathbb{C},1} \subset \mathfrak{G}_{\mathbb{C}}$ denote the connected component of the identity, and let Υ denote the quotient group of components. Then Υ is a free abelian group on $2g$ generators corresponding to $H_1(\Sigma,\mathbb{Z})$ (see [A-B], p. 542). We can find a splitting of the exact sequence $1 \to \mathfrak{G}_{\mathbb{C},1} \to \mathfrak{G}_{\mathbb{C}} \to \Upsilon \to 1$, and this realises $\mathfrak{G}_{\mathbb{C}}$ as a direct product $\mathfrak{G}_{\mathbb{C}} \simeq \mathfrak{G}_{\mathbb{C},1} \times \Upsilon$, with the isomorphism given by $(g,h) \mapsto gh$. For $g \in \mathfrak{G}_{\mathbb{C},1}$, the map $\det g : \Sigma \to \mathbb{C}^*$ is in the identity component of $\mathrm{Map}(\Sigma,\mathbb{C}^*)$. It thus lifts to a map $u \in \mathrm{Map}(\Sigma, \mathbb{C})$, and we can define a character $\chi : \mathfrak{G}_{\mathbb{C},1} \to \mathbb{C}^*$ by $\chi(g) = exp(\int_\Sigma u)$. Then we extend χ to $\mathfrak{G}_{\mathbb{C},1} \times \Upsilon$ by $\chi(g,h) = \chi(g)$. This defines a homomorphism $\mathfrak{G}_{\mathbb{C}} \to \mathbb{C}^*$. We let $\mathfrak{G}_{\mathbb{C}0}$ be the kernel of the character $\chi : \mathfrak{G}_{\mathbb{C}} \to \mathbb{C}^*$, and let $\mathfrak{G}_0 \subset \mathfrak{G}$ be defined by $\mathfrak{G}_0 = \mathfrak{G}_{\mathbb{C}0} \cap \mathfrak{G}$. The remainder of the following proposition follows directly.

Proposition 5.1. (see [BDW], Section 2) *There exists a closed subgroup $\mathfrak{G}^0_{\mathbb{C}} \subset \mathfrak{G}_{\mathbb{C}}$ with quotient $\mathbb{C}^*$ whose Lie algebra is as above. The action of $\mathfrak{G}_0 = \mathfrak{G}^0_{\mathbb{C}} \cap \mathfrak{G}$ on $\mathcal{H}$ is holomorphic and symplectic, and the Ad-invariant moment map for this action is given by (5.2). Moreover, a pair $(\mathcal{E}, \phi) \in \mathcal{H}$ lies in the zero set of Ψ_0 if and only if it admits a solution to the τ-vortex equation for some τ.*

Therefore, the space $\hat{\mathcal{B}} = \Psi_0^{-1}(0)/\mathfrak{G}_0$ is the masterspace we desired. Let $\hat{\mathcal{B}}_0 \subset \hat{\mathcal{B}}$ denote the quotient by $\mathfrak{G}_0$ of the subset of $\Psi_0^{-1}(0)$ where $\mathfrak{G}_0$ acts with at most finite stabilizer. Then by arguments similar to those in Section 4, we have

Proposition 5.2. ([BDW], Proposition 2.3) *$\hat{\mathcal{B}}$ is a compact, Hausdorff topological space. $\hat{\mathcal{B}}_0$ is a Hausdorff, Kähler V-manifold.*

The quotient group $S^1 \simeq \mathfrak{G}/\mathfrak{G}_0$ now acts holomorphically and symplectically on $\hat{\mathcal{B}}_0$ by

$$e^{i\theta}[\mathcal{E}, \phi] = [\mathcal{E}, g_\theta \phi] \,, \tag{5.3}$$

where $g_\theta = \mathrm{diag}(e^{i\theta/R}, \ldots, e^{i\theta/R})$. Moreover, it can now be checked (see [BDW], Proposition 2.17) that a moment map for this circle action is given by

$$f([\mathcal{E}, \phi]) = -2\pi i \left(\frac{\|\phi\|^2}{4\pi R} + \mu(\mathcal{E}) \right) . \tag{5.4}$$

Finally, the principle of reduction in stages makes it clear that for non-critical values of τ, $f^{-1}(\tau)/S^1 \simeq \mathcal{B}_\tau$. Thus, we have achieved our goal of realizing τ as the moment map of a circle action.

§5.2 *Morse flow and birational maps*

Next we analyze the fixed points of the circle action (5.3). These are precisely the critical points of f. For simplicity and for the rest of this section, we shall restrict our attention to the case of rank 2 bundles. In that case, the moment map f on $\hat{\mathcal{B}}_0$ is proper, the circle action is quasi-free, and $\hat{\mathcal{B}}_0$ is a manifold. Also, it follows from the results in Section 3 that the range of τ is the interval $[d/2, d]$, and the critical values between the two extremes are precisely the integers between $d/2$ and d.

Suppose then that $[\mathcal{E}, \phi] \in \hat{\mathcal{B}}_0$ and $e^{i\theta}[\mathcal{E}, \phi] = [\mathcal{E}, \phi]$. Then there must exist a gauge transformation $g \in \mathfrak{G}^0_{\mathbb{C}}$ such that

$$g(\mathcal{E}, \phi) = (g\mathcal{E}, g\phi) = (\mathcal{E}, e^{i\theta/2}\phi)\,.$$

Since $g \in \mathfrak{G}^0_{\mathbb{C}}$, it cannot be a constant scalar endomorphism, and since as mentioned above stable implies simple, the bundle $\mathcal{E}$ must therefore split holomorphically as $\mathcal{E}_\phi \oplus \mathcal{E}_s$, where $\mathcal{E}_\phi$ and $\mathcal{E}_s$ are holomorphic line bundles, $\phi \in H^0(\mathcal{E}_\phi)$, and $g = (e^{i\theta/2}, g_s)$. Again using the fact that $g \in \mathfrak{G}^0_{\mathbb{C}}$ one deduces that $g_s = e^{-i\theta/2}$. Also, since $(\mathcal{E}, \phi)$ gives rise to a solution of the τ-vortex equation, we must have $\deg(\mathcal{E}_s) = \tau$; in particular, such a fixed point can only occur for integral values of τ (for higher rank, critical values can only occur for τ's which are possible slopes of subbundles, i.e. at precisely the non-generic values of τ discussed in Section 3). We summarize this discussion with the following

Proposition 5.3. *Consider the case $R = 2$. Then the critical values of f on $\hat{\mathcal{B}}$, other than the maximum and minimum, occur at precisely all the integer values of τ in $(d/2, d)$. The critical sets Z_τ are of the form*

$$Z_\tau = S^{d-\tau}(X) \times \mathcal{J}_\tau\,,$$

where $S^{d-\tau}(X)$ denotes the $d-\tau$ symmetric product of X, and $\mathcal{J}_\tau$ is the degree τ Jacobian variety of X.

The Kähler metric on $\hat{\mathcal{B}}_0$ and the circle action (5.3) induces a gradient flow

$$\Phi_t : \hat{\mathcal{B}}_0 \times [0, \infty) \longrightarrow \hat{\mathcal{B}}_0$$

defined by the equation

$$\frac{\partial}{\partial t}\Phi_t = -\nabla_{\Phi_t} f\,.$$

It can be verified (see [BDW], Proposition 5.1) that Φ_t is given by

$$\Phi_t[\mathcal{E}, \phi] = [\mathcal{E}, e^{-t/4\pi}\phi]\,. \tag{5.6}$$

The flow is invariant under the circle action, and so if there are no critical values in the interval $[\tau, \tau+\varepsilon]$, then Φ_t descends to give a diffeomorphism of the quotients $\mathcal{B}_\tau \simeq \mathcal{B}_{\tau+\varepsilon}$. Moreover, since the action is holomorphic, this diffeomorphism is actually a biholomorphism. Hence, the only difference between $\mathcal{B}_\tau$ and $\mathcal{B}_{\tau+\varepsilon}$ is in the induced Kähler structure. This is the familiar picture which arises in the Duistermaat-Heckman theorem.

Now let us consider what happens when there exists an intermediary critical value. Let τ be an integer in $(d/2, d)$, and let W^s_τ, W^u_τ denote the stable and unstable manifolds of the critical set Z_τ with respect to the flow (5.6). The two stratifications of $\hat{\mathcal{B}}_0$ obtained from the stable and unstable manifolds correspond to algebraic stratifications which are the analogues of the Seshadri filtration for semistable bundles. For the details of this description we refer to [BDW], Section 4. Here, we simply state the result for rank 2. The set of points in W^u_τ which flow into a point $[\mathcal{E}_\phi \oplus \mathcal{E}_s, \phi] \in Z_\tau$ are the $\mathfrak{G}^0_{\mathbb{C}}$ equivalence classes of pairs $(\mathcal{E}, \phi)$ arising from extensions

$$0 \longrightarrow \mathcal{E}_\phi \longrightarrow \mathcal{E} \longrightarrow \mathcal{E}_s \longrightarrow 0\,,$$

where the section $\phi \in H^0(\mathcal{E})$ is induced from the section of $\mathcal{E}_\phi$ by the injection. The stable manifold W^s_τ consists of equivalence classes of pairs $(\mathcal{E}, \varphi)$ arising from extensions

$$0 \longrightarrow \mathcal{E}_s \longrightarrow \mathcal{E} \longrightarrow \mathcal{E}_\phi \longrightarrow 0\,,$$

where the section $\varphi \in H^0(\mathcal{E})$ projects to $\phi \in H^0(\mathcal{E}_\phi)$.

Now suppose that τ is the only critical value in $[\tau-\varepsilon, \tau+\varepsilon]$. In that case the spaces $\mathcal{B}_{\tau-\varepsilon}$, $\mathcal{B}_{\tau+\varepsilon}$ are symplectic reductions of $\hat{\mathcal{B}}_0$ by the S^1-action. Guillemin and Sternberg studied the appropriate generalization of the Duistermaat-Heckman theorem to this situation [GS]. The result is that the passage from $\mathcal{B}_{\tau-\varepsilon}$ to $\mathcal{B}_{\tau+\varepsilon}$ is via a modification by surgery on a sphere bundle, modulo the circle action. This is more or less Kodaira's description of a blow-up, and it is explained in the context of stable pairs in [T] (also see [D3]). Specifically, we have the following: Let

$$\begin{aligned}\mathbb{P}_\varepsilon(W^+) &= W^s_\tau \cap f^{-1}(\tau+\varepsilon)/S^1\\ \mathbb{P}_\varepsilon(W^-) &= W^u_\tau \cap f^{-1}(\tau-\varepsilon)/S^1\,.\end{aligned}$$

Then we have

Theorem 5.4. ([T], and [BDW], Theorem 6.6) *Suppose that $\tau \in (d/2, d)$ is the only critical value of f in the interval $[\tau - \varepsilon, \tau + \varepsilon]$. Then there is a projective variety $\tilde{\mathcal{B}}_\tau$ and holomorphic maps*

$$\begin{array}{ccccc} & & \tilde{\mathcal{B}}_\tau & & \\ & \rho_- \swarrow & & \searrow \rho_+ & \\ \mathcal{B}_{\tau-\varepsilon} & & & & \mathcal{B}_{\tau+\varepsilon} \end{array}$$

Moreover, for $\tau < d-1$, $\rho_\pm$ are blow-down maps onto the smooth subvarieties $\mathbb{P}_\varepsilon(W^\pm)$. For $\tau = d-1$, ρ_+ is the blow-down map onto $\mathbb{P}_\varepsilon(W^+)$ and ρ_- is the identity.

Theorem 5.4 should in principle apply also in higher rank. The method in [BDW] breaks down because of the existence of singularities in the critical sets which occur for certain values of τ. For these τ, one therefore expects a slightly more complicated modification than the one above.

6.Other Augmented Bundles

In this section we shall comment briefly on other examples of augmented bundles. Some of them had been studied prior to the ones described above; this is the case for parabolic bundles and bundles with a level structure. Other examples are obtained by combining a parabolic structure with one of the augmentations studied above. This leads to the study of parabolic pairs, parabolic Higgs bundles and parabolic triples.

§*6.1 Parabolic bundles*

The notion of parabolic bundle was introduced by Seshadri (cf. [Se1,Se2]) and has been studied extensively in [Se1,MS,Bho1, BhoR] among others. Let I be a finite set of points and $\mathcal{E}$ be a holomorphic vector bundle over a compact Riemann surface X. A *parabolic structure* on $\mathcal{E}$ consists in giving for each point of I a filtration of the fibre $\mathcal{E}_x$:

$$\mathcal{E}_x = F_1(\mathcal{E}_x) \supset \ldots \supset F_{n_x}(\mathcal{E}_x) \supset 0,$$

with "weights"

$$0 \leq \alpha_{x1} < \ldots < \alpha_{xn_x} < 1.$$

For each point we have the *multiplicities*

$$k_{xi} = \dim(F_i(\mathcal{E}_x)/F_{i+1}(\mathcal{E}_x)),$$

and we can define the *parabolic degree* and the *parabolic slope*:

$$\mathrm{pardeg}(\mathcal{E}) = \deg(\mathcal{E}) + \sum_{x \in I}\sum_{i=1}^{n_x} k_{xi}\alpha_{xi},$$

$$\mathrm{par}\,\mu(\mathcal{E}) = \frac{\mathrm{pardeg}(\mathcal{E})}{\mathrm{rank}(\mathcal{E})}.$$

One can define natural morphisms between parabolic bundles. Moreover, every subbundle $\mathcal{E}' \subset \mathcal{E}$ inherits a parabolic structure from that of $\mathcal{E}$, becoming a parabolic subbundle.

Definition 6.1 *A parabolic bundle $\mathcal{E}$ is stable if for every subbundle $\mathcal{E}' \subset \mathcal{E}$*

$$\mathrm{par}\,\mu(\mathcal{E}') < \mathrm{par}\,\mu(\mathcal{E}).$$

These definitions have been generalized by Bhosle to parabolic sheaves on higher dimensional smooth varieties (cf. [Bho2,MY]) and also to parabolic structures defined over divisors of degree greater than 1 [Bho3]. Parabolic bundles can be considered as augmented bundles in which the filtrations defining part of the parabolic structure constitute the augmentation, and the weights are parameters in the definition of stability. The analytic treatment of parabolic bundles requires non-compact Riemann surfaces , but it can be shown that stable parabolic bundles do support special metrics which satisfy equations similar to the Hermitian-Einstein equations (cf. [Bi,MS,P,DW,Si2]). The special structure is reflected now in the fact that the connection of this metric has logarithmic poles at the parabolic points, with coefficients determined by the weights.

One can consider all the augmented bundles studied above in the category of parabolic bundles. Some special cases of pairs and coherent systems on a bundle with parabolic structure have been studied in [Be]. *Parabolic triples*, that is triples in which the bundles are endowed with parabolic structures are considered in [BiGP]. The homomorphism can be either a parabolic morphism or a meromorphic morphism with simple poles at the parabolic points and whose residues respect the parabolic structure in some precise sense. In both cases one can prove a Hitchin-Kobayashi correspondence, and again the metrics involved now have singularities at the parabolic points. Finally, *parabolic Higgs bundles* have been the object of study in [Bi,Y,Si2] for example. These are pairs consisting of a parabolic bundle and a twisted endomorphism, preserving the parabolic structure. The moduli space of parabolic Higgs bundles is relevant for example in connection to the deformations of the moduli space of parabolic bundles. In particular, it contains the cotangent space of the moduli space of parabolic bundles, generalizing the non-parabolic situation (cf. [Hi]).

§6.2 Bundles with a level structure

Let D be a zero-dimensional variety of a compact Riemann surface X with structure sheaf $\mathcal{O}_D$, and let $\mathcal{E}$ be a rank R holomorphic vector bundle on X. A *level structure* on $\mathcal{E}$ consists of a non-zero sheaf morphism

$$\Phi : \mathcal{E} \longrightarrow R.\mathcal{O}_D.$$

A level structure can also be regarded as a k-coherent system with k equal to R (see [LeP1,2] for this point of view). Seshadri (cf. [Se2]) introduced a notion of stability for $(\mathcal{E}, \Phi)$, which does not involve, however, any parameter. The existence of a parametrized notion of stability for a level structure has been detected by Huybrechts and Lehn [HL], who have in fact studied a generalization of a level structure for a smooth projective variety of arbitrary dimension. They consider pairs $(\mathcal{E}, \Phi)$ consisting of a coherent sheaf $\mathcal{E}$ and a homomorphism $\Phi : \mathcal{E} \longrightarrow \mathcal{E}_0$ to a fixed coherent sheaf $\mathcal{E}_0$ (which might be of pure torsion). As mentioned in section 1.5, these pairs can be regarded as specialized triples $(\mathcal{E}_1, \mathcal{E}_2, \Phi)$ in which $\mathcal{E}_1$ is fixed.

References

[AB] Atiyah, M. F. and R. Bott, *The Yang-Mills equations over Riemann surfaces*, Phil. Trans. R. Soc. Lond. A **308** (1982) 523–615.

[Be] A. Bertram, Stable pairs and stable parabolic pairs, preprint.

[BeDW] A. Bertram, G. Daskalopoulos and R. Wentworth, Gromov invariants for holomorphic maps from Riemann surfaces to Grassmannians, preprint.

[Bho1] U. Bhosle, Parabolic vector bundles on curves, *Ark. Mat.* **27** (1989) 15–22.

[Bho2] U. Bhosle, Parabolic sheaves on higher dimensional varieties, *Math. Ann.* **293** (1992) 177-192.

[Bho3] U. Bhosle, Generalised parabolic bundles and applications to torsion-free sheaves on nodal curves, *Ark. Mat.* **30** (1992) 187–215.

[BhoR] U. Bhosle and A. Ramanathan, Moduli of parabolic G-bundles on curves, *Math. Z.* **202** (1984) 161-180.

[Bi] O. Biquard, Fibrés holomorphes et connexions singulières sur une courbe ouverte, Ph.D. Thesis. Ecole Polytechnique, 1991.

[BiGP] O. Biquard and O. García–Prada, in preparation.

[B] S.B. Bradlow, Special metrics and stability for holomorphic bundles with global sections, *J. Diff. Geom.* **33** (1991) 169–214.

[BD1] S.B. Bradlow and G. Daskalopoulos, Moduli of stable pairs for holomorphic bundles over Riemann surfaces, *Int. J. Math.* **2** (1991) 477–513.

[BD2] S.B. Bradlow and G. Daskalopoulos, Moduli of stable pairs for holomorphic bundles over Riemann surfaces II, *Int. J. Math.* **4** (1993) 903-925.

[BDW] S.B. Bradlow, G.D. Daskalopoulos and R. A. Wentworth, Birational equivalences of vortex moduli, preprint.

[BGP] S.B. Bradlow and O. García–Prada, Stable triples, equivariant bundles and dimensional reduction, Orsay preprint 93-83.

[C] K. Corlette, Flat G–bundles with canonical metrics, *J. Diff. Geom.* **28** (1988) 361–382.

[D] G.D. Daskalopoulos, The topology of the space of stable bundles on a compact Riemann surface, *J. Diff. Geom.* **36** (1992) 699-746.

[DU] G.D. Daskalopoulos and K. Uhlenbeck, An application of transversality to the topology of the moduli space of stable bundles, to appear in *Topology.*

[DW] G.D. Daskalopoulos and R. A. Wentworth, Geometric quantization for the moduli space of vector bundles with parabolic structure, preprint.

[D1] S. K. Donaldson, A new proof of a theorem of Narasimhan and Seshadri, *J. Diff. Geom.* **18** (1983) 269–278.

[D2] S.K. Donaldson, Anti-self-dual Yang–Mills connections on a complex algebraic surface and stable vector bundles, *Proc. Lond. Math. Soc.* **3** (1985) 1–26.

[D3] S. K. Donaldson, Instantons in Yang-Mills theory, *Proceedings of the IMA Conference on Geometry and Particle Physics* (F. Tsou, ed.), Oxford University Press, Oxford (1990) 59–75.

[GP] O. García–Prada, Dimensional reduction of stable bundles, vortices and stable pairs, *Int. J. Math.* **5** (1994) 1–52.

[G] D. Gieseker, On moduli of vector bundles on an algebraic surface, *Ann. of Math.* **106** (1977) 45–60.

[Go] P. Gothen, The Betti numbers of the moduli space of stable rank 3 Higgs bundles on a Riemann surface, Warwick Preprint, 71/1993.

[GS] V. Guillemin and S. Sternberg, Birational equivalence in the symplectic category, *Invent. Math.* **97** (1989) 485–522.

[H] N.J. Hitchin, The self-duality equations on a Riemann surface, *Proc. Lond. Math. Soc.* **55** (1987) 59–126.

[HL] D.Huybrechts and M. Lehn, Stable pairs on curves and surfaces, *J.Alg. Geometry*, to appear.

[KN] A. King and P. Newstead, Moduli of Brill–Noether pairs on algebraic curves, preprint, 1994.

[Ki] F. Kirwan, *Cohomology of Quotients in Symplectic and Algebraic Geometry*, Princeton University Press, 1984.

[Ko] S. Kobayashi, *Differential Geometry of Complex Vector Bundles*, Princeton University Press, 1987.

[LeP1] J. Le Potier, Systèmes cohérents et structures de niveau, *Astérisque* **214** (1993).

[LeP2] J. Le Potier, Faisceaux semi-stables et systèmes cohérents, this volume.

[L] M. Lübke, Stability of Einstein–Hermitian vector bundles, *Manuscripta Mathematica* **42** (1983) 245–257.

[M] M. Maruyama, Moduli of stable sheaves I and II, *J. Math. Kyoto Univ.*, **17** (1977) 91–126, and **18** (1978) 557–614.

[MY] M. Maruyama and K. Yokogawa, Moduli of parabolic stable sheaves, *Math. Ann.* **293** (1992) 77–99.

[MS] V.B. Mehta and C.S. Seshadri, Moduli of vector bundles on curves with parabolic structures, *Math. Ann.* **248** (1980) 205–239.

[NS] M.S. Narasimhan and C.S. Seshadri, Stable and unitary bundles on a compact Riemann surface, *Ann. of Math.* **82** (1965) 540–564.

[P] J. Poritz, Parabolic vector bundles and Hermitian-Yang-Mills connections over a Riemann surface, *Int. J. Math.* **4** (1993) 467-501.

[RV] N. Raghavendra and P.A. Vishwanath, Moduli of pairs and generalized theta divisors, *Tohoku Math. J.*, to appear.

[Se1] C.S. Seshadri, Moduli of vector bundles on curves with parabolic strucutures. *Bull. Amer. Math. Soc.* **83** (1977) 124–126.

[Se2] C.S. Seshadri, Fibrés vectoriels sur les courbes algébriques, *Astérisque* **96** (1982).

[Si1] C.T. Simpson, Constructing variations of Hodge structure using Yang–Mills theory and applications to uniformization, *J. Amer. Math. Soc.* **1** (1988) 867–918.

[Si2] C.T. Simpson, Harmonic bundles on noncompact curves, *J. Amer. Math. Soc.* **3** (1990) 713–770.

[Siu] Y.T. Siu, A vanishing theorem for semipositive line bundles over non-Kähler manifolds, *J. Diff. Geom.* **19** (1984) 431–452.

[T] M. Thaddeus, Stable pairs, linear systems and the Verlinde formula, Invent. Math. **117** (1994) 317-353.

[Ti] S. Tiwari, Moduli of holomorphic bundles with global sections over Riemann surfaces, preprint

[UY] K.K. Uhlenbeck and S.T. Yau, On the existence of Hermitian-Yang-Mills connections in stable vector bundles *Comm. Pure. Appl. Math.* **39** (1986) 5257–5293

[Y] K. Yokogawa, Moduli of stable pairs, *J. Math. Kyoto Univ.* **31**–1 (1991) 311–327.

Table 1. Stability Conditions and Related Equations

	Data	Stability Conditions	Special Equations for Metrics	References
Pure Bundle	$\mathcal{E}$	$\mu(\mathcal{E}') < \mu(\mathcal{E})$ if $\mathcal{E}' \subset \mathcal{E}$	$\sqrt{-1}\Lambda F_{\overline{\partial}_E,H} = \mu I$ (Hermitian-Einstein)	[NS] , [Do] , [UY] , [Ko]
Pairs	$(\mathcal{E},\phi)$, $\phi \in H^0(X,\mathcal{E})$	$\mu(\mathcal{E}') < \tau$ if $\mathcal{E}' \subseteq \mathcal{E}$, $\mu(\mathcal{E}/\mathcal{E}_\phi) > \tau$ if $\mathcal{E}_\phi \subseteq \mathcal{E}$ and $\phi \in H^0(X,\mathcal{E}_\phi)$	$\sqrt{-1}\Lambda F_{\overline{\partial}_E,H} + \phi \otimes \phi^* = \tau I$ (τ-Vortex)	[BD1] , [G-P], [T] , [Be]
k-Pairs	$(\mathcal{E},\phi_1,\ldots,\phi_k)$, $\phi_i \in H^0(X,\mathcal{E})$	$\mu(\mathcal{E}') < \tau$ if $\mathcal{E}' \subseteq \mathcal{E}$, $\mu(\mathcal{E}/\mathcal{E}_\phi) > \tau$ if $\mathcal{E}_\phi \subseteq \mathcal{E}$ and $\phi_i \in H^0(X,\mathcal{E}_\phi)$, $1 \leq i \leq k$	$\sqrt{-1}\Lambda F_{\overline{\partial}_E,H} + \Sigma_{i=1}^k \phi_i \otimes \phi_i^* = \tau I$ (k-τ-Vortex)	[BeDW] , [Ti]
Coherent Systems	$(\mathcal{E},V)$, $V \subset H^0(X,\mathcal{E})$	$\mu_\alpha(\mathcal{E}',V') < \mu_\alpha(\mathcal{E},V)$ where $\mu_\alpha(\mathcal{E}',V') = \mu(\mathcal{E}') + \alpha\frac{dim(V')}{rank(\mathcal{E}')}$	$\sqrt{-1}\Lambda F_{\overline{\partial}_E,H} + \Sigma_{i=1}^k \phi_i \otimes \phi_i^* = \tau I$ $< \phi_i,\phi_j >= \alpha I$, $deg(E) + dim(V)\alpha = rank(E)\tau$ where $V = Span\{\phi_1,\ldots,\phi_k\}$ (Orthonormal τ-Vortex)	[LeP], [KN], [RV]
Triples	$(\mathcal{E}_1,\mathcal{E}_2,\Phi)$, $\Phi \in H^0(X,\mathcal{E}_1 \otimes \mathcal{E}_2^*)$	$\theta_\tau(\mathcal{E}_1',\mathcal{E}_2') < 0$, (alt. $\mu_\alpha(\mathcal{E}_1',\mathcal{E}_2') < \mu_\alpha(\mathcal{E}_1,\mathcal{E}_2)$) , where $\theta_\tau(\mathcal{E}_1',\mathcal{E}_2') = (\mu(\mathcal{E}_1' \oplus \mathcal{E}_2') - \tau) - \frac{r_2'}{r_2}\frac{r_1+r_2}{r_1'+r_2'}(\mu(\mathcal{E}_1 \oplus \mathcal{E}_2) - \tau)$ $(\mu_\alpha(\mathcal{E}_1',\mathcal{E}_2') = \mu(\mathcal{E}') + \alpha\frac{r_2'}{r_1'+r_2'})$	$\sqrt{-1}\Lambda F_{\overline{\partial}_E,H_1} + \Phi\Phi^* = \tau I$ $\sqrt{-1}\Lambda F_{\overline{\partial}_E,H_2} - \Phi^*\Phi = \tau' I$ $\deg(E_1) + deg(E_2) = r_1\tau + r_2\tau'$ (Coupled Vortex)	[G-P] [BG-P]
Higgs Bundles	$(\mathcal{E},\Theta)$, $\Theta \in H^0(End\mathcal{E} \otimes K)$	$\mu(\mathcal{E}') < \mu(\mathcal{E})$ for all Θ-invariant subbundles	$\sqrt{-1}\Lambda F_{\overline{\partial}_E,H} + [\Theta,\Theta^*] = 2\pi\mu I$ (Hitchin ASD)	[H], [S] , [C]

Table 2. Infinite Dimensional Picture

	Infinite Dimensional Configuration Space	Holomorphicity Condition	Complex Gauge Group	Group Action
Pure Bundles	$\mathcal{C}$	–	$G_{\mathcal{C}}(E)$	$\overline{\partial}_E \longmapsto g \circ \overline{\partial}_E \circ g^{-1}$
Pairs	$\mathcal{H} \subset \mathcal{C} \times \Omega^0(X,E)$	$\overline{\partial}_E(\phi) = 0$	$G_{\mathcal{C}}(E)$	$\overline{\partial}_E \longmapsto g \circ \overline{\partial}_E \circ g^{-1}$, $\phi \longmapsto g\phi$
k-Pairs	$\mathcal{H}^k \subset \mathcal{C} \times (\Omega^0(X,E))^k$	$\overline{\partial}_E(\phi_i) = 0$ for $1 \le i \le k$	$G_{\mathcal{C}}(E)$	$\overline{\partial}_E \longmapsto g \circ \overline{\partial}_E \circ g^{-1}$, $\phi_i \longmapsto g\phi_i$
Coherent Systems	$\mathcal{H}^{CS} \subset (\mathcal{ST}_k)/GL(k)$	$\overline{\partial}_E(\phi_i) = 0$ for $1 \le i \le k$	$G_{\mathcal{C}}(E)$	$\overline{\partial}_E \longmapsto g \circ \overline{\partial}_E \circ g^{-1}$, $V \longmapsto g(V)$
Triples	$\mathcal{H}^T \subset \mathcal{C}_1 \times \mathcal{C}_2 \times \Omega^0(X, Hom(E_2,E_1))$	$\overline{\partial}_1 \circ \Phi = \Phi \circ \overline{\partial}_2$	$G_{\mathcal{C}}(E_1) \times G_{\mathcal{C}}(E_2)$	$\overline{\partial}_i \longmapsto g_i \circ \overline{\partial}_i \circ g_i^{-1}$, $\Phi \longmapsto g_1 \Phi g_2^{-1}$
Higgs Bundles	$\mathcal{H}^h \subset \mathcal{C} \times \Omega^{1,0}(X,E)$	$\overline{\partial}_E \circ \Theta - \Theta \circ \overline{\partial}_E = 0$	$G_{\mathcal{C}}(E)$	$\overline{\partial}_E \longmapsto g \circ \overline{\partial}_E \circ g^{-1}$, $\Theta \longmapsto g\Theta g^{-1}$

$\mathcal{C} = \{$holomorphic structures on $E\} = \{\overline{\partial}_E : \Omega^0(X,E) \longrightarrow \Omega^{0,1}(X,E)\}$

$\mathcal{ST}_k = \{(\overline{\partial}_E; \phi_1, \phi_2, \ldots, \phi_k) \in \mathcal{C} \times (\Omega^0(X,E))^k$: the sections are linearly independent$\}$

Table 3. Symplectic Descriptions

	Space (M)	Group (G)	Moment Map $(\Psi : M \longrightarrow g)$
Pure Bundles	$\mathcal{C}$	$\mathfrak{G}$	$\Psi(\mathcal{E}) = \Lambda F_{\overline{\partial}_E, H}$
Pairs	$\mathcal{H}$	$\mathfrak{G}$	$\Psi_1(\mathcal{E}, \phi) = \Lambda F_{\overline{\partial}_E, H} - i\, \phi \otimes \phi^*$
k-Pairs	$\mathcal{H}^k$	$\mathfrak{G}$	$\Psi_k(\mathcal{E}, \phi_1, \ldots, \phi_k) = \Lambda F_{\overline{\partial}_E, H} - i\, \Sigma_{i=1}^k \phi_i \otimes \phi_i^*$
Coherent Systems	$\Psi_U^{-1}(-i\alpha I)/U(k)$, where $\Psi_U : \mathcal{H}^k \longrightarrow u(k)$	$\mathfrak{G}$	$\Psi_{CS}(\mathcal{E}, V) = \Psi_k(\mathcal{E}, \phi_1, \ldots, \phi_k)$, where $Span\{\phi_1, \ldots, \phi_k\} = V$ and $\Psi_U(\mathcal{E}, \phi_1, \ldots, \phi_k) = -i\alpha I$
Triples	$\mathcal{H}^T$	$\mathfrak{G}_1 \times \mathfrak{G}_2$	$\Psi_T(\mathcal{E}_1, \mathcal{E}_2, \Phi) = (\Lambda F_{\overline{\partial}_1, H_1} - i\Phi\Phi^*\ , \Lambda F_{\overline{\partial}_2, H_2} + i\Phi^*\Phi\ )$
Higgs Bundles	$\mathcal{H}^h$	$\mathfrak{G}$	$\Psi_h(\mathcal{E}, \Theta) = \Lambda F_{\overline{\partial}_E, H} - i[\Theta, \Theta^*]$

Table 4. Range for Parameters

	Data	Lower and Upper Bounds	Critical Values	Stability $\Leftrightarrow$ Semistability
k-Pairs	$(\mathcal{E}, \phi_1, ..., \phi_k)$ $\phi_i \in H^0(X, \mathcal{E})$	$\frac{d}{R} \leq \tau \leq \frac{d}{R-k}$ Upper bound if $k < R$	$\frac{p}{q} \in Q$ with $q \leq R$	non-critical τ
Coherent Systems	$(\mathcal{E}, V)$ $V \subset H^0(X, \mathcal{E})$	$0 \leq \alpha \leq \frac{d}{R-k}$ Upper bound if $k < R$	$\frac{p}{q} \in Q$ with $q \leq Rk$	non-critical α AND $(R, d) = 1$ OR $(R, k) = 1$
Triples	$(\mathcal{E}_1, \mathcal{E}_2, \Phi)$ $\Phi \in H^0(X, \mathcal{E}_1 \otimes \mathcal{E}_2^*)$	$\mu(\mathcal{E}_1) \leq \tau \leq \mu(\mathcal{E}_1) + \frac{R_2}{\lvert R_1 - R_2 \rvert}(\mu(\mathcal{E}_1) - \mu(\mathcal{E}_2))$ OR $0 \leq \alpha \leq (1 + \frac{R_1+R_2}{\lvert R_1 - R_2 \rvert})(\mu(\mathcal{E}_1) - \mu(\mathcal{E}_2))$ Upper bound if $R_1 \neq R_2$	$\frac{p}{q} \in Q$ with $q \leq R_1 R_2$	non-critical τ AND $(R_1 + R_2, d_1 + d_2) = 1$

ON SURFACES IN $\mathbb{P}^4$ AND 3-FOLDS IN $\mathbb{P}^5$

WOLFRAM DECKER, SORIN POPESCU

CONTENTS

0. INTRODUCTION

We report on some recent progress in the classification of smooth projective varieties with small invariants. This progress is mainly due to the finer study of the adjunction mapping by Reider, Sommese and Van de Ven [So1], [VdV], [Rei], [SV]. Adjunction theory is a powerful tool for determining the type of a given variety. Classically, the adjunction process was introduced by Castelnuovo and Enriques [CE] to study curves on ruled surfaces. The Italian geometers around the turn of the century also started the classification of smooth surfaces in $\mathbb{P}^4$ of low degree. Further classification results are due to Roth [Ro1], who uses the adjunction mapping to get surfaces with smaller invariants already known to him (compare [Ro2] for adjunction theory on 3-folds). Nowadays, through the effort of several mathematicians, a complete classification of smooth surfaces in $\mathbb{P}^4$ and smooth 3-folds in $\mathbb{P}^5$ has been worked out up to degree 10 and 11 respectively. Moreover, in the 3-fold case the classification is almost complete in degree 12. For references see section 7.

One motivation to study these varieties comes from Hartshorne's conjecture [Ha1]. In the case of codimension 2 this suggests that already smooth 4-

folds in $\mathbb{P}^6$ should be complete intersections. Another motivation originates from two mutually corresponding finiteness results. Ellingsrud and Peskine [EP] proved that there are only finitely many families of smooth surfaces in $\mathbb{P}^4$ which are not of general type. However, the question of an exact degree bound d_0 is still open. By [BF] $d_0 \leq 105$. Examples are known only up to degree 15 and one actually believes that $d_0 = 15$. The analogous finiteness result holds for 3-folds in $\mathbb{P}^5$ [BOSS1]. In this case one expects a much higher degree bound. Nevertheless examples had been known so far only up to degree 14 [Ch3]. In this note we present, among other things, three new smooth 3-folds in $\mathbb{P}^5$ of degree $13, 17$ and 18 respectively.

How to construct examples ?

Let us recall that every smooth projective variety of dimension m can be embedded in $\mathbb{P}^{2m+1}$. So, for example, in the surface case we could try to work with general projections from points in $\mathbb{P}^5$. However Severi's theorem [Se] tells us that every non-degenerate smooth surface in $\mathbb{P}^4$ except the Veronese surface is linearly normal. Similarly by Zak's theorem [Za] every non-degenerate smooth 3-fold in $\mathbb{P}^5$ is linearly normal.

There are two other classical construction methods. One is to study linear systems on abstract varieties. This works especially well for rational, abelian and bielliptic surfaces. The other is liaison [PS] starting with a known local complete intersection variety (presumably of lower degree). With a few exceptions these methods failed to produce examples in higher degree. In the case of liaison this is mainly due to the fact that the varieties to be constructed tend to be minimal in their even liaison class (compare [LR]). Whereas, if we consider, for example, linear systems of curves on minimal surfaces, the base points have to be in a special position. Such configurations are hard to find.

In this context a new construction method for surfaces $X \subset \mathbb{P}^4$ (more generally $(n-2)$-folds $X \subset \mathbb{P}^n$) was introduced in [DES] (compare also [Po]). The basic idea is an application of Beilinson's spectral sequence [Bei]: to construct the ideal sheaf $\mathcal{J}_X$ and thus X one has to construct the Hartshorne-Rao modules of X first. Involving corresponding syzygy bundles as suggested by the spectral sequence one finds vector bundles $\mathcal{F}$ and $\mathcal{G}$ on $\mathbb{P}^n$ with $\operatorname{rk} \mathcal{G} = \operatorname{rk} \mathcal{F} + 1$, and a morphism $\varphi \in \operatorname{Hom}(\mathcal{F}, \mathcal{G})$, whose minors define the desired X. From the syzygies of the Hartshorne-Rao modules

one can compute the syzygies of $\mathcal{J}_X$ and so the explicit equations for X. Typically, part of the geometry behind X can already be seen from the syzygies. The smoothness of X can be checked via the implicit function theorem, i.e., by a straightforward computation. Since these computations are very extensive one has to rely on a computer and a computer algebra system. Currently, Macaulay [Mac] is the only system which is powerful enough to handle the computations.
In some cases X is not minimal in its even liaison class, or a minimal element in the complementary even liaison class has low degree and can be identified. In fact, by studying the equations we find examples where X can be reconstructed via liaison from a reducible scheme X' of lower degree. It is hard to find such reducible schemes a priori.

Problem. *Find a geometric construction for all examples constructed via syzygies.* □

Notation. $R = \mathbb{C}[x_0, \ldots, x_n] = \bigoplus_{q \in \mathbb{Z}} S^q V^*$ will be the homogeneous coordinate ring of $\mathbb{P}^n$, so $H^0(\mathcal{O}_{\mathbb{P}^n}(1)) = V^*$. If $X \subset \mathbb{P}^n$ is a fixed smooth subvariety, then d will denote its degree, π its sectional genus, H the hyperplane class and K the canonical class. □

Acknowledgements. Both authors are grateful to Frank-Olaf Schreyer for many helpful conversations. We also thank Mark Gross for interesting discussions. The first author, who lectured in Tokyo and Durham on the topic of this note, would like to thank the Japan Society for the Promotion of Science and the Deutscher Akademischer Austauschdienst for their support and Masaki Maruyama and the University of Kyoto for their hospitality. Last but not least we would like to thank the organizers of the Durham symposium on Vector Bundles in Algebraic Geometry for creating a stimulating atmosphere during this meeting. □

1. Constructions via syzygies

Following [DES] we want to construct a codimension 2 subvariety $X \subset \mathbb{P}^n$ as the determinantal locus of a map between vector bundles. So we are looking for vector bundles $\mathcal{F}$ and $\mathcal{G}$ on $\mathbb{P}^n$ with $\operatorname{rk} \mathcal{F} = f$ and $\operatorname{rk} \mathcal{G} = f + 1$, and a morphism $\varphi \in \operatorname{Hom}(\mathcal{F}, \mathcal{G})$ whose minors vanish in the expected codimension 2. In this case $X = V(\varphi)$ is a locally Cohen-Macaulay subscheme and the

Eagon-Northcott complex [BE]

$$0 \leftarrow \mathcal{O}_X(m) \leftarrow \mathcal{O}(m) \cong \bigwedge^{f} \mathcal{F}^* \otimes \bigwedge^{f+1} \mathcal{G} \leftarrow \mathcal{G} \overset{\varphi}{\leftarrow} \mathcal{F} \leftarrow 0$$

is exact and identifies coker φ with the twisted ideal sheaf

$$\operatorname{coker} \varphi \cong \mathcal{J}_X(m), \qquad m = c_1\mathcal{G} - c_1\mathcal{F}.$$

Furthermore, a mapping cone between the minimal free resolutions of $\mathcal{F}$ and $\mathcal{G}$ is a (not necessarily minimal) free resolution of $\mathcal{J}_X(m)$. So for a given φ we can derive an explicit system of homogeneous equations for its dependency locus X.

Remark 1.1. Let $\varphi_1, \varphi_2 \in \operatorname{Hom}(\mathcal{F}, \mathcal{G})$ be morphisms whose minors vanish in codimension 2. Then $V(\varphi_1)$ and $V(\varphi_2)$ lie in the same irreducible component of the Hilbert scheme (compare e.g. [BB], [BBM], [MDP]). □

To construct a variety with the desired numerical invariants one has to find appropriate $\mathcal{F}$ and $\mathcal{G}$. Clearly $\mathcal{F}$ and $\mathcal{G}$ reflect the structures of the graded finite length R-modules

$$H^i_*\mathcal{J}_X = \bigoplus_{q\in\mathbb{Z}} H^i(\mathbb{P}^n, \mathcal{J}_X(q)), \qquad i = 1, \ldots, \dim X,$$

called the Hartshorne-Rao modules of X. E.g., X is projectively Cohen-Macaulay, i.e., its Hartshorne-Rao modules are trivial, iff $\mathcal{F}$ and $\mathcal{G}$ can be chosen to be direct sums of line bundles. Or compare [Ch2] for the Ω-resolution of a projectively Buchsbaum variety. In this case, in particular, the multiplication maps of the Hartshorne-Rao modules are trivial.

Remark 1.2. Smooth projectively Cohen-Macaulay and smooth projectively Buchsbaum varieties of codimension 2, which are not of general type, are completely classified (compare [Ch3]). □

In any case it is a natural idea to construct the Hartshorne-Rao modules first. Then one may involve corresponding syzygy bundles as direct summands in order to find $\mathcal{F}$ and $\mathcal{G}$. Recall:

Proposition 1.3. *Let* $M = \bigoplus_{q\in\mathbb{Z}} M_q$ *be a graded* R*-module of finite length and let*

$$0 \leftarrow M \leftarrow L_0 \overset{\alpha_1}{\leftarrow} L_1 \leftarrow \ldots \overset{\alpha_{n+1}}{\leftarrow} L_{n+1} \leftarrow 0$$

be its minimal free resolution. Then, for $1 \le i \le n-1$, the sheafified syzygy module

$$\mathcal{F}_i = \mathcal{S}yz_i(M) = (\ker \alpha_i)^\sim = (\mathrm{Im}\ \alpha_{i+1})^\sim$$

is a vector bundle on $\mathbb{P}^n$ with the intermediate cohomology

$$\bigoplus_{q\in\mathbb{Z}} H^j(\mathbb{P}^n, \mathcal{F}_i(q)) = \begin{cases} M & j = i \\ 0 & j \neq i, \quad 1 \le j \le n-1 \end{cases} .$$

Conversely, any vector bundle $\mathcal{F}$ on $\mathbb{P}^n$ with this intermediate cohomology is stably
equivalent with $\mathcal{F}_i$, i.e.,

$$\mathcal{F} \cong \mathcal{F}_i \oplus \mathcal{L}\ , \qquad \mathcal{L} \text{ a direct sum of line bundles.} \qquad \square$$

Example 1.4. Consider $\mathbb{C}$ as a graded R-module sitting in degree 0. The minimal free resolution of $\mathbb{C}(i)$ is the Koszul complex

$$0 \leftarrow \mathbb{C}(i) \leftarrow \bigwedge^0 V^* \otimes R(i) \leftarrow \cdots \leftarrow \bigwedge^{n+1} V^* \otimes R(i-n-1) \leftarrow 0\ .$$

The corresponding syzygy bundles are the twisted bundles of differentials, $\mathcal{S}yz_i(\mathbb{C}(i)) \cong \Omega^i(i)$. It follows from the sheafified Koszul complex, that $\mathrm{Hom}\,(\Omega^i(i), \Omega^j(j)) \cong \bigwedge^{i-j} V$, the isomorphisms being given by contraction (cf. [Bei]). $\square$

Which syzygy bundles should be involved in the construction of $\mathcal{F}$ and $\mathcal{G}$? This can be found out by analyzing Beilinson's spectral sequence for $\mathcal{J}_X(m)$. Recall:

Theorem 1.5. [Bei] *For any coherent sheaf $\mathcal{S}$ on $\mathbb{P}^n$ there is a spectral sequence with E_1-terms*

$$E_1^{pq} = H^q(\mathbb{P}^n, \mathcal{S}(p)) \otimes \Omega^{-p}(-p)$$

converging to $\mathcal{S}$, i.e., $E_\infty^{pq} = 0$ for $p+q \neq 0$ and $\bigoplus E_\infty^{-p,p}$ is the associated graded sheaf of a suitable filtration of $\mathcal{S}$. $\square$

This theorem is often used to construct $\mathcal{S}$ by determining the differentials of the spectral sequence first. A crucial point is that the d_1-differentials

$$\begin{aligned} d_1^{pq} &\in \mathrm{Hom}\,(H^q(\mathbb{P}^n, \mathcal{S}(p)) \otimes \Omega^{-p}(-p),\ H^q(\mathbb{P}^n, \mathcal{S}(p+1)) \otimes \Omega^{-p-1}(-p-1)) \\ &\cong \mathrm{Hom}\,(V^* \otimes H^q(\mathbb{P}^n, \mathcal{S}(p)),\ H^q(\mathbb{P}^n, \mathcal{S}(p+1))) \end{aligned}$$

coincide with the natural multiplication maps. In our case $\mathcal{S} = \mathcal{J}_X(m)$, and we will interpret one part of Beilinson's spectral sequence as the spectral sequence of a vector bundle $\mathcal{F}$, the other part as that of a vector bundle $\mathcal{G}$. The differential between the two parts will give the morphism $\varphi : \mathcal{F} \to \mathcal{G}$ whose cokernel is the desired $\mathcal{J}_X(m)$. The twist m will be mainly n or $n-1$ (compare [DES, 1.7] for the corresponding Beilinson cohomology tables in the surface case).

How to check the smoothness of X ? If the bundle Hom $(\mathcal{F}, \mathcal{G})$ is globally generated and $n \leq 5$, then we know from [Klm], that the generic $\varphi \in$ Hom $(\mathcal{F}, \mathcal{G})$ gives rise to a smooth X. This works well, if X is projectively Cohen-Macaulay. Similarly, if X is projectively Buchsbaum, we may apply [Ch1]. In the general case however, we mostly have to rely on a computer as explained in the introduction.

Example 1.6. We will construct a family of smooth 3-folds $X \subset \mathbb{P}^5$ with the numerical invariants $d = 18$, $\pi = 35$, $\chi(\mathcal{O}_X) = 2$ and $\chi(\mathcal{O}_S) = 26$, where S is a general hyperplane section of X. Let us analyze Beilinson's spectral sequence for $\mathcal{J}_X(4)$. We first need information on the dimensions $h^i\mathcal{J}_X(m)$, $m = -1, \ldots, 4$. In view of Riemann-Roch a plausible Beilinson cohomology table is

i

24					
	1	6	3		

m

Suppose that a smooth 3-fold X with these data exists. Then Beilinson's theorem yields an exact sequence

$$0 \to \mathcal{F} = 24\mathcal{O}(-1) \to \mathcal{G} \to \mathcal{J}_X(4) \to 0\ ,$$

where $\mathcal{G}$ is the cohomology of a monad

$$0 \to \Omega^4(4) \xrightarrow{d_1^{-4,3}} 6\Omega^3(3) \xrightarrow{d_1^{-3,3}} 3\Omega^2(2) \to 0\ .$$

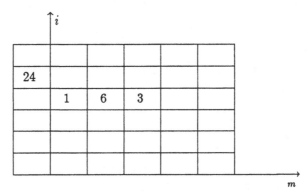

On the other hand, the generic module with Hilbert function $(1,6,3)$ has syzygies of type

$$0 \leftarrow M \leftarrow R(4) \nwarrow 18R(2) \leftarrow 52R(1) \leftarrow 60R \nwarrow \begin{matrix} 24R(-1) \\ \oplus \\ 10R(-2) \end{matrix} \nwarrow 12R(-3) \leftarrow 3R(-4) \leftarrow 0$$

A check on the ranks and the intermediate cohomology of $\mathcal{G}$ and $\mathcal{S}yz_3(M)$ suggests that conversely it is promising to start with $\mathcal{F} = 24\mathcal{O}(-1)$ and $\mathcal{G} = \mathcal{S}yz_3(M)$. Indeed, for the map $\varphi \in \operatorname{Hom}(24\mathcal{O}(-1), \mathcal{S}yz_3(M))$ given by the syzygies, one can check that the minors of φ vanish along a smooth 3-fold X. By construction $\mathcal{J}_X$ has syzygies of type

$$0 \leftarrow \mathcal{J}_X \leftarrow 10\mathcal{O}(-6) \leftarrow 12\mathcal{O}(-7) \leftarrow 3\mathcal{O}(-8) \leftarrow 0 .$$

In particular, X is cut out by 10 sextics. From the syzygies it follows that $\omega_X(1) = \mathcal{E}xt^2(\mathcal{O}_X, \mathcal{O}(-6))(1)$ is a quotient of $24\mathcal{O}$, and since $(K+H)^2 \cdot K = -4$ (compare section 5) we deduce that the Kodaira dimension $\kappa(X) = -\infty$. □

2. Liaison

We recall the definition and some basic results [PS]. Let $X, X' \subset \mathbb{P}^n$ be two locally Cohen-Macaulay subschemes of pure codimension 2 with no irreducible components in common. X and X' are said to be (geometrically) *linked* (r,s), if there exist hypersurfaces V_1 and V_2 of degrees r and s respectively, such that $X \cup X' = V_1 \cap V_2$. Then there are the standard exact sequences

$$0 \to \omega_X \to \mathcal{O}_{V_1 \cap V_2}(r+s-n-1) \to \mathcal{O}_{X'}(r+s-n-1) \to 0 ,$$

$$0 \to \omega_X \to \mathcal{O}_X(r+s-n-1) \to \mathcal{O}_{X \cap X'}(r+s-n-1) \to 0 .$$

The degrees and sectional genera of X and X' are related by

$$d + d' = r \cdot s \qquad \text{and} \qquad \pi - \pi' = \frac{1}{2}(r+s-4)(d-d') ,$$

and

$$\chi(\mathcal{O}_{X'}) = \chi(\mathcal{O}_{V_1 \cap V_2}) - \chi(\mathcal{O}_X(r+s-n-1)) .$$

Under suitable assumptions (e.g., if $H^1(\mathcal{F}(r)) = H^1(\mathcal{F}(s)) = 0$) we may deduce from a given locally free resolution $0 \to \mathcal{F} \to \mathcal{G} \to \mathcal{J}_X \to 0$ of $\mathcal{J}_X$ a resolution

$$0 \to \mathcal{G}^{\check{}}(-r-s) \to \mathcal{F}^{\check{}}(-r-s) \oplus \mathcal{O}(-r) \oplus \mathcal{O}(-s) \to \mathcal{J}_{X'} \to 0$$

of $\mathcal{J}_{X'}$ by taking a mapping cone as in [PS, Prop. 2.5]. Moreover, the Hartshorne-Rao modules of $\mathcal{J}_{X'}$ are $\mathbb{C}$-dual to those of X:

$$(H_*^{n-1-i}\mathcal{J}_X)^* \cong (H_*^i \mathcal{J}_{X'})(n+1-r-s), \qquad i = 1, \ldots, n-2 .$$

Liaison can be used to construct new subvarieties starting from given ones. Hence it is useful to know, under which conditions a residual intersection will be smooth. One result in this direction is a special case of [PS, Prop.4.1]:

Theorem 2.1. *(Peskine-Szpiro). Let $X \subset \mathbb{P}^n$, $n \leq 5$, be a local complete intersection of codimension 2. Let m be a twist such that $\mathcal{J}_X(m)$ is globally generated. Then for every pair $d_1, d_2 \geq m$ there exist forms $f_i \in H^0(\mathcal{J}_X(d_i))$, $i = 1, 2$, such that the corresponding hypersurfaces V_1 and V_2 intersect properly, $V_1 \cap V_2 = X \cup X'$, where*

(i) *X' is a local complete intersection,*
(ii) *X and X' have no common component,*
(iii) *X' is nonsingular outside a set of positive codimension in* Sing X. □

3. Adjunction theory

In this section (X, H) will denote a polarized pair, where X is a smooth, connected, projective variety of dimension $m \geq 2$ and H is a very ample divisor on X. $K = K_X$ will be a canonical divisor on X. Before reviewing the general theory behind the *adjunction map* $\Phi = \Phi_{|K+(m-1)H|}$, we will give an example.

Example 3.1. [Roo] Let $\varphi = (\varphi_{ij})_{\substack{0 \leq i \leq 2 \\ 0 \leq j \leq 3}}$ be a general 3×4-matrix with linear entries in $\mathbb{C}[x_0, \ldots, x_4]$. Then $X = V(\varphi)$ is a smooth surface $X \subset \mathbb{P}^4$ with $d = 6$ and $\pi = 3$. Let H be the hyperplane class of X. By dualizing φ, we obtain the resolution

$$0 \leftarrow \omega_X(1) \leftarrow 3\mathcal{O} \xleftarrow{{}^t\varphi} 4\mathcal{O}(-1) \leftarrow \mathcal{O}(-4) \leftarrow 0 .$$

So $|K+H|$ is base point free, $N = \dim |K+H| = 2$ and we have a well-defined adjunction map $\Phi = \Phi_{|K+H|} : X \to \mathbb{P}^2$. Let y_0, y_1, y_2 be coordinates on $\mathbb{P}^2$. Then graph $(\Phi) \subset \mathbb{P}^4 \times \mathbb{P}^2$ is given by the equations

$$y_0\varphi_{0j}(x) + y_1\varphi_{1j}(x) + y_2\varphi_{2j}(x) = 0\,, \quad j = 0,\dots,3\,.$$

We may rewrite these equations as

$$x_0\psi_{j0}(y) + \cdots + x_4\psi_{j4}(y) = 0\,, \quad j = 0,\dots,3\,,$$

where $\psi = (\psi_{jk})_{\substack{0\le j\le 3\\ 0\le k\le 4}}$ has linear entries in $\mathbb{C}[y_0, y_1, y_2]$. The general fibre of Φ is defined by four independent linear forms in $\mathbb{C}[x_0,\dots,x_4]$. Hence Φ is birational with positive dimensional fibres precisely in the points where ψ drops rank:

$$0 \to 4\mathcal{O}_{\mathbb{P}^2}(-5) \xrightarrow{\psi} 5\mathcal{O}_{\mathbb{P}^2}(-4) \to \mathcal{J}_Z \to 0\,.$$

So $\Phi : X \to \mathbb{P}^2$ expresses X as the blowing up of 10 points in $\mathbb{P}^2$ and X is embedded by the quartics through these points, i.e., by the 4×4-minors of ψ. In other words

$$H \equiv 4L - \sum_{i=1}^{10} E_i$$

(with obvious notations) and X is a Bordiga surface. □

The first general result deals with the existence of the adjunction map. It is a consequence of [Sol], [VdV].

Theorem 3.2. *Let (X,H) and K be as above. Then $|K+(m-1)H|$ is base point free unless*

(i) $(X, \mathcal{O}_X(H)) \cong (\mathbb{P}^m, \mathcal{O}_{\mathbb{P}^m}(1))$ *or* $(\mathbb{P}^2, \mathcal{O}_{\mathbb{P}^2}(2))$,

(ii) $(X, \mathcal{O}_X(H)) \cong (Q, \mathcal{O}_Q(1))$, *where $Q \subset \mathbb{P}^{m+1}$ is a smooth hyperquadric,*

(iii) $(X, \mathcal{O}_X(H))$ *is a scroll over a smooth curve.* □

If $|K+(m-1)H|$ is base point free, then we denote by

$$\begin{array}{ccc} X & \xrightarrow{\Phi} & \mathbb{P}^N \\ & r\searrow \quad \nearrow s & \\ & X' & \end{array}$$

the Stein factorization of the adjunction map Φ. X' is normal, r is connected and s is finite.

Theorem 3.3. [So2] *Let (X, H) and K be as above and suppose that $|K+(m-1)H|$ is base point free. Then there are the following possibilities:*

(i) $\dim \Phi(X) = 0$, *and* $K \equiv -(m-1)H$, *i.e.,* X *is Fano of index* $(m-1)$,

(ii) $\dim \Phi(X) = 1$, *and the general fibre of* r *is a smooth quadric* Q *such that* H *induces* $\mathcal{O}_Q(1)$,

(iii) $\dim \Phi(X) = 2 < m$ *and* r *exhibits* X *as a scroll over a smooth surface,*

(iv) $\dim \Phi(X) = m$. □

If $\dim \Phi(X) = m$ we write $L' = r_*(H)$, $K' = K_{X'}$ and $H' = K'+(m-1)L'$. The next result tells us, that in this case r contracts precisely the linear $\mathbb{P}^{m-1} \subset X$ with normal bundle $\mathcal{O}_{\mathbb{P}^{m-1}}(-1)$ (necessarily disjoint).

Theorem 3.4. [So1],[So2] *Suppose that* $\dim \Phi(X) = m$. *Then* $r : X \to X'$ *is the blowing up of a finite number of points on the smooth projective variety* X'. L' *and* H' *are ample and*

$$r^*(H') \equiv K + (m-1)H . \qquad \square$$

In the above situation (X', L') is called the *first reduction* of (X, H) [So5]. When is s an embedding ? The answer is given by

Theorem 3.5. [SV] *Suppose that* $\dim \Phi(X) = m$. *Then* H' *is very ample, unless* X *is a surface and*

(i) $X = \mathbb{P}^2(p_1, \ldots, p_7)$ *and* $H \equiv 6L - \sum_{i=1}^{7} 2E_i$ *(the Geiser involution),*

(ii) $X = \mathbb{P}^2(p_1, \ldots, p_8)$ *and* $H \equiv 6L - \sum_{i=1}^{7} 2E_i - E_8$,

(iii) $X = \mathbb{P}^2(p_1, \ldots, p_8)$ *and* $H \equiv 9L - \sum_{i=1}^{8} 3E_i$ *(the Bertini involution),*

(iv) $X = \mathbb{P}(\mathcal{E})$, *where* $\mathcal{E}$ *is an indecomposable rank 2 bundle over an elliptic curve, and* $H \equiv 3B$, *where* B *is a section with* $B^2 = 1$ *on* X

□

For surfaces the *adjunction process*, i.e., the study of $|K+H|$, $|K'+H'|$ etc., will finally lead to a minimal model. For 3-folds $X \subset \mathbb{P}^5$ the situation is quite different. In this case it is often successful to study $|K+H|$ instead of $|K+2H|$. Compare section 5 for details and applications of further general results of adjunction theory.

4. Surfaces in $\mathbb{P}^4$

In this section X will denote a smooth non-degenerate surface in $\mathbb{P}^4$ and $d = H^2$ its degree, $\pi = \frac{1}{2}H \cdot (K + H) + 1$ its sectional genus and $\chi = \chi(\mathcal{O}_X) = 1 - q + p_g$ its Euler characteristic.
K^2 may be computed from the double point formula (cf. [Ha2, Appendix A, 4.1.3.])

$$d^2 - 10d - 5H \cdot K - 2K^2 + 12\chi = 0 \ .$$

In order to classify surfaces of a given degree, one first has to work out a finite list of admissible numerical invariants. One may apply Halphen's upper bound for π [GP] in connection with the lifting theorem of Roth [Ro1, p.152] and the following classification results:

Theorem 4.1. [Ro1], [Au]. *Let X be contained in a hyperquadric $V^2 \subset \mathbb{P}^4$. Then $\pi = 1 + [d(d-4)/4]$ and X is either the complete intersection of V^2 with another hypersurface, or X is linked to a plane in the complete intersection of V^2 with another hypersurface.* □

Theorem 4.2. [Ko], [Au]. *Let X be contained in an irreducible cubic hypersurface $V^3 \subset \mathbb{P}^4$. Then either X is projectively Cohen-Macaulay and linked on V^3 to an irreducible scheme of degree ≤ 3, or X is linked on V^3 to a Veronese surface, or to a quintic elliptic scroll.* □

Corollary 4.3. *If X is contained in a cubic hypersurface and if $d \geq 9$, then X is of general type.* □

To derive a lower bound for π and bounds for χ we may use Severi's Theorem [Se] together with Riemann-Roch, the Hodge index theorem, the Enriques-Kodaira classification and adjunction theory. In the context of section 3 we note:

Theorem 4.4. [Au], [La]. *If X is a scroll, then X is a rational cubic or an elliptic quintic scroll.* □

Theorem 4.5. [BR], [ES]. *If X is a conic bundle, then X is a Del Pezzo surface of degree 4, or a Castelnuovo surface.* □

Corollary 4.6. *If $d \geq 6$, then the adjunction map Φ is defined and $(K + H)^2 > 0$, i.e., $\dim \Phi(X) = 2$.* □

Once the numerical invariants are fixed, we use the information on the dimensions $h^i\mathcal{J}_X(m)$ provided by Riemann-Roch and [DES, 1.7]. In some cases more information on the dimensions and the structures of the Hartshorne-Rao modules may be obtained by studying the relations between the multi-secants to X, the plane curves on X and the syzygies of $\mathcal{J}_X$ (compare [PR]). This information is helpful for construction and classification purposes. In other cases one has to go through the adjunction process to analyze, how a given X fits into the Enriques-Kodaira classification. In any case it is crucial to know the number of exceptional lines on X. Le Barz' 6-secant formula [LB] tells us, that the number of 6-secants to X (if finite) plus the number of exceptional lines equals a polynomial expression $N_6 = N_6(d, \pi, \chi)$ (if X does not contain a line with self-intersection ≥ 0). This fits well with the ideas of section 1. Once having constructed X explicitly , we can compute the 6-secants easily. For examples we refer to [DES],[Po].

With the following example we would like to demonstrate, that the construction via syzygies is not always as straightforward as in Example 1.6.

Example 4.9. [Po] Let us construct a family of smooth surfaces $X \subset \mathbb{P}^4$ with $d = 11$, $\pi = 11$ and $\chi = 3$. In view of [DES, 1.7] a plausible Beilinson cohomology table for $\mathcal{J}_X(4)$ is

i

2				
	1			
		1	4	3

m

Suppose that a smooth surface X with these data exists. Then Beilinson's theorem yields a resolution of type

$$0 \to \mathcal{F} = 2\mathcal{O}(-1) \oplus \Omega^3(3) \xrightarrow{\varphi} \mathcal{G} \to \mathcal{J}_X(4) \to 0\ ,$$

where $\mathcal{G}$ is the cohomology of a monad

$$0 \to \Omega^2(2) \xrightarrow{d_1^{-2,1}} 4\Omega^1(1) \xrightarrow{d_1^{-1,1}} 3\mathcal{O} \to 0\ .$$

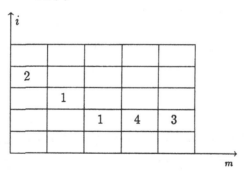

Arguing as in example 1.6, we conversely choose $\mathcal{G} = \mathcal{S}yz_1(M)$, where M is a module with Hilbert function $(1, 4, 3)$ and a minimal free presentation of type

$$0 \leftarrow M \leftarrow S(2) \overset{(\alpha,\beta)}{\longleftarrow} S(1) \oplus 7S\ .$$

So M is the tensor product of the Koszul complex given by the linear form α and the module M' presented by β. We may assume that $\alpha = x_4$ and that M' is a module over $R' = \mathbb{C}[x_0, \ldots, x_3]$. M' has the same Hilbert function as M, namely $(1,4,3)$. The general such M' has syzygies of type

$$0 \leftarrow M' \leftarrow R'(2) \overset{\beta}{\longleftarrow} 7R' \longleftarrow \begin{array}{c} 8R'(-1) \\ \oplus \\ (3+a)R'(-2) \end{array} \longleftarrow \begin{array}{c} aR'(-2) \\ \oplus \\ 8R'(-3) \end{array} \overset{{}^t\gamma}{\longleftarrow} 3R'(-4) \leftarrow 0$$

with $a = 0$. It is easy to see, that in this case no morphism $\varphi \in \operatorname{Hom}(\mathcal{F}, \mathcal{G})$ is injective. The trick for the construction of X is to choose β special in order to obtain some extra syzygies and thus a larger space $\operatorname{Hom}(\mathcal{F}, \mathcal{G})$. We will construct a module M' with the above type of syzygies and $a = 1$. Equivalently, we will construct the $\mathbb{C}$-dual module M'^* by defining its presentation matrix $\gamma = (\gamma_1, \gamma_2)$. Choose four general lines $L_1, \ldots, L_4$ in the hyperplane $V(x_4)$, denote by ϵ the presentation matrix in the direct sum of the four Koszul complexes built on these lines and let δ be a general 3×4-matrix with entries in $\mathbb{C}$. Then ϵ and thus also $\gamma_1 = \delta\epsilon$ has four linear 1-syzygies. Let γ_2 be given by 3 general quadrics. Then γ presents an artinian module as desired. With these choices the generic $\varphi \in \operatorname{Hom}(\mathcal{F}, \mathcal{G})$ yields a smooth surface X cut out by 8 quintics and 4 sextics. In general it is a plausible guess and in many cases true that the number of 6-secants to a surface in $\mathbb{P}^4$ is precisely the number of sextic generators of its homogeneous ideal. Indeed, in our case it is easy to see that $L_1, \ldots, L_4$ are precisely the 6-secants to X [Po, Proposition 3.32]. Le Barz' 6-secant formula gives $N_6(11, 11, 3) = 5$. Hence there is one exceptional line on X. One can show that there are no other exceptional curves [Po, Proposition 3.31]. Since $K^2 = 1$ by the double point formula X is of general type. □

In some cases it is quite subtle to construct artinian modules with the desired graded Betti numbers. From this point of view the most difficult surfaces are the abelian and bielliptic surfaces known so far [ADHPR2]. These are also the surfaces with the most beautiful geometry behind (compare [ADHPR1] and Hulek' s article on the Horrocks-Mumford bundle in this volume). The link between the geometry and the syzygies is provided by the distribution of the 2- and 3-torsion points on the Heisenberg invariant elliptic normal curves in $\mathbb{P}^4$. In turn, these curves are related to the Horrocks-Mumford bundle. Our knowledge on this bundle has influenced the construction of further families of surfaces (compare [ADHPR1, Thm. 32], [DES, 2.5], [Po, 4.1 and 7.4]).

5. 3-FOLDS IN $\mathbb{P}^5$

In this section X will denote a smooth, non-degenerate 3-fold in $\mathbb{P}^5$, S a general hyperplane section, $d = H^3$ its degree and $\pi = \frac{1}{2}H^2 \cdot (K + 2H) + 1$ its sectional genus.
We have two double point formulae, one for X,

$$K^3 = -5d^2 + d(2\pi + 25) + 24(\pi - 1) - 36\chi(\mathcal{O}_S) - 24\chi(\mathcal{O}_X),$$

and one for S, which may be rewritten as

$$H \cdot K^2 = \frac{1}{2}d(d+1) - 9(\pi - 1) + 6\chi(\mathcal{O}_S)$$

(compare e.g. [Ok2]). So the basic invariants of X are $d, \pi, \chi(\mathcal{O}_X)$ and $\chi(\mathcal{O}_S)$. Equivalently one may consider the pluridegrees

$$d_i = (K+H)^i \cdot H^{3-i} = c_{2+i}(\mathcal{J}_X(5)), \qquad i = 0, \ldots, 3,$$

introduced in [BBS]. By Zak's theorem [Za] X is linearly normal. Moreover $h^1(X, \mathcal{O}_X) = 0$ by Barth-Larsen-Lefschetz [BL]. In particular S is linearly normal and regular. Clearly X is projectively Cohen-Macaulay iff S has this property. So by studying S we obtain from Theorem 4.1 and Theorem 4.2:

Proposition 5.1. *Let X be contained in a cubic hypersurface. Then X is projectively Cohen-Macaulay. In particular X is of general type if $d \geq 13$.* □

To work out a finite list of admissible invariants for a given degree one may again start with Halphen's upper bound for π. Further tools are a congruence obtained from Riemann-Roch [BSS2, 0.11], the inequalities deduced from the semipositivity of $\mathcal{N}_{X/\mathbb{P}^5}(-1)$ [BOSS1, Proposition 2.2] and adjunction theory. In the context of section 3 we recall some classification results. $X_1, \ldots, X_{30}$ will denote the 3-folds listed in table 7.3. The first result follows from Theorem 4.4 and Theorem 4.5 (compare also [BOSS2]).

Proposition 5.2.

(i) *If X is a scroll over a smooth curve, then $X = X_1$ is a Segre cubic scroll.*

(ii) *If X is a Fano 3-fold of index 2, then $X = X_2$ is a complete intersection of two quadric hypersurfaces.*

(iii) *If X is a quadric bundle over a smooth curve, then $X = X_3$ is a Castelnuovo 3-fold.* □

Theorem 5.3. [Ott]. *If X is a scroll over a smooth surface, then X is one of the following:*

(i) a *Segre scroll $X = X_1$,*

(ii) a *Bordiga scroll $X = X_4$,*

(iii) a *Palatini scroll $X = X_6$,*

(iv) a *scroll $X = X_{11}$ over a K3 surface.* □

From now on we suppose that X is none of the exceptional 3-folds above. Then the adjunction map Φ is defined and the connected morphism r of its Stein factorization contracts the linear $\mathbb{P}^2 \subset X$ with normal bundle $\mathcal{O}_{\mathbb{P}^2}(-1)$ to points.

Proposition 5.4. [BSS2] *r is an isomorphism unless $X = X_7$.* □

From now on we suppose that $X \neq X_7$. Then X coincides with its first reduction.

The next step in adjunction theory is to study $K + H$. This is big and nef unless X is one of the special varieties listed in [So5]. In our case these are classified:

Theorem 5.5. [BOSS2]. *$K + H$ is big and nef unless*

(i) *(X, H) is a Fano 3-fold of index 1. Then $X = X_5$ is a complete intersection of type $(2, 3)$.*

(ii) *(X, H) is a Del Pezzo fibration over a smooth curve. Then $X = X_8$ or $X = X_9$.*

(iii) *(X, H) is a conic bundle over a surface. Then $X = X_{12}$ or $X = X_{20}$.* □

From now on we suppose that $K + H$ is big and nef. Then S is of general type and minimal. Therefore X is called to be of *log-general type* [BSS1].

In this case further numerical information is provided by the generalized Hodge index theorem [BBS, Lemma 1.1] and the parity relations [BBS, Lemma 1.4].
From the Kawamata-Shokurov base point free theorem (see [KMM, §3]) we know that for some $m > 0$ the linear system $|m(K+H)|$ gives rise to a morphism, say $\Psi : X \to X''$. For m large enough we can assume that Ψ has connected fibers and normal image. We write $L'' = \Psi_*(H)$, $K'' = K_{X''}$ and $H'' = K'' + L''$. Then L'' and H'' are ample and

$$\Psi^*(H'') \equiv K + H$$

(cf. [BFS, (0.2.6)]). (X'', L'') is called the *second reduction* of (X, H) [So4],[BFS].

Proposition 5.6. [BSS3, Corollary 1.3] *If $d \neq 10$ and $d \neq 13$, then X'' is smooth and Ψ is an isomorphism outside a disjoint union $\mathcal{C}$ of smooth curves. Let C be an irreducible component of $\mathcal{C}$ and let $D := \Psi^{-1}(C)$. Then the restriction Ψ_D of Ψ to D is a $\mathbb{P}^1$-bundle $\Psi_D : D \to C$ and $\mathcal{N}^X_{D|F} \cong \mathcal{O}_{\mathbb{P}^1}(-1)$ for any fiber F of Ψ_D. In fact, Ψ is simply the blowing up along $\mathcal{C}$.* □

Remark 5.7. i) If $d = 10$, then there is exactly one case where X'' is not smooth. Namely, for $X = X_{16}$ the second reduction morphism Ψ, which is defined by $|K + H|$, contracts the quadric surface K to a singular point p. Moreover, $\Psi(X) \subset \mathbb{P}^6$ is a complete intersection of type $(2,2,3)$, while X is the projection from p of $\Psi(X)$ (see also section 7).
ii) From [BSS3, (0.5.1) and (1.1)] and [Ed, (3.1.3)] it follows that in case $d = 13$ the second reduction is singular iff there exist on X divisors $D \cong \mathbb{P}^2$, with $\mathcal{N}^X_D \cong \mathcal{O}_{\mathbb{P}^2}(-2)$, which are contracted to points. We are not aware of any such example. □

Example 5.8. Let $X \subset \mathbb{P}^5$ be a smooth 3-fold with $d = 11$ and $\pi = 14$. Then $\chi(\mathcal{O}_S) = 8$ and $\chi(\mathcal{O}_X) = 0$ (compare [BSS2]). Every smooth surface in $\mathbb{P}^4$ with the same invariants as S is linked $(4,4)$ to a Castelnuovo surface [Po, Prop. 3.70]. In particular S and hence X are projectively Cohen-Macaulay with syzygies of type

$$0 \to 2\mathcal{O}(-5) \oplus \mathcal{O}(-6) \xrightarrow{\varphi} 4\mathcal{O}(-4) \to \mathcal{J}_X \to 0\ .$$

Consequently $X = X_{18}$ is linked $(4,4)$ to a Castelnuovo 3-fold. Conversely this shows the existence of 3-folds of type X_{18} [BSS2].

What kind of 3-fold is X ?
From the invariants we compute the Kodaira dimension $\kappa(X) = 0$. In order to show that X is a blown up Calabi-Yau 3-fold we study $|K + H|$. By dualizing φ we obtain the resolution

$$0 \leftarrow \omega_X(1) \leftarrow \mathcal{O}(1) \oplus 2\mathcal{O} \overset{t_\varphi}{\leftarrow} 4\mathcal{O}(-1) \leftarrow \mathcal{O}(-5) \leftarrow 0 .$$

So $|K + H|$ is base point free, $\dim |K + H| = 7$, and we have a well-defined map $\Psi_{|K+H|} : X \to X''$, where X'' is a 3-fold in $\mathbb{P}^7$. Moreover $h^0(\mathcal{O}_S(K_S - H_S)) = h^2(\mathcal{O}_S(1)) = 1$ by Riemann-Roch and Severi's theorem, thus S is minimal and there exists a rigid curve $D \in |K_S - H_S|$ with $H_S \cdot D = 4$, $p_a(D) = 0$. In particular, $|K_S| = |D + H_S|$ defines an embedding outside the support of D and maps the divisor D onto a line L in $\mathbb{P}^6 = \mathbb{P}(H^0(\mathcal{O}_S(K_S)))$. It follows that $\Psi = \Psi_{|K+H|} : X \to X'' \subset \mathbb{P}^7$ coincides with the second reduction morphism. Moreover, by Proposition 5.6, X'' is smooth, K is a smooth rational scroll $\mathbb{P}^1 \times \mathbb{P}^1 \overset{(1,2)}{\hookrightarrow} \mathbb{P}^5$, which is contracted by Ψ to the line $L \subset X''$, while X is exactly the blow up of X'' along this line. Riemann-Roch gives $\chi(\mathcal{O}_X(2H + 2K)) = 32$, hence $h^0(\mathcal{J}_{X''}(2)) \geq h^0(\mathcal{O}_{\mathbb{P}^7}(2)) - h^0(\mathcal{O}_X(2H + 2K)) = 4$. In other words, X'' lies on 4 linearly independent hyperquadrics. In fact, as one can check, $X'' \subset \mathbb{P}^7$ is the complete intersection $\Sigma_{(2,2,2,2)}$ of 4 hyperquadrics. Conversely, let L be a line in $\mathbb{P}^7$ and $\Sigma_{(2,2,2,2)} \subset \mathbb{P}^7$ a smooth complete intersection of 4 hyperquadrics containing L. Then a general projection $X = \mathrm{proj}_L \Sigma_{(2,2,2,2)} \subset \mathbb{P}^5$ is a 3-fold of type X_{18}. □

Remark 5.9. Similarly, [Po, Proposition 3.59] yields an easy proof for the uniqueness of the examples of smooth 3-folds with $d = 11$ and $\pi = 13$ constructed in [BSS2]. The uniqueness for the other two families with $d = 11$ in [BSS2] is clear from [GP]. □

The construction via syzygies of all smooth 3-folds $X \subset \mathbb{P}^5$ known so far is straightforward. Nevertheless, it is sometimes quite subtle to determine the structure of X. We will give examples of this kind in the next section.

6. Examples: Two families of birational Calabi-Yau 3-folds in $\mathbb{P}^5$

In this section we will construct and study a family of smooth 3-folds $X \subset \mathbb{P}^5$

with $d = 17$, $\pi = 32$, $\chi(\mathcal{O}_X) = 0$ and $\chi(\mathcal{O}_S) = 24$. We will also describe a family of smooth 3-folds $X' \subset \mathbb{P}^5$ obtained via linkage $X' \underset{(5,6)}{\sim} X$.

Let us first explain how to construct X via syzygies. In view of Riemann-Roch the following is a plausible Beilinson cohomology table for $\mathcal{J}_X(5)$:

i ↑

1					
	4	2			
				2	

→ m

Suppose that a 3-fold X with these data exists. Then Beilinson's theorem yields an exact sequence

$$0 \to \mathcal{F} = \mathcal{O}(-1) \oplus 4\Omega^4(4) \xrightarrow{\varphi} \mathcal{G} = 2\Omega^3(3) \oplus 2\mathcal{O} \to \mathcal{J}_X(5) \to 0. \qquad (6.1)$$

Conversely, as one can check, the minors of a generic $\varphi \in \operatorname{Hom}(\mathcal{F}, \mathcal{G})$ vanish along a smooth 3-fold X as desired. By construction, $\mathcal{J}_X$ has syzygies of type

$$0 \leftarrow \mathcal{J}_X \leftarrow 2\mathcal{O}(-5) \oplus 5\mathcal{O}(-6) \leftarrow 8\mathcal{O}(-7) \leftarrow 2\mathcal{O}(-8) \leftarrow 0\,. \qquad (6.2)$$

What kind of 3-fold is X ? From the syzygies we see that X can be linked $(5,5)$ to a 3-fold Z of degree 8. It is not too hard to identify the scheme Z.

Starting conversely with Z we will reconstruct X and study its geometry. Z can be described as follows: Let $Y = \mathbb{P}^1 \times \mathbb{P}^2 \overset{(1,1)}{\hookrightarrow} \mathbb{P}^5$ be a Segre cubic scroll and let $L_1, \ldots, L_5$ be five general lines in $\mathbb{P}^2$. Then for $i = 1, \ldots, 5$ the quadric $Q_i = \mathbb{P}^1 \times L_i \overset{(1,1)}{\hookrightarrow} \mathbb{P}^5$ is contained in Y and spans a linear subspace $\Pi_i \subset \mathbb{P}^5$ of dimension 3. Clearly, $\Pi_i \cap Y = Q_i$ (Y is cut out by quadrics) and $\Pi_i \cap \Pi_j = \mathbb{P}^1 \times \{p_{ij}\}$, where $\{p_{ij}\} = L_i \cap L_j$ for $i < j$. Hence the scheme

$$Z := Y \cup \Pi_1 \cup \cdots \cup \Pi_5$$

is locally Cohen-Macaulay, and moreover a local complete intersection outside the lines $L_{ij} := \mathbb{P}^1 \times \{p_{ij}\}$. Write $Z_k = Y \cup \bigcup_{i=1}^{k} \Pi_i$, $k = 0, \ldots, 5$. Then $Z_0 = Y$ and $Z_5 = Z$. From the exact sequences

$$0 \to \mathcal{J}_{Z_{k-1}}(m-1) \to \mathcal{J}_{Z_k}(m) \to \mathcal{J}_{Z_{k-1} \cap \mathbb{P}^4, \mathbb{P}^4}(m-1) \to 0, \qquad (6.3)$$

where $\mathbb{P}^4 \subset \mathbb{P}^5$ is a general hyperplane through Π_k, we deduce that $h^0\mathcal{J}_Z(3) = 0$, $h^0\mathcal{J}_Z(4) = 1$ and $h^0\mathcal{J}_Z(5) = 26$, and that $\mathcal{J}_Z(5)$ is globally generated.

Proposition 6.4. *Let X be linked to Z in the complete intersection of two general quintic hypersurfaces containing Z. Then X is smooth, it contains the lines L_{ij} and $\mathcal{J}_X(5)$ has a resolution of type* (6.1).

Proof. By a variant of Theorem 2.1 (compare [PS]) X is smooth outside the lines L_{ij}. By using the exact sequences (6.3) we see that the general quintic hypersurface through Z contains the first infinitesimal neighborhood of L_{ij}, which is a multiplicity 5 structure on such a line. Higher infinitesimal neighborhoods are not contained in the general quintic hypersurface through Z. Moreover the tangent cone to Z at a point $p \in L_{ij}$ is $\Pi_i \cup \Pi_j \cup \mathrm{T}_p Y$, and $\Pi_i \cap \mathrm{T}_p Y = \mathrm{T}_p Q_i$. Now a local computation shows that indeed X is smooth along and contains the lines L_{ij}. That $\mathcal{J}_X(5)$ has a Beilinson cohomology table as above follows via liaison from the exact sequences (6.3). □

Remark 6.5. (i) By dualizing (6.1) we find that $\omega_X(1)$ has a presentation of type

$$0 \leftarrow \omega_X(1) \leftarrow \mathcal{O}(1) \oplus 18\mathcal{O} \leftarrow 50\mathcal{O}(-1) \leftarrow \cdots .$$

Thus $|K+H|$ is base point free and $\dim |K+H| = 24$.

(ii) From the double point formulae we compute

$$H^2 \cdot K = 28, \qquad H \cdot K^2 = 18 \quad \text{and} \quad K^3 = -52 .$$

In particular $(K+H) \cdot K^2 = -34$, hence $\kappa(X) \le 1$. In fact, as we will see later, X is a birational Calabi-Yau 3-fold. □

We use in the sequel the above liaison to describe the geometry of X:

Lemma 6.6. *Each linear subspace Π_i intersects X along a smooth sextic surface S_i. A general element in the residual pencil $|H - S_i|$ is a smooth* *blown-up $K3$ surface of degree 11, sectional genus 12, which is embedded in its corresponding $\mathbb{P}^4$ by a linear system of type*

$$|H_{min} - 2E_1 - \sum_{i=2}^{10} E_i| \,, \quad with \quad H_{min}^2 = 24 \,.$$

Proof. It follows from the standard liaison exact sequences that X meets Π_i along a divisor in the class $4H_{\Pi_i} - K_{\Pi_i} - Q_i$, hence along a sextic surface S_i, which is smooth for general choices in the liaison. For the second statement in the lemma, we observe that a general element in $|H - S_i|$ is linked $(4,4)$ inside the hyperplane H to the configuration of planes $P_i \cup \bigcup_{j \neq i}(H \cap \Pi_j)$, where P_i is the plane residual to Q_i in the intersection $H \cap Y$. Therefore the lemma follows from the following:

Proposition 6.7. [Po] *Let T be a configuration $T = P \cup P_1 \cup P_2 \cup P_3 \cup P_4 \subset \mathbb{P}^4$, where P is a plane, while P_i, $i = 1, \ldots, 4$, is a plane meeting P along a line, such that no three of the lines have common intersection points. Then T can be linked in the complete intersection of two general quartic hypersurfaces to a smooth, non-minimal $K3$ surface $S \subset \mathbb{P}^4$ with $d = 11$ and $\pi = 12$, embedded by a linear system*

$$H_S \equiv H_{min} - 2E_1 - \sum_{i=2}^{10} E_i \,, \quad H_{min}^2 = 24 \,.$$

Moreover, P meets S along the exceptional conic E_1 and an extra scheme of length 6, while each intersection $P_i \cap S$ is a plane quintic curve. Residual to it there is a base point free pencil of elliptic space curves of degree 6. □

Lemma 6.8. *$|H - S_i|$ is a base point free pencil, $i = 1, \ldots, 5$.*

Proof. Let H_i denote a general hyperplane containing Π_i, and let K_i be the surface residual to S_i in $H_i \cap X$. Then $S_i \cap K_i \subset \Pi_i \cap K_i$, and in fact equality holds since deg $S_i \cap K_i = 32 - \pi(S_i) - \pi(K_i) + 1 = 11$. Thus if $C_i = S_i \cap K_i$, then $C_i^2 = 2p_a(C_i) - 2 - K_{S_i} \cdot C_i = 0$, where the intersection numbers are computed on S_i, and the lemma follows. □

As a corollary, we deduce that X is an elliptic 3-fold, namely

Corollary 6.9. *For all $i \neq j$, the linear system $|H - S_i| \boxtimes |H - S_j|$ induces an elliptic fibration*

$$\varphi_{|H-S_i| \boxtimes |H-S_j|} : X \to \mathbb{P}^1 \times \mathbb{P}^1,$$

with elliptic space curves of degree 6 as fibres.

Proof. Fix a general point in $\mathbb{P}^1 \times \mathbb{P}^1$, i.e., two general hyperplanes, H_i containing Π_i and H_j containing Π_j, and denote as above by K_i and K_j the residual surfaces to S_i and S_j respectively. By Proposition 6.7, $H_i \cap \Pi_j \cap K_i$ and $H_j \cap \Pi_i \cap K_j$ are plane quintic curves, hence $K_i \cap K_j$ is an elliptic space curve of degree 6, namely the residual to $H_i \cap \Pi_j \cap K_i$ in $H_j \cap K_i$, or equivalently the residual to $H_j \cap \Pi_i \cap K_j$ in $H_i \cap K_j$. □

By liaison we deduce that X meets the Segre scroll Y along a surface T_2 in the class $4H_Y - K_Y - \sum_{i=1}^{5} Q_i$, thus along a conic bundle of degree 10 and sectional genus 6. Moreover, the standard liaison exact sequences yield on X the linear equivalence

$$4H - K \equiv T_2 + \sum_{i=1}^{5} S_i. \tag{6.10}$$

We study in the sequel the structure of the map defined by the composition of the cartesian product of the 5 pencils $|H - S_i|$, $i = 1, \ldots, 5$, with the Segre embedding to $\mathbb{P}^{31}$:

$$\Upsilon = \Upsilon_{|5H - \sum_{i=1}^{5} S_i|} : X \to \mathbb{P}^1 \times \mathbb{P}^1 \times \mathbb{P}^1 \times \mathbb{P}^1 \times \mathbb{P}^1 \hookrightarrow \mathbb{P}^{31}. \tag{6.11}$$

Lemma 6.12. *The canonical divisor K of X has two components* T_1 *and* T_2. T_1 *is a scroll of degree 18 and sectional genus 10, while* T_2 *is the above conic bundle of degree 10 and sectional genus 6.*

Proof. Let, as in the proof of Lemma 6.8, K_i be a general element in the pencil $|H - S_i|$. We recall that $(H - S_i)^2 = 0$, thus $K\ |_{K_i} \equiv K + (H - S_i)\ |_{H-S_i} \equiv K_{H-S_i}$. In other words, K meets a $K3$ surface K_i along its canonical divisor, namely, by Proposition 6.7, along 9 exceptional lines and one exceptional conic in the plane residual to Q_i in $H \cap Y$. In conclusion, the exceptional conics sweep the conic bundle T_2, which is thus a component of K, while the exceptional lines on the K_i's are rulings of a scroll T_1 of degree $H^2 \cdot K - 10 = 18$. Since $\omega_{S_i} = \mathcal{O}_{S_i}(2)$ and $S_i \cap S_j = L_{ij}$, (6.10) restricted to S_i yields $\mathrm{T}_2 \cap S_i \equiv 2H_{S_i} - \sum_{j \neq i} L_{ij}$. On the other side from (6.10) again we infer:

$$(H - S_i)\ |_{S_i} + (2H - \sum_{j \neq i} S_j)\ |_{S_i} + 2H\ |_{S_i} \equiv \mathrm{T}_2\ |_{S_i} + H\ |_{S_i} + K\ |_{S_i}.$$

Thus, since $(H - S_i)\ |_{S_i} \equiv H_{K_i}$, it follows that K intersects S_i along a hyperplane section of X. We deduce that T_1 must intersect S_i along a curve of degree 9 and genus 10, which in turn must be a section of this scroll since it meets the exceptional lines of K_i in one point. In other words, T_1 is a scroll of degree 18 and genus 10. □

Lemma 6.13.

i) *The linear system $|K + H|$ defines a birational morphism $\Psi = \Psi_{|K+H|} : X \to \Psi(X) \subset \mathbb{P}^{23}$, which contracts the scroll T_1 to a curve of degree 27. Moreover, X is the blowing up of $\Psi(X)$ along this curve.*

ii) *The morphism $\Upsilon = \Upsilon_{|5H - \sum_{i=1}^{5} S_i|} : X \to \Upsilon(X) \subset \mathbb{P}^{31}$, induced by $|H + K + T_2|$, contracts the conic bundle T_2 to a curve and is birational on its image.*

Proof. Let K_i be a general element of the pencil $|H - S_i|$. Part i) follows easily since $|K + H|$ induces on K_i the adjunction morphism $\Phi_i = \Phi_{|H_{K_i} + K_{K_i}|} : K_i \to \Phi_i(K_i) \subset \mathbb{P}^{12}$, which is birational and blows down only the 9 exceptional lines $K \cap K_i$. A similar argument works for part ii) since $|5H - \sum_{i=1}^{5} S_i| = |H + K + T_2|$ restricts to K_i as the map onto the second adjoint surface given by the adjunction process. □

We can show now that X is a non minimal Calabi-Yau 3-fold, namely:

Proposition 6.14. *The morphism*

$$\Upsilon = \Upsilon_{|5H - \sum_{i=1}^{5} S_i|} : X \to \mathbb{P}^1 \times \mathbb{P}^1 \times \mathbb{P}^1 \times \mathbb{P}^1 \times \mathbb{P}^1 \hookrightarrow \mathbb{P}^{31}$$

is birational on its image and contracts only the canonical divisor of X to a curve. Moreover, the image $\Upsilon(X)$ is a smooth complete intersection of type $(1,1,1,1,1)^2$ in $\times_{i=1}^{5} \mathbb{P}^1$, hence a minimal Calabi-Yau 3-fold in $\mathbb{P}^{29}$.

Proof. The smoothness of $\Upsilon(X)$ follows from the fact that the iterated adjunction morphisms for K_i blow down only the (-1)-lines and (-1)-conics onto the minimal model of K_i. To see further that $\Upsilon(X)$ is a complete intersection of the type claimed we need to compute some intersection numbers. By Lemma 6.8, $(H - S_i)^2 = 0$, thus $(H - S_i)^2 \cdot H = 0$ and $(H - S_i)^2 \cdot S_j = 0$, which yields $H \cdot S_i^2 = -5$, $S_i^2 \cdot S_j = -4$, for $i \neq j$, and $S_i^3 = -16$. Moreover $S_i \cdot S_j \cdot S_k = 0$, for $i \neq j \neq k$, $i \neq k$, since $\Pi_i \cap \Pi_j \cap \Pi_k \subset L_{ij} \cap L_{ik} \cap L_{jk} = \emptyset$, and so we deduce that $\deg \Upsilon(X) = (5H - \sum_{i=1}^{5} S_i)^3 = 120$. On the other side, $\deg \times_{i=1}^{5} \mathbb{P}^1 = 5! = 120$ in $\mathbb{P}^{31}$, while $\Upsilon(X)$ spans only a $\mathbb{P}^{29}$ since $h^0(\mathcal{O}_X(H+K+T_2)) \leq h^0(\mathcal{O}_X(H+2K)) = \chi(\mathcal{O}_X(H+2K)) = 30$. The proposition follows. □

Proposition 6.15. *Let V^5 and V^6 be general hypersurfaces of degrees 5 and 6 resp. containing X. Then X can be linked in the complete intersection of V^5 and V^6 to a smooth 3-fold $X' \subset \mathbb{P}^5$. X' has invariants $d' = 13$, $\pi' = 18$, $\chi(\mathcal{O}_{X'}) = 0$, $\chi(\mathcal{O}_{S'}) = 10$ and $p_g(X') = h^0(\mathcal{J}_X(5)) - 1 = 1$. Hence $(H')^2 \cdot K' = 8$, $H' \cdot (K')^2 = -2$ and $(K')^3 = -4$ by the double point formulae.*

Proof. Smoothness follows from a Bertini argument since, on V^5, X is cut out by sextic hypersurfaces (compare 2.1 and [PS]). The numerical information follows from the standard liaison exact sequences. □

Corollary 6.16. *X' is the degeneracy locus of a morphism*

$$0 \to \mathcal{O}(-1) \oplus 2\Omega^2(2) \to 4\Omega^1(1) \oplus 2\mathcal{O} \to \mathcal{J}_{X'}(5) \to 0 \, .$$

Hence $\mathcal{J}_{X'}$ has syzygies of type

$$0 \leftarrow \mathcal{J}_{X'}(5) \leftarrow 2\mathcal{O} \oplus 19\mathcal{O}(-1) \leftarrow 50\mathcal{O}(-2) \leftarrow 48\mathcal{O}(-3) \leftarrow 22\mathcal{O}(-4) \leftarrow 4\mathcal{O}(-5) \leftarrow 0 \, .$$

Proof. This follows from (6.1) via liaison or by applying Beilinson's theorem. □

What type of 3-fold is X'?

Proposition 6.17. *K' is a smooth scroll of degree 8, sectional genus 3 over a plane quartic curve. Moreover, the Segre scroll $Y = \mathbb{P}^1 \times \mathbb{P}^2$ meets X' along the scroll K' and a curve of degree 9, arithmetic genus 4 on the scroll T_2.*

Proof. From general liaison arguments it follows that Z intersects X' along the canonical divisor of X' plus may be something of bigger codimension. On the other side, $V^6 \cap \Pi_i = S_i$ since $\Pi_i \cap X = S_i$ for all i. We deduce that the 2-dimensional part of the scheme theoretical intersection $Y \cap X'$ is exactly K'. Now Pic (Y) is generated by the classes $P = [\{\text{point}\} \times \mathbb{P}^2]$ and $Q = [\mathbb{P}^1 \times \mathbb{P}^1]$, and $P^2 = 0$, $Q^3 = 0$, $Q^2 \cdot P = 1$. Then the scroll T_2 is of class $4H_Y - K_Y - \sum_{i=1}^{5} Q_i \equiv 6P + 2Q$. But K' is residual to T_2 in $Y \cap V^6$. Moreover, T_2 is cut out on Y outside the Π_i's by the sextic hypersurfaces through X. It follows that K' is smooth for a general choice of the liaison, and that K' is of class $6H_Y - T_2 \equiv 4Q$. In particular, K' is a scroll over a plane quartic curve and has the claimed invariants. Outside K', X' can meet the scroll Y only inside $T_2 \subset Y \cap X$. The proposition follows now since $X' \cap T_2 \equiv (5H_X - K_X) \cdot T_2$ is a curve of degree 41, arithmetic genus 80, with a component of degree 32 on the scroll K'. □

Proposition 6.18.

i) $|H' + K'|$ *is base point free and big.*

ii) $\Psi' = \Psi_{|H'+K'|} : X' \to \mathbb{P}^9$ *is birational on its image* $M = \Psi'(X')$, *which is a smooth Calabi-Yau 3-fold, with* deg $M = 27$, $\pi(M) = 28$ *and* $c_3(M) = -64$.

iii) Ψ' *contracts the scroll* K' *to a curve of degree 6 and genus 3, and is an isomorphism outside this scroll. Moreover,* X' *is the blow up of* M *along this curve.*

Proof. i) From the syzygies we see that $\omega_{X'}$ is a quotient of $\mathcal{O} \oplus 4\mathcal{O}(-1)$, thus $|H' + K'|$ is base point free and big since $(H' + K')^3 = 27$. Moreover dim $|H' + K'| = 9$

ii) From the liaison exact sequence

$$0 \to \mathcal{J}_{V^5 \cap V^6}(6) \to \mathcal{J}_X(6) \to \omega_{X'}(1) \to 0$$

we deduce that the map $\Psi' : X' \to \mathbb{P}^9$ is in fact the composition of the restriction to X' of the rational morphism $\Xi : \mathbb{P}^5 \dashrightarrow \mathbb{P}^{16} = \mathbb{P}(H^0(\mathcal{J}_X(6))$ given by the sextic hypersurfaces through X, with a projection from $\mathbb{P}^{16} \dashrightarrow \mathbb{P}^9$ along $\mathbb{P}^6 = \mathbb{P}(H^0(\mathcal{J}_{V^5 \cap V^6}(6))$. Thus in order to show that Ψ' is birational on its image it is enough to check that Ξ is birational on its image and that the projection $\mathbb{P}^{16} \dashrightarrow \mathbb{P}^9$ is generic. But one checks easily that 5 general sextic hypersurfaces through X meet in exactly one point outside X. In particular, it follows that $\Psi' : X' \to M$ coincides with the second reduction map of X'.

iii) Since $(H' + K') \cdot K' = 0$, Ψ' contracts the scroll K'. Its image is isomorphic to the plane quartic curve, which is the base of the scroll K'. From Remark 5.7 it follows that M is smooth and Ψ' is an isomorphism outside the scroll K', unless there are divisors $D \cong \mathbb{P}^2 \subset X'$ with $\mathcal{N}_D^{X'} \cong \mathcal{O}_{\mathbb{P}^2}(-2)$, which are contracted to singular points on M. Assume that such a divisor D exists. Then a general hyperplane section S' of X' contains a (-2)-line L. But on S', $h^0(\mathcal{O}_{S'}(K_{S'} - H_{S'})) = h^3(\mathcal{O}_{X'}) = 1$, so if $D_{S'} \in |K_{S'} - H_{S'}|$, then $K_{S'} \cdot L = 0 = 1 + D_{S'} \cdot L$, and thus L must be a component of $D_{S'}$. On the other side $D_{S'}$ is an irreducible hyperplane section of the smooth scroll $K_{X'}$, and therefore contains no such line as a component. □

7. Overview

In this section we collect some information on the families of smooth non general type surfaces in $\mathbb{P}^4$ and 3-folds in $\mathbb{P}^5$ known to us.

Table 7.1. Known families of smooth non general type surfaces in $\mathbb{P}^4$

	Enriques-Kodaira Classification						
degree	rational	ruled irrat.	Enriques	$K3$	abelian	bielliptic	elliptic
$d \leq 4$	6			1			
$d = 5$	1 [Ca]	1 [Seg]					
$d = 6$ [Io1],[Ok1]	1 [Bo],[Ve] [Wh]			1			
$d = 7$ [Io1],[Ok3]	1 [Io1],[Ok3]			1 [Ro1]			1 [Ba]
$d = 8$ [Io2],[Ok4]	2 [Ok4],[Al1]			1 [Ok4]			1 [Ba]
$d = 9$ [AR]	2 [Al1],[Al2]		1 [Cos],[CV]	1 [Ro1]			1 [AR]
$d = 10$ [Ra],[PR]	3 [DES],[Ra]		1 [DES],[Br]	2 [Ra],[Po]	1 [Co],[HM] [HL],[Ram]	1 [Ser] [ADHPR1]	2 [Ra]
$d = 11$ [Po]	3+2 [DES],[Po]		1 [DES]	5 [DES],[Po]			1 [Po]
$d = 12$				1 [DES]			3 [Po]
$d = 13$			2 [DES],[Po]	1 [Po]			
$d = 14$				1 [Po]			
$d = 15$					2 [HM],[Po]	1 [ADHPR1]	

Remark 7.2. (i) The classification of smooth surfaces in $\mathbb{P}^4$ is complete up to degree 10, and there is a partial classification in degree 11. In the first column of Table 7.1 we refer to the papers, where one can find the classification results. In the other columns we indicate the number of families known and the corresponding references. The classification up to degree 5 is classical. More information can be found in [DES, Appendix B], [Po, Appendix] and [ADHPR2].

(ii) Two families of rational surfaces of degree 11 are due to Schreyer (unpublished).

(iii) One of the families of K3 surfaces of degree 11 has been first constructed by Ranestad (compare [Po, Proposition 3.41]). □

Table 7.3. Known families of smooth, non-degenerate, non general type 3-folds in $\mathbb{P}^5$

X	d	π	p_g	$\chi(\mathcal{O}_X)$	$\chi(\mathcal{O}_S)$	$\kappa(X)$	liaison	type	classification	ref.
X_1	3	0	0	1	1	$-\infty$	$X_1 \overset{(2,2)}{\sim} \mathbb{P}^3$	Segre embedd. of $\mathbf{P}^1 \times \mathbf{P}^2$	rational scroll	
X_2	4	1	0	1	1	$-\infty$	$X_2 = \Sigma_{(2,2)}$	Fano 3-fold of index 2	rational	[Kl]
X_3	5	2	0	1	1	$-\infty$	$X_3 \overset{(2,3)}{\sim} \mathbb{P}^3$	Castelnuovo 3-fold; quadric fibration over $\mathbf{P}^1$ via $\lvert K+2H\rvert$	rational	[Io1],[Ok2]
X_4	6	3	0	1	1	$-\infty$	$X_4 \overset{(3,3)}{\sim} X_1$	Bordiga 3-fold; $X_4 = \mathbf{P}(\mathcal{E})$, $\mathcal{E}$ rk 2 vb on $\mathbf{P}^2$ $c_1 = 4$, $c_2 = 10$, via $\lvert K+2H\rvert$	rational scroll	[Io1],[Ok2]
X_5	6	4	0	1	2	$-\infty$	$X_5 = \Sigma_{(2,3)}$	Fano 3-fold of index 1	unirational not rational	[Io1],[Ok2] [En],[Fa1]
X_6	7	4	0	1	1	$-\infty$	X_6 proj. Buchsbaum	$X_6 = \mathbb{P}(\mathcal{E})$, $\mathcal{E}$ rk 2 vb on $\mathbb{P}^2(x_1,\ldots x_6)$; via $\lvert K+2H\rvert$	rational scroll	[Io1],[Ok3] [Pa]
X_7	7	5	0	1	2	$-\infty$	$X_7 \overset{(3,3)}{\sim} \Sigma_{(1,2)}$	$X_7 = \Sigma_{(2,2,2)}(x_0)$, the blowing up of a c.i. $\Sigma_{(2,2,2)} \subset \mathbf{P}^6$; via $\lvert K+2H\rvert$	blown up Fano of index 1	[Io1],[Ok3]
X_8	7	6	0	1	3	$-\infty$	$X_8 \overset{(2,4)}{\sim} \mathbb{P}^3$	Del Pezzo fibration over $\mathbf{P}^1$, with gen. fibre $\mathbf{P}^2(x_1,\ldots x_6)$; via $\lvert K+H\rvert$	rational	[Io1],[Ok3]
X_9	8	7	0	1	3	$-\infty$	$X_9 \overset{(3,3)}{\sim} \mathbb{P}^3$	Del Pezzo fibration over $\mathbf{P}^1$; gen. fibre c.i. $(2,2)$ in $\mathbf{P}^4$; via $\lvert K+H\rvert$	rational	[Io2]
X_{10}	8	9	1	0	6	0	$X_{10} = \Sigma_{(2,4)}$	minimal Calabi-Yau 3-fold		
X_{11}	9	8	0	2	2	$-\infty$	X_{11} proj. Buchsbaum	$\mathbf{P}^1$ bundle over a minimal $K3$ surface $S \subset \mathbf{P}^8$; via $\lvert K+H\rvert$	scroll, not rational	[Ch3]
X_{12}	9	9	0	1	4	$-\infty$	$X_{12} \overset{(3,4)}{\sim} X_1$	conic bundle over $\mathbf{P}^2$, via $\lvert K+H\rvert$	rational	[BSS1]
X_{13}	9	10	1	0	6	0	$X_{13} = \Sigma_{(3,3)}$	minimal Calabi-Yau 3-fold		
X_{14}	9	12	2	-1	9	1	$X_{14} \overset{(2,5)}{\sim} \mathbb{P}^3$	minimal $K3$ fibration over $\mathbf{P}^1$, via $\lvert K\rvert$; $4K + 3H > 0$. log-gen type		[BSS1]
X_{15}	10	11	0	1	5	$-\infty$	$X_{15} \overset{(4,4)}{\sim} X_4$	log-general type; $\lvert K+H\rvert$ is birational onto $\mathbf{P}^3$	rational	[BSS1]
X_{16}	10	12	1	0	7	0	$X_{16} \overset{(3,4)}{\sim} \Sigma_{(1,2)}$	$X_{16} = \mathrm{proj}_p \Sigma_{(2,2,3)}$, with $\Sigma_{(2,2,3)} \subset \mathbf{P}^6$ c.i. singular at p; via the inverse of $\lvert K+H\rvert$	birational Calabi-Yau; $H^2K = 2$	[BSS1]
X_{17}	11	13	0	1	6	$-\infty$	$X_{17} \overset{(4,5)}{\sim} X_{11}$	blown up Fano 3-fold; $\lvert K+H\rvert$ is birational onto a hypercubic in $\mathbf{P}^4$;	unirational not rational	[Ch3] [BSS2]
X_{18}	11	14	1	0	8	0	$X_{18} \overset{(4,4)}{\sim} X_3$	$X_{18} = \mathrm{proj}_L \Sigma_{(2,2,2,2)}$, with $L \subset \Sigma_{(2,2,2,2)} \subset \mathbf{P}^7$ smooth c.i. and L a line; via $\lvert K+H\rvert$	birational Calabi-Yau; $H^2K = 4$	[BSS2]
X_{19}	11	15	2	-1	10	1	$X_{19} \overset{(3,4)}{\sim} \mathbb{P}^3$	minimal $K3$ fibration over $\mathbf{P}^1$ via $\lvert K\rvert$; fibres are $(2,3)$ c.i. in $\mathbf{P}^4$		[BSS2]

X	d	π	p_g	$\chi(\mathcal{O}_X)$	$\chi(\mathcal{O}_S)$	$\kappa(X)$	liaison	type	classification	ref
X_{20}	12	15	0	2	6	$-\infty$	$X_{20} \overset{(5,5)}{\sim} X_{25}$	conic bundle over a $K3$ quartic surface $S \subset \mathbb{P}^3$, via $\|K+H\|$	not rational	[BOSS2]
X_{21}	12	15	0	1	7	$-\infty$	$X_{21} \overset{(5,5)}{\sim} X_{15} \cup X_1$	$\|K+H\|$ defines a birational map onto a Bordiga $X_4 \subset \mathbb{P}^5$	rational	[Ed]
X_{22}	12	16	1	0	9	0	$X_{22} \overset{(5,5)}{\sim} X_{27}$	$\|K+H\|$ defines a birational morphism onto $\mathbb{P}(R_2) \cap Bl_{p_2}\mathbb{P}^8$; see [Ch3] for details	birational Calabi-Yau; $H^2K = 6$	[Ch3]
X_{23}	12	17	2	-1	11	1	$X_{23} \overset{(4,5)}{\sim} X_9$	$\|K\|$ defines a $K3$ fibration over $\mathbb{P}^1$; the fibres are $\Sigma_{(2,2,2)}$ in $\mathbb{P}^5$	minimal	[Ed]
X_{24}	12	18	3	-2	13	2	$X_{24} \overset{(3,6)}{\sim} X_4$	$\|K\|$ defines an elliptic fibration (in plane cubics) over $\mathbb{P}^2$	minimal elliptic	[Ed]
X_{25}	13	18	0	1	9	$-\infty$	$X_{25} \overset{(5,5)}{\sim} X_{20}$	log-gen type		[BOSS2]
X_{26}	13	18	1	0	10	0	$X_{26} \overset{(5,6)}{\sim} X_{29}$	$\|K+H\|$ defines a birational map onto a Calabi-Yau 3-fold $Y \subset \mathbb{P}^9$ with $c_3(Y) = -64$	blown up Calabi-Yau	
X_{27}	13	19	0	1	11	$-\infty$	$X_{27} \overset{(4,5)}{\sim} X_6$	$\|K+H\|$ defines a birational map onto $\mathbb{G}(1,5) \cap \mathbb{P}^9$; birat. to cubic 3-fold in $\mathbb{P}^4$ [Fa2]	unirational not rational	[Is] [Ch3]
X_{28}	14	22	1	0	14	0	$X_{28} \overset{(5,5)}{\sim} X_{17}$	$\|K+H\|$ defines a birational map onto $\mathbb{G}(1,6) \cap \mathbb{P}^{13}$;	birational Calabi-Yau	[Ch3]
X_{29}	17	32	1	0	24	0	$X_{29} \overset{(5,5)}{\sim} X_1 \cup \bigcup_{i=1}^{5} \mathbb{P}^3$	$K = K_1 + K_2$, $H^2K_1 = 10$; $\|H+K+K_1\|$ birat. onto a c.i. $(1,1,1,1,1)^2$ in $\times_{i=1}^{5} \mathbb{P}^1 \subset \mathbb{P}^{31}$	birational Calabi-Yau (elliptic)	
X_{30}	18	35	0	2	26	$-\infty$		log-general type	not rational	

Remark 7.4. (i) The classification of smooth 3-folds in $\mathbb{P}^5$ is complete up to degree 11 and almost complete in degree 12.
(ii) Some of the information in Table 7.3 is new. It can be obtained along the lines of section 5 (compare Example 5.8).
(iii) In order to construct X_{21} via a liaison $X_{21} \overset{(5,5)}{\sim} X_{15} \cup X_1$, one has to choose X_{15} and X_1 in a special position. $\square$

References

[ADHPR1] Aure, A.B., Decker, W., Hulek, K., Popescu, S., Ranestad, K., *The Geometry of Bielliptic Surfaces in* $\mathbb{P}^4$, Int. J. of Math. (to appear).

[ADHPR2] ———, *Syzygies of abelian and bielliptic surfaces in* $\mathbb{P}^4$, in preparation.

[Al1] Alexander, J., *Surfaces rationelles non-speciales dans* $\mathbb{P}^4$, Math. Z. **200** (1988), 87–110.

[Al2] ———, *Speciality one rational surfaces in* $\mathbb{P}^4$, Complex Projective Geometry, Proceedings Bergen-Trieste, London Math. Soc., LNS **179**, 1992, pp. 1–23.

[AR] Aure, A.B., Ranestad, K., *The smooth surfaces of degree 9 in* $\mathbb{P}^4$, Complex Projective Geometry, Proceedings Bergen-Trieste, London Math. Soc., LNS **179**, 1992, pp. 32–46.

[Au] Aure, A.B., *On surfaces in projective 4-space*, Thesis, Oslo 1987.

[Ba] Baker, H. F., *Principles of Geometry*, vol. VI, Cambridge University Press, 1933.

[Bau] Bauer, I., *Projektionen von glatten Flächen in den* $\mathbb{P}^4$, Dissertation, Bonn, 1992.

[BB] Ballico, E., Bolondi, G., *The variety of module structures*, Arch. der Math. **54** (1990), 397–408.

[BBM] Ballico, E., Bolondi, G., Migliore, J.C., *The Lazarsfeld-Rao problem for liaison classes of two-codimensional subschemes of* $\mathbb{P}^n$, Amer. J. of Math. **113** (1991), no. 1, 117–128.

[BBS] Beltrametti, M.C., Biancofiore A., Sommese, A.J., *Projective n-folds of log-general type, I*, Transactions of the AMS **314** (1989), no. 2, 825–849.

[BE] Buchsbaum, D.A., Eisenbud, D., *Generic free resolutions and a family of generically perfect ideals*, Adv. Math. **18** (1975), 245–301.

[Bei] Beilinson, A., *Coherent sheaves on* $\mathbb{P}^N$ *and problems of linear algebra*, Funct. Anal. Appl. **12** (1978), 214-216.

[BF] Braun, R., Fløystad, G., *A bound for the degree of surfaces in* $\mathbb{P}^4$ *not of general type*, Preprint Bayreuth/Bergen (1993).

[BFS] Beltrametti, M.C., Fania, M.L., Sommese, A.J., *On the adjunction theoretic classification of projective varieties*, Math. Ann. **290** (1991), 31–62.

[BL] Barth, W., Larsen M.E., *On the homotopy types of complex projective manifolds*, Math. Scand. **30** (1972), 88–94.

[BM] Bolondi, G., Migliore, J.C., *The structure of an even liaison class*, Trans. AMS. **316** (1989), no. 1, 1–38.

[Bo] Bordiga, G., *La superficie del 6d ordine, con 10 rette, nello spazio* $\mathbb{P}^4$*; e le sue proiezioni nello spazio ordinario*, Atti. Accad. Naz. Lincei. Mem., (4), **IV** (1887), 182.

[BOSS1] Braun, R., Ottaviani, G., Schneider, M, Schreyer, F.-O., *Boundedness for non general type 3-folds in* $\mathbb{P}^5$, Complex Analysis and Geometry, Plenum Press (to appear).

[BOSS2] ———, *Classification of log-special 3-folds in* $\mathbb{P}^5$, Preprint Bayreuth (1992).

[BR] Braun, R., Ranestad, K., *Conic bundles in projective fourspace*, Preprint Bayreuth (1993).

[Br] Brivio, S., *Smooth Enriques surfaces in* $\mathbb{P}^4$ *and exceptional bundles*, Math. Z. **213** (1993), 509–521.

[BSS1] Beltrametti, M.C., Schneider, M., Sommese, A.J., *Threefolds of degree 9 and 10 in* $\mathbb{P}^5$, Math. Ann. **288** (1990), 413–444.

[BSS2] ———, *Threefolds of degree 11 in* $\mathbb{P}^5$, Complex Projective Geometry, Proceedings Bergen-Trieste, London Math. Soc., LNS **179**, 1992, pp. 59–80.

[BSS3] ———, *Some special properties of the adjunction theory for 3-folds in* $\mathbb{P}^5$, Preprint Bayreuth, 1993.

[Ca] Castelnuovo, G., *Sulle superficie algebriche le cui sezioni sono curve iperellittiche*, Rend. Palermo **IV** (1890), 73–88.

[CE] Castelnuovo, G., Enriques, F., *Sur quelques resultats nouveaux dans la theorie des surfaces algebriques*, Note V, Théorie des Fonctions Algebriques de

Deux Variables Indépendantes, by Picard, E. and Simart, G., Chelsea Pub Co., Bronx, New York, 1971.

[Ch1] Chang, M.-C., *A filtered Bertini-type theorem*, J. reine angew. Math. **397** (1989), 214–219.

[Ch2] ———, *Characterization of arithmetically Buchsbaum subschemes of codimension 2 in* $\mathbb{P}^n$, J. Diff. Geometry **31** (1990), 323–341.

[Ch3] ———, *Classification of Buchsbaum subvarieties of codimension 2 in projective space*, J. reine angew. Math. **401** (1989), 101–112.

[Co] Comessatti, A., *Sulle superficie die Jacobi semplicimente singolari*, Tipografia della Roma Accad. dei Lincei, Roma 1919.

[Cos] Cossec, F., *On the Picard group of Enriques surfaces*, Math. Ann. **271** (1985), 577–600.

[CV] Conte, A., Verra, A., *Reye constructions for nodal Enriques surfaces*, Preprint **129**, Genova 1990.

[DES] Decker, W., Ein, L., Schreyer, F.-O., *Construction of surfaces in* $\mathbb{P}^4$, J. of Algebraic Geometry **2** (1993), 185–237.

[Ed] Edelmann, G., *3-Mannigfaltigkeiten im* $\mathbb{P}^5$ *vom Grad 12*, Thesis, Bayreuth 1993.

[EP] Ellingsrud. G., Peskine, C., *Sur les surfaces lisse de* $\mathbb{P}_4$, Inv. Math. **95** (1989), 1–12.

[ES] Ellia, Ph., Sacchiero, G., *Surfaces lisses de* $\mathbb{P}^4$ *reglées en coniques*, talk at the conference "Projective Varieties", Trieste, June 1989.

[Fa1] Fano, G., *Osservazioni sopra alcune varietà non razionali aventi tutti i generi nulli*, Atti Acc. Torino **50** (1915), 1067–1072.

[Fa2] ———, *Sulle sezioni spaziali della varietà Grassmanniana della rette spazio a cinque dimensioni*, Rend. R. Accad. Lincei **11, no. 6** (1930), 329–356.

[GP] Gruson, L., Peskine, Ch., *Genre des courbes de l'espace projectif*, Algebraic Geometry, Trømso 1977, LNM, vol. 687, Springer, Berlin, Heidelberg, New York, Tokyo, 1978, pp. 31–59.

[Ha1] Hartshorne, R., *Varieties of small codimension in projective space*, Bull. A.M.S. **80** (1974), 1017–1032.

[Ha2] ———, *Algebraic geometry*, Springer, Berlin, Heidelberg, New York, Tokyo, 1977.

[HL] Hulek, K., Lange, H., *Examples of abelian surfaces in* $\mathbb{P}^4$, J. reine und angew. Math. **363** (1985), 201–216.

[HM] Horrocks, G., Mumford, D., *A rank 2 vector bundle on* $\mathbb{P}^4$ *with 15,000 symmetries*, Topology **12** (1973), 63–81.

[Hu] Hulek, K., *Projective geometry of elliptic curves*, Astérisque **137**.

[Io1] Ionescu, P., *Embedded projective varieties of small invariants*, Proceedings of the week of algebraic geometry, Bucharest,1982, LNM, vol. 1056, Springer, Berlin, Heidelberg, New York, Tokyo, 1984.

[Io2] ———, *Embedded projective varieties of small invariants II*, Rev. Roumaine Math. Pures Appl. **31** (1986), 539-544.

[Io3] ———, *Generalized adjunction and applications*, Math. Proc. Cambridge Phil. Soc. **9** (1986), 457–472.

[Is] Iskovskih, V. A., *Anticanonical models of three-dimensional algebraic varieties*, J. Soviet Math. **13, nr. 4** (1980), 815–868.

[Kle] Klein, F., *Zur Theorie der Liniencomplexe des ersten und zweiten Grades*, Math. Ann. **2** (1870), 198–226.

[Klm] Kleiman, S., *Geometry on grassmanians and applications to splitting bundles and smoothing cycles*, Publ. Math. I.H.E.S. **36** (1969), 281–297.

[KMM] Kawamata, Y., Matsuda, K., Matsuki, K., *Introduction to the minimal model problem*, Algebraic Geometry, Sendai 1985, Advanced Studies in Pure Math., vol. 10, 1987, pp. 283–360.

[Ko] Koelblen, L., *Surfaces de $\mathbb{P}_4$ tracées sur une hypersurface cubique*, J. reine und angew. Math. **433** (1992), 113–141.

[La] Lanteri, A., *On the existence of scrolls in $\mathbb{P}^4$*, Lincei-Rend. Sc. fis.mat.e.nat **LXIX** (1980), 223–227.

[LB] Le Barz, P., *Formules pour les multisecantes des surfaces*, C.R. Acad. Sc. Paris **292, Serie I** (1981), 797–799.

[LR] Lazarsfeld, R., Rao, P., *Linkage of general curves of large degree*, Algebraic Geometry - open problems, Ravello 1982, LNM, vol. 997, Springer, Berlin, Heidelberg, New York, Tokyo, 1983, pp. 267–289.

[Mac] Bayer, D., Stillman, M., *Macaulay: A system for computation in algebraic geometry and commutative algebra*, Source and object code available for Unix and Macintosh computers. Contact the authors, or download from **zariski.harvard.edu** via anonymous ftp..

[MDP] Martin-Deschamps, M., Perrin, D., *Sur la classification des courbes gauches*, Astérisque **184-185** (1990).

[Ok1] Okonek, C., *Moduli reflexiver Garben und Flächen von kleinem Grad in $\mathbb{P}^4$*, Math. Z. **184** (1983), 549–572.

[Ok2] ———, *3-Mannigfaltigkeiten in $\mathbb{P}^5$ und ihre zugehörigen stabilen Garben*, Manuscripta Math. **38** (1982), 175–199.

[Ok3] ———, *Über 2-codimensionale Untermannigfaltigkeiten vom Grad 7 in $\mathbb{P}^4$ and $\mathbb{P}^5$*, Math. Z. **187** (1984), 209–219.

[Ok4] ———, *Flächen vom Grad 8 im $\mathbb{P}^4$*, Math. Z. **191** (1986), 207–223.

[Ok5] ———, *On codimension 2 submanifolds in $\mathbb{P}^4$ and $\mathbb{P}^5$*, Math. Gottingensis **50** (1986).

[Ott] Ottaviani, G., *3-Folds in $\mathbb{P}^5$ which are scrolls*, Annali Sc. Norm. Sup. Pisa (to appear).

[Pa] Palatini, F., *Sui sistemi lineari di complessi lineari di rette nello spazio a cinque dimensioni*, Atti Ist. Veneto **60** (1900), 371–383.

[Pe] Peskine, Ch., *Hilbert polynomials of smooth surfaces in $\mathbb{P}^4$. Comments*, Preprint Paris 1993.

[Po] Popescu, S., *On smooth surfaces of degree ≥ 11 in $\mathbb{P}^4$*, Dissertation, Saarbrücken, 1993.

[PR] Popescu, S., Ranestad, K., *Surfaces of degree 10 in projective four-space via linear systems and linkage*, Preprint Saarbrücken/Oslo 1993.

[PS] Peskine, Ch., Szpiro, L., *Liaison des variétés algébriques I*, Invent. Math. **26** (1974), 271–302.

[Ra] Ranestad, K., *On smooth surfaces of degree ten in the projective fourspace*, Thesis, Univ. of Oslo, 1988.

[Ram] Ramanan, S., *Ample divisors on abelian surfaces*, Proc. London Math. Soc. **51** (1985), 231–245.

[Rei] Reider, I., *Vector bundles of rank 2 linear systems on algebraic surfaces*, Ann. Math. **127** (1988), 309–316.

[Ro1] Roth, L., *On the projective classification of surfaces*, Proc. of London Math. Soc. **42** (1937), 142–170.

[Ro2] ———, *Algebraic Threefolds*, Springer, Berlin, Göttingen, Heidelberg, 1955.

[Roo] Room, T. G., *A General configuration in Space of any Number of Dimensions Analogous to the Double-Six of Lines in Ordinary Space*, Proc. Royal Soc. London **CXI, Series A** (1926), 386–404; *The geometry of determinantal loci*, Cambridge University Press, Cambridge, 1938.

[Sch] Schneider, M., *3-folds in $\mathbb{P}^5$: classification in low degree and finiteness results*, Geometry of complex projective varieties, Cetraro, June 1990, Seminars and Conferences, vol. 9, Mediterranean Press, 1993, pp. 275–289.

[Se] Severi, F., *Intorno ai punti doppi improri di una superficie generale dello spazio ai quattro dimensioni, e a suoi punti tripli apparenti*, Rend. Circ. Math., Palermo **15** (1901), 33–51.

[Seg] Segre, C., Rend. Palermo **II** (1888), 42–52.

[Ser] Serrano, F., *Divisors of bielleptic surfaces and embeddings in* $\mathbb{P}^4$, Math. Z. **203** (1990), 527–533.

[So1] Sommese, A.J., *Hyperplane sections of projective surfaces I. The adjunction mapping*, Duke Math. J. **46** (1979), 377–401.

[So2] ———, *On hyperplane sections*, Algebraic Geometry, Proceedings, Chicago Circle Conference, 1980, LNM, vol. 862, Springer, Berlin, Heidelberg, New York, Tokyo, 1981, pp. 232–271.

[So3] ———, *On the minimality of hyperplane sections of projective threefolds*, J. reine und angew. Math. **329** (1981), 16–41.

[So4] ———, *Configurations of -2 rational curves on hyperplane sections of projective threefolds*, Classification of Algebraic and Analytic Manifolds, Katata Symposium 1982, Prog. Math., vol. 39, Birkhäuser, Basel, 1983, pp. 465–497.

[So5] ———, *On the adjunction theoretic structure of projective varieties*, Complex Analysis and Algebraic Geometry, Proceedings Göttingen 1985, LNM, vol. 1194, Springer, Berlin, Heidelberg, New York, Tokyo, 1986, pp. 175–213.

[SV] Sommese, A.J., Van de Ven, A., *On the adjunction mapping*, Math. Ann. **278** (1987), 593–603.

[VdV] Van de Ven, A., *On the 2-connectedness of very ample divisors on a surface*, Duke Math. J. **46** (1979), 403–407.

[Ve] Veronese, G., *Behandlung der projektivischen Verhältnisse der Räume von verschiedenen Dimensionen durch das Princip des Projicirens und Schneidens*, Math. Ann. **XIX** (1882), 161–234.

[Wh] White, F.P., *The projective generation of curves and surfaces in space of four dimensions*, Proc. Camb. Phil. Soc. **21** (1922), 216–227.

[Za] Zak, F. L., *Projections of algebraic varieties*, Math. USSR Sbornik **44** (1983), 535–544.

Wolfram Decker

Fachbereich Mathematik, Universität des Saarlandes, D 66041 Saarbrücken, Germany

E-mail address: decker@math.uni-sb.de

Sorin Popescu

Fachbereich Mathematik, Universität des Saarlandes, D 66041 Saarbrücken, Germany

E-mail address: popescu@math.uni-sb.de

Exceptional bundles and moduli spaces of stable sheaves on $\mathbb{P}_n$

J.-M. Drézet

1 Introduction

In this paper I try to show how the exceptional bundles can be useful to study vector bundles on projective spaces. The exceptional bundles appeared in [5], and they were used to describe the ranks and Chern classes of semi-stable sheaves. In [1] the generalized Beilinson spectral sequence, built with exceptional bundles, was defined, and it was used in [2] and [3] to describe some moduli spaces of semi-stable sheaves on $\mathbb{P}_2$. The general notion of exceptional bundle and helix, on $\mathbb{P}_n$ and many other varieties, is due mainly to A.L. Gorodentsev and A.N. Rudakov (cf. [7] , [14]). A.N. Rudakov described completely in [12] the exceptional bundles on $\mathbb{P}_1 \times \mathbb{P}_1$, and used them in [13] to describe the ranks and Chern classes of semi-stable sheaves on this variety. The exceptional vector bundles on $\mathbb{P}_3$ have been studied (cf [4], [10], [11]) but they have not yet been used to describe semi-stable sheaves on $\mathbb{P}_3$. On higher $\mathbb{P}_n$ almost nothing is known.

In the second part of this paper, new invariants of coherent sheaves of non-zero rank are defined. In some cases they are more convenient than the Chern classes.

In the third part the exceptional bundles and helices are defined, and their basic properties are given.

In the fourth part, I define some useful hypersurfaces in the space of invariants of coherent sheaves on $\mathbb{P}_n$. On $\mathbb{P}_2$, this space is $\mathbb{R}^2$, with coordinates (μ, Δ), where μ is the *slope* and Δ the *discriminant* of coherent sheaves, as defined in [5]. On $\mathbb{P}_n$, the space of invariants is $\mathbb{R}^n$, and the coordinates are the invariants defined in the second part.

In the fifth part, the description of ranks and Chern classes of semi-stable sheaves on $\mathbb{P}_2$ is recalled. The ranks and Chern classes of semi-stable sheaves on $\mathbb{P}_3$ are not known, and in this case I can only try to formulate the problem correctly, using the notions of the fourth part.

In the sixth part, some partial results are given on the description of the simplest moduli spaces of semi-stable sheaves on $\mathbb{P}_n$. A moduli space is simple when the corresponding point in the space of invariants belongs to many hypersurfaces defined in part 4 (in this case a suitable generalized Beilinson spectral sequence applied to the sheaves of this moduli space is supposed to degenerate). In the case of $\mathbb{P}_n$, $n \geq 3$, many questions remain open.

2 Logarithmic invariants

Let X be a projective smooth algebraic variety of dimension n, E a vector bundle (or coherent sheaf) on X, of rank $r > 0$. The *logarithmic invariants* $\Delta_i(E) \in A^i(E) \otimes \mathbb{Q}$ of E are defined formally by the following formula :

$$log(ch(E)) = log(r) + \sum_{i=1}^{n} (-1)^{i+1} \Delta_i(E),$$

where $ch(E)$ is the Chern character of E. For example, we have

$$\Delta_1(E) = \frac{c_1}{r} \ , \ \Delta_2(E) = \frac{1}{r}(c_2 - \frac{r-1}{2r} c_1^2),$$

$$\Delta_3(E) = \frac{1}{r}(\frac{c_3}{2} + c_1 c_2(\frac{1}{r} - \frac{1}{2}) + c_1^3(\frac{1}{3r^2} - \frac{1}{2r} + \frac{1}{6})),$$

$$\begin{aligned} \Delta_4(E) &= \frac{1}{r}(\frac{c_4}{6} + c_2^2(\frac{1}{2r} - \frac{1}{12}) + c_1 c_3(\frac{1}{2r} - \frac{1}{6}) \\ &\quad + c_1^2 c_2(\frac{1}{r^2} - \frac{1}{r} + \frac{1}{6}) + c_1^4(\frac{1}{4r^3} - \frac{1}{2r^2} + \frac{7}{24r} - \frac{1}{24}), \end{aligned}$$

where for $1 \leq i \leq n$, c_i is the i-th Chern class of E. The first invariant is the *slope* and the second the *discriminant* of E.

Proposition 2.1 *Let L be a line bundle, E,F vector bundles on X. Then*

1. $\Delta_1(L) = c_1(L)$ *and* $\Delta_i(L) = 0$ *if* $i > 1$.
2. $\Delta_i(E\otimes F) = \Delta_i(E) + \Delta_i(F)$ *if* $1 \leq i \leq n$. *Thus* $\Delta_i(E\otimes L) = \Delta_i(E)$ *if* $2 \leq i \leq n$.
3. $\Delta_i(E^*) = (-1)^i\Delta_i(E)$ *if* $1 \leq i \leq n$.

This is clear from the definition of the Δ_i's.

Since $ch(E)/r$ is a polynomial in $\Delta_1(E), \dots, \Delta_n(E)$, the Riemann-Roch theorem on X can be written in the following way :

$$\frac{\chi(E)}{r} = P(\Delta_1(E), ..., \Delta_n(E)),$$

where P is a polynomial with rational coefficients that depends only on X. If X is a surface with fundamental class K, we have

$$P(\Delta_1, \Delta_2) = \frac{\Delta_1(\Delta_1 - K)}{2} + \chi(\mathcal{O}_X) - \Delta_2.$$

If X is a threefold with fundamental class K, and if c_2 is the second Chern class of the tangent bundle of X, we have

$$P(\Delta_1, \Delta_2, \Delta_3) = \Delta_3 - \Delta_1\Delta_2 + \frac{1}{2}K\Delta_2 + \frac{1}{6}\Delta_1^3 - \frac{1}{4}K\Delta_1^2 + \frac{1}{12}(K^2 + c_2)\Delta_1 + \chi(\mathcal{O}_X).$$

In particular, for $X = \mathbb{P}_3$,

$$P(\Delta_1, \Delta_2, \Delta_3) = \Delta_3 - \Delta_1\Delta_2 - 2\Delta_2 + \binom{\Delta_1 + 3}{3}.$$

For $X = \mathbb{P}_4$,

$$\begin{aligned} P(\Delta_1, \Delta_2, \Delta_3, \Delta_4) &= -\Delta_4 + \Delta_1\Delta_3 + \frac{1}{2}\Delta_2(\Delta_2 - \Delta_1^2) + \frac{5}{2}(\Delta_3 - \Delta_1\Delta_2) \\ &\quad + \binom{\Delta_1 + 4}{4}. \end{aligned}$$

In the case of $\mathbb{P}_n$, let Γ be the hyperplane of elements of rank 0 in the Grothendieck group $K(\mathbb{P}_n)$. Then we have a surjective map

$$(\Delta_1, \dots, \Delta_n) \ : \ K(\mathbb{P}_n)\backslash\Gamma \longrightarrow \mathbb{Q}.$$

Two elements of $K(\mathbb{P}_n)\backslash\Gamma$ are in the same fibre of this map if and only if they are collinear.

3 Exceptional bundles

3.1 Definition of exceptional bundles

Let E be an algebraic vector bundle on a smooth projective irreducible algebraic variety X. Then E is called *exceptional* if $H^i(X, Ad(E)) = 0$ for every i. If X is one of the varieties considered here (a projective space or a smooth quadric surface) then E is exceptional if and only if E is *simple* (i.e. the only endomorphisms of E are the homotheties) and $Ext^i(E,E) = 0$ for every $i \geq 1$.

For example, on $\mathbb{P}_n$ the line bundles are exceptional. So is the tangent bundle. In general, if E is an exceptional bundle and L a line on X, then $E \otimes L$ is also exceptional.

3.2 Helices

Suppose that $X = \mathbb{P}_n$, with $n \geq 2$. An infinite sequence $(E_i)_{i\in\mathbb{Z}}$ of exceptional bundles is called *exceptional* if the following three conditions are satisfied :

1. The sequence is *periodical*, i.e. for all $i \in \mathbb{Z}$ we have

$$E_{i+n+1} \simeq E_i(n+1).$$

2. There exists an integer i_0 such that for $i_0 \leq i < j \leq i_0 + n$ we have

$$Ext^k(E_i, E_j) = 0 \ \text{ if } k > 0,$$
$$Ext^k(E_j, E_i) = 0 \ \text{ for all } k.$$

3. For every integer j, the canonical morphism

$$ev_j : E_{j-1} \otimes Hom(E_{j-1}, E_j) \longrightarrow E_j$$
$$(\text{resp. } ev_j^* : E_{j-1} \longrightarrow E_j \otimes Hom(E_{j-1}, E_j)^*)$$

is surjective (resp. injective).

If $\sigma = (E_i)_{i\in\mathbb{Z}}$ is a sequence of exceptional bundles, let $\tau(\sigma)$ denote the sequence $(E'_i)_{i\in\mathbb{Z}}$, where $E'_i = E_{i-1}$ for each i. Suppose that σ satisfies condition 1. Then any subsequence $(E_i, \ldots, E_{i+n})$ is called a *foundation* or a *basis* of σ. Suppose that σ is exceptional. Then it is not difficult to see that condition 2 above is verified for every integer i_0, and that $Ker(ev_j)$ and $Coker(ev_j^*)$ are exceptional bundles. We can thus define two new sequences of exceptional bundles, associated to σ and $j \ mod\,(n+1)$. The first sequence

$$L_j(\sigma) = (E'_i)_{i\in\mathbb{Z}}$$

is defined by :

$$\begin{aligned} E'_i &= E_i \text{ if } i \neq j-1 \ (mod\ n+1) \ \ and \ \ i \neq j \ (mod\ n+1), \\ E'_{j-1+k(n+1)} &= Ker(ev_j)(k(n+1)), \\ E'_{j+k(n+1)} &= E_{j-1+k(n+1)}, \end{aligned}$$

for all k. The second sequence $R_{j-1}(\sigma)$ is defined in the same way, by replacing in σ each pair $(E_{j-1+k(n+1)}, E_{j+k(n+1)})$ by

$$(E_{j+k(n+1)}, Coker(ev_j^*)(k(n+1))).$$

The sequence $L_j(\sigma)$ is called the *left mutation* of σ at E_j and $R_{j-1}(\sigma)$ the *right mutation* of σ at E_{j-1}. For these two sequences, conditions 1 and 2 above are satisfied.

Suppose that condition 3 is also satisfied for $L_j(\sigma)$, i.e. that it is an exceptional sequence. Then it has a foundation of type

$$(E_{j-1}, E_{j+1}, \ldots, E_{j+n-1}, F_1),$$

where F_1 is an exceptional bundle. It is possible to define

$$L_j^2(\sigma) = L_{j-1} \circ L_j(\sigma).$$

Suppose that this is again an exceptional sequence. Then it has a foundation of type

$$(E_{j-1}, E_j, \ldots, E_{j+n-2}, F_2, E_{j+n-1}),$$

where F_2 is an exceptional bundle. It is then possible to define

$$L_j^3(\sigma) = L_{j-2} \circ L_j^2(\sigma).$$

If this process can be continued, we can define the exceptional sequence $L_j^k(\sigma)$, for $1 \leq k \leq n$, which has, if $k \leq n-1$, a foundation of type

$$(E_{j-1}, E_{j+1}, \ldots, E_{j+n-k}, F_k, E_{j+n-k+1}, \ldots, E_{j+n-1}),$$

where F_k is an exceptional bundle. In particular, $L^{n-1}(\sigma)$ has a foundation of type

$$(E_{j-1}, E_{j+1}, F_{n-1}, E_{j+2}, \ldots, E_{j+n-1}),$$

so $L_j^n(\sigma)$ has a foundation of type

$$(E_{j-1}, F_n, E_{j+1}, \ldots, E_{j+n-1}).$$

The exceptional bundle $F_k(-n)$, for $1 \leq k \leq n$, is denoted by $L^{(k)}E_j$, and $L^{(0)}E_j = E_j$.

A *helix* is an exceptional sequence σ such that for every integer j, the sequences $L_j^k(\sigma)$ are defined, for $1 \leq k \leq n$, and such that

$$L_j^n(\sigma) = \tau(\sigma).$$

The last condition means that $F_n \simeq E_j$.

The helices have an interesting property: any left or right mutation of a helix is a helix. So it is possible to define infinitely many helices and exceptional bundles simply by making successive mutations of one helix. The simplest helix is the sequence $(\mathcal{O}(i))_{i \in \mathbb{Z}}$. The helices that can be obtained by successive mutations of this helix are called *constructive helices* and the corresponding exceptional bundles are the *constructive exceptional bundles.* All helices and vector bundles on $\mathbb{P}_2$ are constructive (see [1, 5, 7]), and so are all helices on $\mathbb{P}_3$ (see [11]).

All the mutation transformations defined above can be expressed in terms of τ and L_0 only. For example, we have $L_j = \tau^i \circ L_0 \circ \tau^{-j}$. There are some relations among τ and L_0 :

$$L_0^n = \tau,$$
$$L_0 \circ \tau \circ L_0 \circ \tau^{-1} \circ L_0 \circ \tau = \tau \circ L_0 \circ \tau^{-1} \circ L_0 \circ \tau \circ L_0,$$
$$L_0 \circ \tau^n = \tau^n \circ L_0,$$
$$L_0 \circ \tau^i \circ L_0 \circ \tau^{-i} = \tau^i \circ L_0 \circ \tau^{-i} \circ L_0 \quad \text{if} \quad 2 \leq i \leq n-1.$$

3.3 Generalized Beilinson spectral sequence

Let $(E_0, \dots, E_n)$ be a foundation of a constructive helix. Then the sequence

$$(L^{(n)}E_n, L^{(n-1)}E_{n-1}, \dots, L^{(1)}E_1, E_0)$$

is a foundation of the helix $L_n \circ L_{n-1} \circ \dots \circ L_1(\sigma)$. There exists a canonical resolution of the diagonal Δ of $\mathbb{P}_n \times \mathbb{P}_n$:

$$0 \longrightarrow L^{(n)}E_n \boxtimes E_n^* \longrightarrow L^{(n-1)}E_{n-1} \boxtimes E_{n-1}^* \longrightarrow \dots$$
$$\longrightarrow \dots L^{(1)}E_1 \boxtimes E_1^* \longrightarrow E_0 \boxtimes E_0^* \xrightarrow{\phi} \mathcal{O}_\Delta \longrightarrow 0,$$

where ϕ is the trace morphism. It follows easily that for every coherent sheaf $\mathcal{E}$ on $\mathbb{P}_n$ there exists a spectral sequence $E_r^{p,q}$ of coherent sheaves on $\mathbb{P}_n$, converging to $\mathcal{E}$ in degree 0 and to zero in other degrees, such that the only possibly non-zero $E_1^{p,q}$ terms are

$$E_1^{p,q} = E_{-p}^* \otimes H^q(\mathcal{E} \otimes L^{(-p)}E_{-p}),$$

for $-n \leq p \leq 0$, $0 \leq q \leq n$. The morphisms $d_1^{p,q}$ come from the morphisms in the preceding resolution of $\mathcal{O}_\Delta$. This spectral sequence is called the *generalized Beilinson spectral sequence* associated to $\mathcal{E}$ and the foundation $(E_0, \ldots, E_n)$. If this foundation is $(\mathcal{O}(i))_{0\leq i\leq n}$, the generalized Beilinson spectral sequence if of course the ordinary Beilinson spectral sequence.

From the generalized Beilinson spectral sequence one can deduce the *generalized Beilinson complex*

$$0 \longrightarrow X_{-n} \longrightarrow X_{-n+1} \longrightarrow \ldots \longrightarrow X_{-1} \longrightarrow X_0 \longrightarrow X_1 \longrightarrow \ldots \longrightarrow X_n \longrightarrow 0$$

where for $-n \leq k \leq n$

$$X_k = \bigoplus_{p+q=k} E_1^{p,q}.$$

This complex is exact in non-zero degrees and its cohomology in degree 0 is isomorphic to $\mathcal{E}$.

4 The geometry associated to exceptional bundles

4.1 The space of invariants and its canonical hypersurface

Consider the space $\mathbb{R}^n$, with coordinates $(\Delta_1, \ldots, \Delta_n)$. Then to each coherent sheaf $\mathcal{E}$ on $\mathbb{P}_n$ with non-zero rank one associates the point

$$(\Delta_1(\mathcal{E}), \ldots, \Delta_n(\mathcal{E}))$$

of $\mathbb{R}^n$, which will be also denoted by $\mathcal{E}$. Recall that there exists a polynomial P in n variables with rational coefficients, such that for every coherent sheaf $\mathcal{E}$ on $\mathbb{P}_n$ with non-zero rank we have

$$\chi(\mathcal{E}) = rk(\mathcal{E}).P(\Delta_1(\mathcal{E}), \ldots, \Delta_n(\mathcal{E})).$$

The hypersurface H of $\mathbb{R}^n$ defined by the equation

$$P(0, 2\Delta_2, 0, 2\Delta_4, \ldots) = 0$$

is called the *canonical hypersurface.* If $\mathcal{E}$ is a stable sheaf on $\mathbb{P}_n$, then $\mathcal{E}$ belongs to the halfspace

$$P(0, 2\Delta_2, 0, 2\Delta_4, ...) < 0$$

if and only if the expected dimension of the moduli space of semi-stable sheaves that contains $\mathcal{E}$ is strictly positive.

For example, on $\mathbb{P}_2$, the equation of H is

$$\Delta_2 = \frac{1}{2}.$$

On $\mathbb{P}_3$ it is

$$\Delta_2 = \frac{1}{4},$$

and on $\mathbb{P}_4$

$$\Delta_4 = \Delta_2^2 - \frac{35}{6}\Delta_2.$$

Question 1 *Is the expected dimension of a moduli space of semi-stable sheaves on $\mathbb{P}_n$ always nonnegative ?*

4.2 Hypersurfaces associated to exceptional bundles and limit hypersurfaces

Let E be an exceptional bundle on $\mathbb{P}_n$. Then to E one associates the hypersurface $S(E)$ of $\mathbb{R}^n$ defined by the equation

$$P(\Delta_1(E) - \Delta_1, \ldots, \Delta_n(E) - \Delta_n) \;=\; 0.$$

It contains the points corresponding to sheaves $\mathcal{E}$ such that

$$\chi(\mathcal{E}, E) \;=\; \sum_{i=0}^{n}(-1)^i dim(Ext^i(\mathcal{E}, E)) \;=\; 0.$$

To define the *limit hypersurfaces* we need to consider an *exceptional pair*, i.e. a pair (E_0, E_1) of exceptional bundles that can be inserted as a pair of consecutive elements in some helix. Then it follows from the definition of a helix that there exists a sequence $(F_i)_{i\in\mathbb{Z}}$ of exceptional bundles such that $F_0 = E_0$, $F_1 = E_1$, and for every integer i we have an exact sequence

$$0 \longrightarrow F_{i-1} \longrightarrow F_i \otimes Hom(F_i, F_{i+1}) \longrightarrow F_{i+1} \longrightarrow 0.$$

Then the hypersurfaces $S(F_i)$ have a limit when i tends to $+\infty$ or $-\infty$. It is possible to define more complicated limit hypersurfaces (limits of limits, and so on). Let $C(E)$ denote the intersection of $S(E)$ and the canonical hypersurface H.

In the case of $\mathbb{P}_2$, the curve $S(E)$ is a parabola in $\mathbb{R}^2$, of equation

$$\Delta_2 \;=\; \frac{1}{2}(\Delta_1(E) - \Delta_1)^2 + \frac{3}{2}(\Delta_1(E) - \Delta_1) + \frac{1}{2} + \frac{1}{2rk(E)^2},$$

and $C(E)$ consists of two points on the line H. The limit points on H coincide with the non-limit points.

In the case of $\mathbb{P}_3$, the equation of the surface $S(E)$ is

$$\Delta_3 \;=\; -\frac{1}{6}z^3 + (\Delta_2 + \frac{5}{12} - \frac{1}{4rk(E)^2})z + \Delta_3(E),$$

with $z = \Delta_1 - \Delta_1(E) - 2$. The curve $C(E)$ is obtained by taking $\Delta_2 = 1/4$ in the preceding equation. In this case, the limit curves are distinct from the non-limit ones. Some other surfaces and curves may be interesting in the case of $\mathbb{P}_3$: the images of the preceding ones by the translations

$$(\Delta_1, \Delta_2, \Delta_3) \longrightarrow (\Delta_1, \Delta_2, \Delta_3 + k),$$

where k is an integer. In [11], Nogin proved that the semi-orthogonal bases of $K(\mathbb{P}_3)$ are the sequences

$$([E_0] \otimes \alpha^k, [E_1] \otimes \alpha^k, [E_2] \otimes \alpha^k, [E_3] \otimes \alpha^k),$$

(E_0, E_1, E_2, E_3) beeing a foundation of a helix, α the class of the ideal sheaf of a point and k an integer. The multiplication by α^k corresponds to the preceding translation in $\mathbb{R}^3$.

4.3 The case of $\mathbb{P}_1 \times \mathbb{P}_1$

The space of invariants is here $\mathbb{R}^3$ with coordinates (a, b, Δ_2), a, b beeing the two coordinates of Δ_1. The equation of H is

$$\Delta_2 = \frac{1}{2}.$$

The surfaces $S(E)$ are quadrics, and the corresponding conics $C(E)$ have been used in [12]. It is also possible here to define the notion of limit surface (or curve). This case is similar to the case of $\mathbb{P}_3$ (cf. [4]).

5 Existence theorems

5.1 The existence theorem on $\mathbb{P}_2$

Let $\mathcal{E}$ be a stable coherent sheaf on $\mathbb{P}_2$, not exceptional, and E an exceptional bundle such that $rk(E) < rk(\mathcal{E})$ and $\mid \Delta_1(E) - \Delta_1(\mathcal{E}) \mid \leq 1$. Then we have

$$\chi(E, \mathcal{E}) \leq 0 \text{ if } \Delta_1(\mathcal{E}) \leq \Delta_1(E),$$

and

$$\chi(\mathcal{E}, E) \leq 0 \text{ if } \Delta_1(\mathcal{E}) > \Delta_1(E) .$$

The first condition means that the point $\mathcal{E}$ in $\mathbb{R}^2$ is over the curve $S(E^*(-3))$[1] and the second that it is over the curve $S(E)$. Conversely the following is proved in [5]:

[1]this means that $\Delta_2(E)$ is greater than or equal to the Δ_2 coordinate of the point of $S(E^*(-3))$ whose first coordinate is $\Delta_1(E)$.

Theorem 5.1 *Let $q = (\Delta_1, \Delta_2)$ be a point in $\mathbb{Q}^2$. Suppose that for every exceptional bundle E such that*

$$\mid \Delta_1(E) - \Delta_1 \mid \leq 1$$

the point q is over $S(E^*(-3))$ if $\Delta_1(\mathcal{E}) \leq \Delta_1(E)$, and over $S(E)$ if $\Delta_1(\mathcal{E}) > \Delta_1(E)$. Then for every triple (r, c_1, c_2) of integers, with $r > 0$ such that $\Delta_i(r, c_1, c_2) = \Delta_i$ for i=1,2, there exists a stable vector bundle of rank r and Chern classes c_1, c_2.

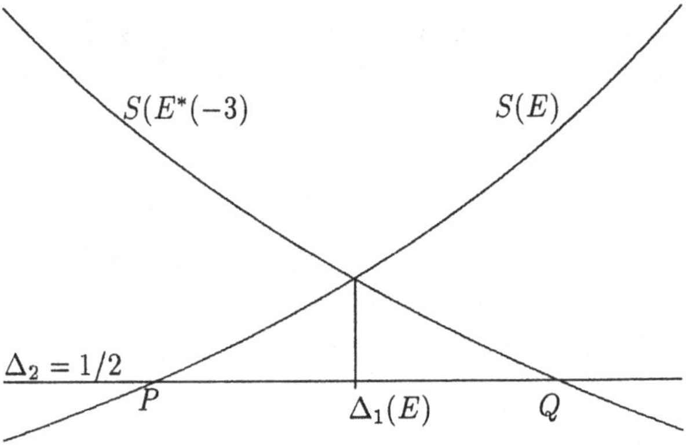

The coordinates of the point P (resp. Q) above are $(\Delta_1(E) - x_E, 1/2)$ (resp. $(\Delta_1(E) + x_E, 1/2)$), where x_E is the smallest root of the equation

$$x^2 + 3x + \frac{1}{3rg(E)^2} = 0.$$

Let $I_E =]\Delta_1(E) - x_E, \Delta_1(E) + x_E[$, and $\mathcal{E}xc$ be the set of isomorphism classes of exceptional bundles on $\mathbb{P}_2$. Let $M(r, c_1, c_2)$ denote the moduli space of semi-stable coherent sheaves on $\mathbb{P}_2$, of rank r and Chern classes c_1, c_2. The preceding theorem can be improved and one obtains easily the final form of the existence theorem on $\mathbb{P}_2$:

Theorem 5.2 1. *The family of intervals $(I_E)_{E \in \mathcal{E}xc}$ is a partition of $\mathbb{Q}$.*

2. *There exists a unique mapping*

$$\delta : \mathbb{Q} \longrightarrow \mathbb{Q}$$

such that for all integers r, c_1, c_2 with $r \geq 1$ one has

$$dim(M(r, c_1, c_2)) \geq 0 \Longleftrightarrow \Delta_2 \geq \delta(\Delta_1),$$

(with $\Delta_1 = \frac{c_1}{r}$ and $\Delta_2 = \frac{1}{r}(c_2 - \frac{r-1}{2r}c_1^2)$.)

3. If E in an exceptional bundle on $\mathbb{P}_2$, δ in given on $]\Delta_1(E)-x_E, \Delta_1(E)]$ by $S(E^*(-3))$ and on $[\Delta_1(E), \Delta_1(E)+x_E[$ by $S(E)$.

5.2 The existence theorem on $\mathbb{P}_1 \times \mathbb{P}_1$

A.N. Rudakov has proved in [13] a result analogous to theorem 5.1, using the exceptional bundles on $\mathbb{P}_1 \times \mathbb{P}_1$, but it seems more difficult than in the case of $\mathbb{P}_2$ to obtain the analogous result to theorem 5.2.

Question 2 *It is easy to deduce from Rudakov's result that there exists a surface S in $\mathbb{R}^3$, defined by an equation*

$$\Delta_2 = f(\Delta_1),$$

such that for all integers r, a, b, c_2 with $r \geq 1$, there exists a stable non-exceptional coherent sheaf on $\mathbb{P}_1 \times \mathbb{P}_1$ of rank r and Chern classes $(a, b), c_2$ if and only if the associated point in $\mathbb{R}^3$ (whose coordinates are the corresponding Δ_1, Δ_2) is over[2] S . It would be interesting to give a description of S. It should be made of pieces of the surfaces $S(E)$ or perhaps of the limit surfaces defined in section 4.3 .

5.3 The existence theorem on $\mathbb{P}_n$, $n \geq 3$

In this case, almost nothing is known, except for rank-2 stable reflexive sheaves on $\mathbb{P}_3$ (cf. [8]).

Question 3 *Is there a surface S in $\mathbb{R}^3$, of equation*

$$\Delta_2 = f(\Delta_1, \Delta_3),$$

such that for all integers r, c_1, c_2, c_3 with $r \geq 1$, the moduli space $M(r, c_1, c_2, c_3)$ of semi-stable sheaves on $\mathbb{P}_3$ of rank r and Chern classes c_1, c_2, c_3 has a positive dimension if and only if the associated point of $\mathbb{R}^3$ (whose coordinates are the corresponding $\Delta_1, \Delta_2, \Delta_3$) is over[3] S ?

In particular, do the gaps *in c_3 found in [8] can be filled if one allows non-reflexive rank-2 stable sheaves ?*

If it exists, is S made of pieces of the $S(E)$ and the limit surfaces ?

[2] this means that $\Delta_2 \geq f(\Delta_1)$.

[3] this means that $\Delta_2 \geq f(\Delta_1, \Delta_3)$.

6 Descriptions of moduli spaces of semi--stable sheaves using exceptional bundles

6.1 The case of $\mathbb{P}_2$

Let E be an exceptional bundle on $\mathbb{P}_2$, and Δ_1 a rational number such that $\Delta_1(E) - x_E < \Delta_1 \leq \Delta_1(E)$. There exists exceptional bundles F,G such that (E, F, G) is a foundation of a helix. Then to study moduli spaces of semi-stable sheaves $M(r, c_1, c_2)$ such that $c_1/r = \Delta_1$ it is convenient to use the Beilinson spectral sequence associated to $(G^*(3), F^*(3), E^*(3))$. We obtain a good description of $M(r, c_1, c_2)$ if $\Delta_2 = \delta(\Delta_1)$, i.e. if the point of $\mathbb{R}^2$ corresponding to $M(r, c_1, c_2)$ lies on the curve $S(E^*(-3))$ (or more generally in some cases where $M(r, c_1, c_2)$ is *extremal*, i.e if $dim(M(r, c_1, c_2)) > 0$ and $dim(M(r, c_1, c_2 - 1)) \leq 0$).

Suppose that $\Delta_2 = \delta(\Delta_1)$. Let H be the exceptional bundle cokernel of the canonical map

$$F \longrightarrow G \otimes Hom(F, G)^*,$$

and

$$m = -\chi(\mathcal{E} \otimes H^*), \; k = -\chi(\mathcal{E} \otimes G^*),$$

where $\mathcal{E}$ is a coherent sheaf of rank r and Chern classes c_1, c_2. Then $m > 0$ and $k > 0$. If $\mathcal{E}$ is semi-stable then the only two non zero $E_1^{p,q}$ terms in the Beilinson spectral sequence associated to (G^*, F^*, E^*) and $\mathcal{E}$ are $F(-3) \otimes H^0(\mathcal{E} \otimes H^*(3))$ and $G(-3) \otimes H^0(\mathcal{E} \otimes G^*(3))$. So the spectral sequence degenerates and we have an exact sequence

$$0 \longrightarrow F(-3) \otimes \mathbb{C}^m \longrightarrow G(-3) \otimes \mathbb{C}^k \longrightarrow \mathcal{E} \longrightarrow 0\,.$$

Consider now the vector space

$$Hom(F \otimes \mathbb{C}^m, G \otimes \mathbb{C}^k) \;=\; L(Hom(F, G)^* \otimes \mathbb{C}^m, \mathbb{C}^k),$$

with the obvious action of the reductive group

$$G_0 = (GL(m) \times GL(k))/\mathbb{C}^*.$$

This action can be linearized in an obvious way, so we have the notion of semi-stable (or stable) point of $\mathbb{P}(W)$. A non-zero element of W will be called *semi-stable* (resp. *stable*) if its image in $\mathbb{P}(W)$ is. Let $q = dim(W)$ and

$$N(q, m, k) \;=\; \mathbb{P}(W)^{ss}/G_0\,.$$

which is a projective variety. The following result is proved in [2] :

Theorem 6.1 *1. Let α be a non-zero element of W, f the corresponding morphism of vector bundles. Then f is injective as a morphism of* *sheaves, and coker(f) is semi-stable (resp. stable) if and only if α is semi-stable (resp. stable).*

2. *The map $f \longmapsto coker(f)$ defines an isomorphism*

$$N(q,m,k) \simeq M(r,c_1,c_2).$$

There is a similar result for some other extremal moduli spaces of semi-stable sheaves (cf. [3]). In this case we have to consider morphisms of the following type

$$(E(-3)\otimes \mathbb{C}^h)\oplus(F(-3)\otimes \mathbb{C}^m) \longrightarrow G(-3)\otimes \mathbb{C}^k,$$

and the group acting on the space of such morphisms is non reductive.

There is a canonical isomorphism

$$N(q,m,k) \simeq N(q,k,qk-m).$$

Hence to $N(q,k,qk-m)$ is associated another moduli space $M(r',c_1',c_2')$ which is canonically isomorphic to $M(r,c_1,c_2)$, with

$$\Delta_1(E) - x_E < \frac{c_1'}{r'} < \frac{c_1}{r} \leq \Delta_1(E).$$

Finally, we obtain an infinite sequence of moduli spaces of semi-stable sheaves all isomorphic to $M(r,c_1,c_2)$. This phenomenon does not occur for moduli spaces such that $\Delta_2 > \delta(\Delta_1)$. Here are some examples of descriptions of moduli spaces obtained with the preceding theorem :

$$M(1,0,1) \simeq M(3,-1,2) \simeq M(8,-3,8) \simeq \mathbb{P}_2 \ ,$$

$$M(4,-2,4) \simeq M(24,-10,60) \simeq M(140,-58,1740) \simeq \mathbb{P}_5 \ ,$$

$$M(4,-1,3) \simeq M(11,-4,13) \simeq M(29,-11,73) \simeq N(3,2,3) \ .$$

Question 4 *It is also possible to study more complicated moduli spaces (non extremal ones) using some foundation of a helix. We can for example obtain descriptions of moduli spaces by monads. What is the best choice for the generalized Beilinson spectral sequence? Or are there foundations of helices that would lead to more interesting monads than those obtained from the classical Beilinson spectral sequence?*

6.2 The case of $\mathbb{P}_n$, $n \geq 3$

If we want to use exceptional bundles to study a moduli space of semi-stable sheaves on $\mathbb{P}_n$, we have to choose a foundation of helix such that the Beilinson spectral sequence associated to it and to sheaves in this moduli space is as simple as possible. This is the case of course if many $E_1^{p,q}$ are zero. Let $(E_0, \ldots, E_n)$ be a foundation of helix, $r, c_1, \ldots, c_n$ integers with $r \geq 1$, $\mathcal{E}$ a coherent sheaf on $\mathbb{P}_n$ of rank r and Chern classes $c_1, \ldots, c_n$, and $E_r^{p,q}$ the Beilinson spectral sequence associated to $(E_0, \ldots, E_n)$ and $\mathcal{E}$. Then, if $0 \leq i \leq n$, we can hope that the terms $E_1^{p,-i}$ will be zero for all p only if $\chi(\mathcal{E} \otimes L^{(i)}E_i) = 0$. This means that the point corresponding to $\mathcal{E}$ in the space of invariants belongs to the hypersurface $S(L^{(i)}E_i)$. Of course this is not sufficient.

Suppose for example that all the $E_1^{p,-i}$, for $0 \leq i \leq n-2$, vanish, and that $E_1^{n,-n} = 0$. Then we have an exact sequence

$$0 \longrightarrow E_n^* \otimes H^{n-1}(\mathcal{E} \otimes L^{(n)}E_n) \longrightarrow E_{n-1}^* \otimes H^{n-1}(\mathcal{E} \otimes L^{(n-1)}E_{n-1}) \longrightarrow \mathcal{E} \longrightarrow 0.$$

In general the vanishing of the cohomology groups necessary to obtain the above exact sequence are very hard to verify.

Let E,F be exceptional bundles on $\mathbb{P}_n$ which are consecutive terms in some helix, and m,k two positive integers. We want to study morphisms

$$E \otimes \mathbb{C}^m \longrightarrow F \otimes \mathbb{C}^k.$$

Let

$$W \;=\; Hom(E \otimes \mathbb{C}^m, F \otimes \mathbb{C}^k) \;=\; L(\mathbb{C}^m \otimes Hom(E,F)^*, \mathbb{C}^k)\;.$$

on which acts the reductive group G_0. Recall the characterization of semi-stable and stable points of W:

Proposition 6.2 *Let α be a non-zero element of W. Then α is semi-stable (resp. stable) if and only if for every non-zero subspace $H \subset \mathbb{C}^m$, if*

$$K = \alpha(Hom(E,F)^* \otimes H),$$

we have

$$\frac{dim(K)}{dim(H)} \geq \frac{k}{m} \quad (resp. >)$$

Let $q = dim(Hom(E,F))$. Then we have $dim(N(q,m,k)) > 0$ if and only if

$$x_q < \frac{m}{k} < \frac{1}{x_q},$$

where x_q is the smallest root of the equation

$$X^2 - qX + 1 \;=\; 0.$$

Suppose that there exist an injective morphism of sheaves

$$E \otimes \mathbb{C}^m \longrightarrow F \otimes \mathbb{C}^k .$$

Let $\mathcal{E}$ be its cokernel. Then it is easy to see that the preceding inequalities are verified if and only if the expected dimension of the moduli space of semi-stable sheaves with the same invariants as $\mathcal{E}$ is positive.

The first problem is the injectivity of stable maps.

Proposition 6.3 *Let x be a point of $\mathbb{P}_n$. Suppose that the canonical map*

$$ev_x : E_x \otimes Hom(E, F) \longrightarrow F_x$$

is stable (for the action of $(GL(E_x) \times GL(F_x))/\mathbb{C}^*$). Suppose that

$$\frac{k}{m} \geq \chi(E^* \otimes F) - \frac{rk(F)}{rk(E)}.$$

Then the morphism of vector bundles associated to a semi-stable element of W is injective on the complement of a finite set, and injective if the preceding inequality is strict.

The proof uses the same arguments as on $\mathbb{P}_2$. Of course, the stability of ev_x is independent of x.

Question 5 *Is ev_x always stable ?*

The answer is yes on $\mathbb{P}_2$.

If we allow $\frac{n}{m}$ to be smaller than the bound in the preceding proposition, it may happen that a semi-stable morphism is non injective on some subvariety of $\mathbb{P}_n$. For example let (E_0, E_1, E_2, E_3) de a foundation of some helix on $\mathbb{P}_3$. It follows easily from the generalized Beilinson spectral sequence that if C is a smooth curve in $\mathbb{P}_3$, and $\mathcal{F}$ a vector bundle on C such that

$$H^0(L^{(3)}E_3 \otimes i_*\mathcal{F}) = H^1(L^{(3)}E_3 \otimes i_*\mathcal{F}) = 0,$$

(where i is the inclusion of C in $\mathbb{P}_3$) then there exists an exact sequence

$$0 \longrightarrow H^0(L^{(2)}E_2 \otimes i_*\mathcal{F}) \otimes E_2^* \longrightarrow H^0(L^{(1)}E_1 \otimes i_*\mathcal{F}) \otimes E_1^* \longrightarrow \cdots$$
$$\cdots \longrightarrow H^0(E_0 \otimes i_*\mathcal{F}) \otimes E_0^* \longrightarrow i_*\mathcal{F} \longrightarrow 0.$$

In this case, we get a morphism

$$H^0(L^{(2)}E_2 \otimes i_*\mathcal{F}) \otimes E_2^* \longrightarrow H^0(L^{(1)}E_1 \otimes i_*\mathcal{F}) \otimes E_1^*$$

which is non injective along C. For example, this happens if C is a degree 19 curve not on a cubic, $\mathcal{F}$ a general line bundle on C of degree $g+2$ on C (where g is the genus of C), and

$$(E_0, E_1, E_2, E_3) = (\mathcal{O}, Q(3), Q^*(4), Q_3^*(4)),$$

where Q (resp. Q_3) is the cokernel of the canonical morphism

$$\mathcal{O}(-1) \longrightarrow \mathcal{O} \otimes H^0(\mathcal{O}(1))^* \quad (\text{resp. } \mathcal{O}(-3) \longrightarrow \mathcal{O} \otimes H^0(\mathcal{O}(3))^*).$$

In this case we have a morphism

$$Q(-4) \otimes \mathbb{C}^{10} \longrightarrow Q^*(-3) \otimes \mathbb{C}^{11}$$

which is non injective along C.

Question 6 *What is the smallest number z such that if $\frac{k}{m} > z$, then every semi-stable morphism is injective on a nonempty open subset of $\mathbb{P}_n$?*

The next problem is the relation between the (semi-)stability of morphisms and the (semi-)stability of cokernels. On $\mathbb{P}_n$, $n \geq 3$, no general result is known. The only non trivial case where the problem is completely solved is on $\mathbb{P}_3$, with $E = \mathcal{O}(-2)$, $F = \mathcal{O}(-1)$, $m = 2$, $k = 4$. In this case, R.M. Miro-Roig and G. Trautmann have proved in [9] that the (semi-)stability of the map is equivalent to the (semi-)stability of the cokernel, and it follows that the moduli space $M(2,0,2,4)$ is isomorphic to $N(4,2,4)$.

Question 7 *In which cases is there an equivalence between the (semi-) stability of morphisms and the (semi-)stability of cokernels ?*

Suppose that every (semi-)stable morphism is injective on a nonempty open subset of $\mathbb{P}_n$, and that m and k are relatively prime (so $N(q,m,k)$ is smooth). Then there exists a *universal cokernel* on $N(q,m,k)\times\mathbb{P}_n$, i.e a coherent sheaf $\mathcal{F}$ on $N(q,m,k)\times\mathbb{P}_n$, flat on $N(q,m,k)$, such that for every stable morphism α, if $\pi(\alpha)$ denotes its image in $N(q,m,k)$, then $\mathcal{F}_{\pi(\alpha)}$ is isomorphic to $coker(\alpha)$ (for every closed point y in $N(q,m,k)$, $\mathcal{F}_y$ denotes the restriction of $\mathcal{F}$ to $\{y\}\times\mathbb{P}_n$). This family of sheaves on $\mathbb{P}_n$ is a *universal deformation* at each point of $N(q,m,k)$. It is also *injective*, i.e. if y, y' are distinct points of $N(q,m,k)$ then the sheaves $\mathcal{F}_y$, $\mathcal{F}_{y'}$ on $\mathbb{P}_n$ are not isomorphic.

Question 8 *Let S be a smooth projective variety, F a coherent sheaf on $S\times\mathbb{P}_n$, flat on S, such that for every closed point s, $\mathcal{F}_s$ has no torsion, $\mathcal{F}$ is a universal deformation of $\mathcal{F}_s$, and such that if s, s' are distinct points of S, the sheaves $\mathcal{F}_s$ and $\mathcal{F}_{s'}$ are not isomorphic. Does it follow that S is a component of a moduli space of stable sheaves on $\mathbb{P}_n$, and $\mathcal{F}$ the universal sheaf ?*

References

[1] Drézet, J.-M. *Fibrés exceptionnels et suite spectrale de Beilinson généralisée sur* $\mathbb{P}_2(\mathbb{C})$. Math. Ann. 275 (1986), 25-48.

[2] Drézet, J.-M. *Fibrés exceptionnels et variétés de modules de faisceaux semi-stables sur* $\mathbb{P}_2(\mathbb{C})$. Journ. Reine Angew. Math. 380 (1987), 14-58.

[3] Drézet J.-M. *Variétés de modules extrémales de faisceaux semi-stables sur* $\mathbb{P}_2(\mathbb{C})$. Math. Ann. 290 (1991), 727-770.

[4] Drézet, J.-M. *Sur les équations vérifiées par les invariants des fibrés exceptionnels.* Preprint Paris VII (1993).

[5] Drézet, J.-M., Le Potier, J. *Fibrés stables et fibrés exceptionnels sur* $\mathbb{P}_2$. Ann. Ec. Norm. Sup. 18 (1985), 193-244.

[6] Gorodentsev, A.L. *Exceptional bundles on surfaces with a moving anti-canonical class.* Math. USSR Izvestiya. AMS transl. 33 (1989), 67-83.

[7] Gorodentsev, A.L., Rudakov, A.N. *Exceptional vector bundles on projective spaces.* Duke Math. Journ. 54 (1987), 115-130.

[8] Miro-Roig, R.M. *Gaps in Chern Classes of Rank 2 Stable Reflexive Sheaves.* Math. Ann. 270 (1985), 317-323.

[9] Miro-Roig, R.M., Trautmann, G. *The Moduli Scheme* $M(0,2,4)$ *over* $\mathbb{P}_3$. Preprint Univ. Kaiserslautern.

[10] Nogin, D.Y. *Helices of period 4 and Markov type equations.* To appear in Math. USSR Izvestiya.

[11] Nogin D.Y. *Helices on some Fano threefolds : constructivity of semi-orthogonal bases of* K_0. Preprint.

[12] Rudakov A.N. *Exceptional vector bundles on a quadric.* Math. USSR Izvestiya. AMS transl. 33 (1989), 115-138.

[13] Rudakov A.N. *A description of Chern classes of semi-stable sheaves on a quadric surface.* Preprint.

[14] Rudakov A.N. et al. *Helices and Vector Bundles : seminaire Rudakov.* London Math. Soc. Lect. Note Series 148. Cambridge Univ. Press (1990).

FLOER HOMOLOGY AND ALGEBRAIC GEOMETRY

S. K. DONALDSON

The Mathematical Institute, Oxford

§1. INTRODUCTION

Let X be a smooth, oriented, closed 4-manifold whose intersection form is not negative-definite. A choice of Riemannian metric allows us to construct moduli spaces of Yang-Mills instanton connections on bundles over X, and these can be used to define differential-topological invariants of the 4-manifold which have been studied, and applied to concrete problems, quite extensively over the last few years. One of the main themes in this work is the calculation of the invariants by "cut and paste" techniques: if X is the union of two manifolds-with-boundary X_1, X_2, meeting along a 3-manifold $Y \subset X$, then one aims to define "relative" invariants for the pieces X_1, X_2, together with a procedure for calculating the invariants of X from the relative invariants. There is now a substantial theory—based on techniques from differential geometry and global analysis—for doing this, which has been developed by a number of authors. One prominent development is the notion of *Floer homology*: at least in favourable situations one knows how to associate "Floer homology groups "$HF_*(Y)$ to the (oriented) 3-manifold Y so that the 4-manifold with boundary X_1 defines an invariant $\Psi(X_1) \in HF_*(Y)$. The other piece, X_2, defines an invariant $\Psi(X_2) \in HF_*(\overline{Y})$, where $\overline{Y}$ denotes Y with the opposite orientation. There is a dual pairing

$$HF_*(Y) \otimes HF_*(\overline{Y}) \to \mathbf{Z}$$

and the invariant of X is expresssed by a "gluing formula" of the shape

$$\Psi(X) =< \Psi(X_1), \Psi(X_2 > .$$

We will go in to this into more detail in the next section We should say that various authors, particularly Taubes and Morgan, Mrowka and Ruberman, have developed rather different approaches to these questions, and many of the problems that arise in practice can be handled by direct methods without explicit reference to Floer's theory. For brevity, however. we will sometimes refer to all of these differential geometric ideas as "Floer homology" techniques. The common threads running through all the differential geometric work in this area are first the use of a family of Riemannian metrics on X in which the neck around Y is stretched out into a long tube, and second the importance of the *representation variety* of flat connections over Y.

Now suppose that X is the 4-manifold underlying a complex algebraic surface. If we choose a Kähler metric on X, the instanton moduli spaces may be identified with moduli spaces of stable holomorphic bundles, and the invariants can be defined in a purely algebro-geometric way (to do this in detail requires some technical work involving compactifications [L],[M]). This gives another route by which one can calculate invariants, which has also been developed quite extensively. The interaction between this algebro-geometric approach and the "Floer homology" approach sketched above is one of the interesting aspects of the subject, and leads to the following state of affairs. The Floer homology approach, based on differential geometry, allows the calculation of invariants for algebraic surfaces which can be viewed as entirely algebraic entities. Moreover, the kind of decompositions $X_1 \cup_Y X_2$ that arise often have an algebro-geometric interpretation. For example, Y might be the link of an isolated surface singularity and X_1 might be the Milnor fibre of the singularity, or a resolution. Another kind of example arises when X is a small smoothing of a singular space with a normal crossing where two components meet in a common curve. In these kind of situations one might expect that there should be an algebro-geometric version of the Floer homology theory, perhaps equivalent to the differential geometric one, but based entirely on algebro-geometric techniques. The development of this conjectural theory seems to be an interesting problem in algebraic geometry; an answer would be conceptually satisfactory and might also break new ground: for example in algebraic geometry one can study moduli spaces of bundles over varieties of any dimension—not just surfaces—so one might hope that there is a similarly general "Floer theory". This line of thought might fit in with a suggestion of Arnold [A], about a possible analogue of the Casson invariant for contact manifolds.

The remarks above are intended as a general motivation and background for this article, but we must state at once that we will not make any systematic progress in the direction which is suggested; the development of an algebro-geometric analogue of the Floer theory. Rather we will discuss a number of topics which might eventually be seen as fitting into such a theory. In §2 we will recall the main definitions in the subject in more detail and then review the "Fukaya-Floer" construction developed by the author and P.J. Braam in [BD2]. In §3 we will illustrate the ideas with some very simple calculations for elliptic surfaces, motivated by recent work of R. Friedman [F]. In §4 we will discuss the relation of this Fukaya-Floer theory with bundles over a product of curves and with the "quantum cohomology" product for moduli spaces of stable bundles over curves; as another simple illustration of the theory we calculate this product in the case of genus 2.

§2. Floer homology and the Fukaya-Floer Theory

We begin by recalling the definition of the 4-manifold invariants in more detail. Let $E \to X$ be a $U(2)$ bundle and fix a connection on $\Lambda^2 E$. We write M_E for the moduli space of "U(2)-instantons" on E: connections A with curvature $F(A)$ such that $F + *F$ is a multiple of the identity and which induce the given connection on $\Lambda^2 E$. Here $*$ is the Hodge $*$-operator defined by a given Riemannian metric on X. The moduli space has a "formal dimension"

$$\dim M_E = 2(4c_2(E) - c_1(E)^2) - 3(1 - b_1(X) + b^+(X)),$$

where b_1 is the first Betti number and b^+ is the rank of the positive part of the intersection form. We suppose that $b^+ - b_1$ is odd so the dimension of the moduli space is even, $2d$ say. Let $M_E^* \subset M_E$ be the subset of *irreducible* connections. There is a universal $SO(3) = PU(2)$ bundle $\xi \to M_E^* \times X$, which yields a Pontryagin class in $H^4(M_E^* \times X)$. We define $\mu : H_2(X) \to H^2(M_E^*)$ by $\mu(\alpha) = -\frac{1}{4}p_1(\xi)/\alpha$, and a class $\nu \in H^4(M_E^*)$ by $\nu = -\frac{1}{4}p_1(\xi)/pt$. Now the main idea is that if there are no reducible connections in the moduli space these cohomology classes are defined over all of M_E and we should form pairings:

$$\Psi_{a,b,\alpha}(X) = < \mu(\alpha)^a \nu^b, [M] >,$$

where $a + 2b = d$. (In our discussion b will normally be 0, in which case we may omit it from the notation.) The main issues that have to be addressed here are the following.

(1) To avoid reducible connections. The case when $c_1(E)$ is "odd" (i.e. not divisible by 2 in $H^2(X)/\text{Torsion}$) works better here, just as—in algebraic geometry—the stable bundle theory is easier in the "co-prime" case.
(2) To choose a generic metric, or otherwise perturb the problem, so that the moduli space is a manifold of the correct dimension.
(3) To compactify the moduli space suitably, or to define representatives of the cohomology classes such that $\mu(\alpha)^a \nu^b$ has compact support and can be evaluated on the moduli space.

These issues are now well-understood (see [MM] for a recent treatment) so suffice it to say here that if $b^+(X) \geq 1$ one can define numbers $\Psi_{a,b,\alpha}(X)$ in the above fashion, and if $b^+ > 1$ these are independent of the metric on X and so give differential topological invariants of the 4-manifold. If $b^+ = 1$ these numbers depend on the metric only through the "periods" of its self-dual harmonic form [KM]. Letting α vary, the invariants are *polynomial functions* on $H_2(X)$.

Now suppose that $X = X_1 \cup_Y X_2$, as in the introduction. In discussing the Floer homology groups we will concentrate on the case when $c_1(E)$ is odd on Y, so there are no (projectively) flat reducible connections over Y. We also suppose that all these flat connections are isolated and "non-degenerate", so the representation variety $R(Y)$ is a finite set. For each $\rho \in R$ we can define an index $\lambda(\rho) \in \mathbf{Z}/8$. Let ρ_+, ρ_- be two such flat connections. The central objects in the Floer theory are the moduli spaces $M(\rho_-, \rho_+)$ of instantons over the tube $Y \times \mathbf{R}$ with limits $\rho_\pm$ at $\pm\infty$, and the dimension of $M(\rho_-, \rho_+)$ is $\lambda(\rho_+) - \lambda(\rho_-)$. More precisely, this moduli space falls into a collection of components $M^p(\rho_-, \rho+)$, labelled by a relative second Chern class, and the component M^p has (formal) dimension p, equal to $\lambda(\rho_+) - \lambda(\rho_-)$ modulo 8. The translations of the tube act in an obvious way on the moduli spaces so we get "reduced" moduli spaces $M^p/\mathbf{R}$ of dimension one less.

Now Floer's theory proceeds as follows. If $\lambda(\rho_+) - \lambda(\rho_-) = 1$ the space $M^1(\rho_-, \rho_+)/\mathbf{R}$ is a finite collection of points and we define a number $n(\rho_+, \rho_-)$ by counting these points with appropriate signs. For each $\lambda \in \mathbf{Z}/8$ we let C_λ be the $\mathbf{Q}$-vector space (we will work with rational coefficients but this is not really essential) generated by points $\rho \in R$ with $\lambda(\rho) = \lambda$. We then interpret the numbers $n(\rho_-, \rho_+)$ as the matrix elements of a collection of maps:

$$\partial : C_\lambda \to C_{\lambda-1},$$

$$\partial(<\rho_->) = \sum n(\rho_-, \rho_+) <\rho_+> .$$

The crux of the theory is that $\partial^2 = 0$, so we have associated $\mathbf{Z}/8$-graded *Floer homology groups* $HF_*(Y)$. Changing the orientation of Y essentially reverses the grading on the chain conmplex and we get a dual pairing between $HF_*(Y)$ and $HF_*(\overline{Y})$ induced by the obvious pairing on the chain groups.

We now come to the invariants for the manifolds-with-boundary X_1, X_2. We choose complete metrics on the interiors of these manifolds which contain half-infinite cylinders $Y \times [0,\infty), Y \times (-\infty, 0]$ as their "ends", and by an "instanton on X_i" we mean a (finite energy) solution of the equations over the corresponding non-compact manifold. Any finite energy solution has a flat limit along the cylinder, and for each $\rho \in R(Y)$ we can form moduli spaces $M(X_i, \rho)$ of instantons with this limit; again, these moduli spaces will have components of various dimensions. If $M(X_1, \rho)$ has a component $M^{2d}(X_1, \rho)$ of dimension $2d$ we can mimic the discussion for closed 4-manifolds and define, for any $\alpha \in H_2(X_1)$ a number

$$\psi_{a,b,\alpha}(X_1, \rho) = <\mu(\alpha)^a \nu^b, [M^{2d}(X_1, \rho)]>,$$

where $a + 2b = d$. We will sometimes abbreviate this to $\psi(X_1, \rho)$. The next step is to define, for fixed a, b, α, a chain

$$\psi(X_1) = \sum_\rho \psi(X_1, \rho) \ <\rho>$$

in $CF_*(Y)$. One shows that this is a cycle in the chain complex, and that its homology class

$$\Psi(X_1) = \Psi_{a,b,\alpha}(X_1) \in HF_*(Y)$$

is independent of the choice of metrics. Likewise we get a Floer homology class $\Psi_{a',b',\alpha'}(X_2)$ in $HF_*(\overline{Y})$ and the gluing formula is

$$\Psi_{n,b+b',\alpha+\alpha'}(X) = \sum_{\substack{a,a' \\ a+a'=n}} \binom{n}{a} \Psi_{a,b,\alpha}(X_1)\ \Psi_{a',b',\alpha'}(X_2)$$

This describes the part of the X-invariant which lies in $s^*(H^2(X_1) + H^2(X_2))$ inside the full polynomial algebra $s^*(H^2(X))$. One point here is that if $b^+(X) = 1$ then the left hand side of the formula is not a priori defined as an invariant of X, because of the dependence on the periods of the harmonic form, while all the terms in the right hand side are (differential) topological invariants. The explanantion is that in this case the formula should be read as giving the invariants for X in the "chamber" corresponding to the metrics with a very long neck.

The things that need to be done to complete the theory are:

(1) To extend the gluing formulae to all of $s^*(H^2(X))$, that is to classes $\alpha \in H_2(X)$ which come from relative classes $\alpha_i \in H_2(X_i, Y)$ with the same boundary in $H_1(Y)$.

(2) To deal with the problems that arise from *reducible connnections.*

These points, individually, are quite well understood. A prototype case for the second aspect is that of a connected sum $X = X_1 \sharp X_2$: in this case Y is just the 3-sphere. If $b^+ > 1$ for each of X_1, X_2 the the invariants of X vanish in this situation; if one of the factors, X_2 say, has a negative definite intersection form then one encounters formulae expressing the instanton invariants of X in terms of the similar invariants of X_1 and the abelian reducible connections over X_2. This is straightforward in low dimensional cases, but the general case leads into difficult questions to do with the compactification which have been tackled succesfully by Orzvath [O]. (The corresponding discussion in algebraic geometry, when $X_2 = \overline{\mathbf{CP}}^2$, involves comparing moduli spaces of bundles on a surface with those on its blow-up.) For splittings by general 3-manifolds Y one can define *equivariant Floer homology groups* [AB] and gluing formulae which are valid in low dimensional cases; needing complicated correction terms in general for just the same reasons as in the case of connected sums.

Turning to the second aspect: a complete theoretical treatment here is available in terms of certain "Fukaya-Floer groups" [BD2], which we shall now review. Let γ be a loop in Y and form a complex $CFF_* = CFF_*(Y, \gamma)$ with a double grading:

$$CFF_i = \tilde{CF}_i(Y) \oplus \tilde{CF}_{i-2} \oplus \tilde{CF}_{i-4} \oplus \dots.$$

Here $\tilde{CF}_*$ is the $\mathbf{Z}$-graded complex defined by the $\mathbf{Z}/8$- graded complex CF_*, in the obvious way. There is a differential $\underline{\partial} : CFF_i \to CFF_{i-1}$, which has one component given by the ordinary Floer differential

$$\partial_0 : \tilde{CF}_{i-j} \to \tilde{CF}_{i-j-1},$$

but also other components $\underline{\partial} = \sum_{n\geq 0} \partial_n$:

$$\partial_n : \tilde{CF}_{i-j} \to \tilde{CF}_{i-j-2n-1}.$$

The differential ∂_n is defined by its matrix entries, as for ∂_0. For ρ_+, ρ_- with index difference $2n+1$ we consider the $2n$-dimensional moduli space $M^{2n+1}(\rho_-, \rho_+)/\mathbf{R}$, which we represent as a submanifold of $M(\rho_-, \rho_+)$ by fixing a "centre of mass" along the tube. We can define a 2-dimensional cohomology class $\mu(\gamma \times \mathbf{R})$ over $M^{2n+1}/\mathbf{R}$ by using a relative version of the construction for ordinary classes in H_2. The ρ_-, ρ_+ matrix entry of ∂_n is obtained by "evaluating" $\mu(\gamma \times \mathbf{R})^n$ on $M^{2n+1}(\rho_-, \rho_+)/\mathbf{R}$. This construction gives a filtered complex $(CFF_*, \underline{\partial})$ with cohomology $HFF_*(Y, \gamma)$ and a spectral sequence:

$$HF_*(Y) \otimes H_*(\mathbf{CP}^\infty) \Rightarrow HFF_*(Y, \gamma)$$

The groups HFF_* depend only on the homology class of γ in $H_1(Y)$ (or, more precisely, a choice of homology between two such loops induces a natural isomorphism between the Fukaya-Floer groups). The first differential in the above spectral sequence is

$$d_3 : HF_i \to HF_{i-3},$$

and this can be identified with a cup product in Floer homology. In general if A is a class in $H_p(Y)$, for $p = 0, 1, 2$, there is a corresponding class $\mu(A)$ in the $(4-p)$-dimensional cohomology of the space of connections over Y. The Floer homology groups can be regarded formally as homology or cohomology groups of this space of connections in "half" the (infinite) dimension of the space, and there is an analogue of the usual cup product with $\mu(A)$ mapping the Floer homology of Y to itself and shifting degree by $-(4-p)$ (this becomes $+(4-p)$ if one chooses to work with Floer cohomology, which is a trivial change of notation). The differential d_3 is this cup product associated to the class $A = \gamma$ in $H_1(Y)$.

The gluing construction in this theory is a natural extension of that in the ordinary case described above. If α_1 is a relative classes in $H_2(X_1, Y)$ with the boundary γ in $H_1(Y)$ we define cohomology classes $\mu(\alpha_1)$ over moduli spaces of instantons over X_1 with fixed limits, and then define numbers by evaluating powers of these on the moduli spaces. These numbers give the coefficients of a cycle in CFF_* and the corresponding homology class $\Psi_{\alpha_1}(X_1)$

in $HFF_*(Y,\gamma)$ is independent of choices. If α_2 is a similar relative class in $H_2(X_2,Y)$, with the same boundary γ, we get an invariant in $HFF_*(\overline{Y},\gamma)$. Now there is a class $\alpha \in H_2(X)$ given by gluing α_1 and α_2, and the main formula is an expression for $\Psi_{a,\alpha}(X)$ via a sequence of bilinear pairings on the Fukaya-Floer groups $\sigma_a : HFF_*(Y) \otimes HFF_*(\overline{Y}) \to \mathbf{Q}$:

$$\Psi_{a,\alpha}(X) = \sigma_a(\Psi_{\alpha_1}(X_1), \Psi_{\alpha_2}(X_2)).$$

§3. Elliptic surfaces

In this section we will apply the theory above to make some concrete calculations, inspired by the work of Friedman [F]. We consider the family of manifolds $S_n, n \geq 1$ underlying simply-connected elliptic surfaces, without multiple fibres, and with $p_g = n-1$. It is well known that there is a unique deformation class of such surfaces, for each n, [FM]. We will think of the manifolds in the following way. The surface S_1 is the rational elliptic surface, obtained by blowing up the plane at the 9 intersection points of two general cubic curves and the elliptic fibration $\pi_1 : S_1 \to \mathbf{CP}^1$ is induced by the pencil of cubics generated by these two. Then S_n, with an elliptic fibration $\pi_n : S_n \to \mathbf{CP}^1$ is given by pulling back π_1 under a degree n map from $\mathbf{CP}^1$ to $\mathbf{CP}^1$. If we allow this branched cover of $\mathbf{CP}^1$ to degenerate we can easily see that S_n can also be obtained as a small smoothing of the singular surface given by gluing n copies of S_1 along pairs of fibres. From the topological point of view, S_n is the "fibrewise connected sum" of n copies of S_1 Thus if we let W be a neighbourhood of a smooth fibre, i.e. the manifold-with-boundary $D^2 \times T^2$, and Z_n be the complement of W in S_n we can write

$$S_{n-1} = Z_{n-1} \cup_Y W, \quad S_n = Z_{n-1} \cup_Y Z_1,$$

where the gluing is performed across the 3-torus Y. We want to compute the invariants defined by $U(2)$ bundles E_n over S_n, where $c_1(E_n)$ is *odd*—say 1—on the fibre.

Now the Fukaya-Floer theory becomes very simple in this situation: indeed so simple that the complicated apparatus described in the previous section is scarcely necessary. We need to consider the projectively flat bundles over $Y = S^1 \times T^2$ with first Chern class dual to the S^1 factor. If we restrict to the 2-torus there is just one such connection, up to equivalence, corresponding to the well known fact in algebraic geometry that there is just one stable bundle with odd degree and fixed determinant over a curve of genus 1. Going to the 3-torus we get *two* points in our representation variety, because there are two choices ± 1 of the holonomy around the S^1 factor. Call these points ρ^+, ρ^- respectively. It is also easy to see that the index difference between these points is 4—corresponding to the general fact that the $SO(3)$ version of Floer homology is naturally $\mathbf{Z}/4$-graded;

see [BD1] for example. Thus, with a suitable normalisation of the grading, the Floer complex $\tilde{CF}_*(Y)$ in this situation has one generator in each dimension $4j$, and there is no room for any differentials so this also gives the Floer groups. Likewise, when we go to the Fukaya-Floer theory, with the homology class γ corresponding to the S^1 factor in $Y = S^1 \times T^2$, there are still no differentials, for reasons of parity, and the Fukaya-Floer groups have a free set of generators ρ_k^+ in degree $2k$, ρ_k^- in degree $2k+4$.

We want to discuss invariants defined by 4-dimensional moduli spaces over closed manifolds and in this case the Fukaya-Floer gluing theory boils down to the following very simple set-up. Let $X = X_1 \cup_Y X_2$ be a decomposition of a 4-manifold across a copy of the 3-torus Y and E be a bundle over X with $\dim M(E) = 4$, and which restricts to the bundle considered above on Y. Let α be a class in $H_2(X)$ which is formed from relative classes α_i in X_i with common boundary the class γ specified above. Then, after possibly interchanging X_1 and X_2, the relevant moduli spaces are:

(1) 4-dimensional moduli spaces $M^4(X_1, \rho_+), M^4(X_2, \rho_-)$ which give, by pairing with classes $\mu(\alpha_i)^2$, numbers $\Psi_2(X_1), \Psi_2(X_2)$ say.
(2) 0-dimensional moduli spaces $M^0(X_1, \rho_-), M^0(X_2, \rho_+)$ which give, by "counting points", numbers $\Psi_0(X_1), \Psi_0(X_2)$ say.

The gluing formula is just

$$< \mu(\alpha)^2, [M_E] >= \Psi_2(X_1)\Psi_0(X_2) + \Psi_0(X_1)\Psi_2(X_2).$$

We will also need the formula appropriate to 0-dimensional moduli spaces, and of course this is simpler, having the shape: $\Psi_0(X) = \Psi_0(X_1)\Psi_0(X_2)$.

With this theory in place we will now proceed with our calculation. (Since this is intended mainly as an illustration of the general theory we will not dwell too much on one technical point that enters: the proper treatment of signs and orientations.) There will be four pieces of geometric input into the calculations, i.e. calculations of invariants in special cases. The first is just for the invariant $\Psi_0(W)$ defined by the zero-dimensional moduli spaces over the tubular neighbourhood W. Each of the connections $\rho^\pm$ over the boundary 3-torus bounds a unique projectively flat connection over W, since the holonomy on the circle linking the torus is in the centre ± 1. Then there is a simple general principle, derived from the Chern-Weil theory, that tells us that if an instanton moduli space contains any flat connections then it contains *only* flat connections (see [BD1], for example). Thus the moduli spaces over W consist of just one point, and one finds that the invariants are 1. This gives us a very simple gluing formula in the following general situation: suppose that V_1 and V_2 are manifolds each containing a copy of the torus neighbourhood W, and we form a new manifold $V_1 \sharp V_2$—a kind of generalised fibrewise connected sum—by cutting out each of the copies of W and gluing the resulting boundaries. (So in this notation S_n is the multiple fibre sum $S_1 \sharp \ldots \sharp S_1$.) Then applying the gluing formula for the zero dimensional invariants times we get the rather obvious rule for Ψ_0:

$$\Psi_0(V_1 \sharp V_2) = \Psi_0(V_1)\Psi_0(V_2).$$

The next two pieces of geometric input are calculations for the surface S_2. This surface is a $K3$ surface, and as such the invariants have been calculated in a variety of ways [OG],[FS], the most direct being to use the algebro-geometric work of Mukai [Mu] (compare [DK]). In any case we will assume known the facts that

$$\Psi_0(S_2) = 1$$

and

$$\Psi_{2,\beta}(S_2) = \beta.\beta \text{ for } \beta \in H_2(S_2).$$

The first of these, and the multiplicativity property, immediately tells us that

$$\Psi_0(S_n) = 1 \text{ for all } n.$$

Now let us look at the 4-dimensional invariants $\Psi_{4,\beta}(S_n)$. If the Ψ_0-invariants of V_1, V_2 are 1 the gluing formula tells us that if $\beta \in H_2(V_1 \sharp V_2)$ is the sum of terms $\beta_i \in H_2(V_i)$, each of which can be represented *outside* the copy of W, then

$$\Psi_{2,\beta}(V_1 \sharp V_2) = \Psi_{2,\beta_1}(V_1) + \Psi_{2,\beta_2}(V_2).$$

We deduce easily from this, and the calculation for S_2, that $\Psi_{2,\beta}(S_n)$ is $\beta.\beta$, for any n and for classes β which can be represented outside a tubular neighbourhood, i.e. for classes where the intersection $\beta.F$ with the fibre is zero. The remaining problem is to deal with the classes which intersect the fibre, and it suffices to consider a "horizontal" class h in $H_2(S_n; \mathbf{Q})$ with $h.h = 0$, $h.F = 1$. In the general picture we consider classes $a_i \in H_2(V_i)$ which meet the boundary of W in compatible loops, and form a class $a \in H_2(V)$. Then three applications of the Fukaya-Floer gluing formula, together with the discussion above of the 0-dimensional invariants, gives:

$$\Psi_{2,a}(V) = \Psi_0(V_1)\Psi_{2,a_2}(V_2) + \Psi_0(V_2)\Psi_{2,a_1}(V_1) - 2\Psi_{2,H}(W)\ \Psi_0(V_1)\Psi_0(V_2)$$

where H is the class in $H_2(W, \partial W)$ represented by a small transverse disc. Let us write $\Psi_{2,H}(W) = \lambda$. Applying this formula to our elliptic surfaces we get:

$$\Psi_2(S_{m+n}) = \Psi_2(S_m) + \Psi_2(S_n) - 2\lambda,$$

and hence

$$\Psi_2(S_n) = (c - 2\lambda)n + 2\lambda,$$

for some constant c. Then, knowing that the invariant vanishes for $n = 2$, we have

$$\Psi_{2,h}(S_n) = (2 - n)\lambda,$$

for the horizontal class h. Putting this together with the other case we get, for any $\alpha \in H_2(S_n)$,

$$\Psi_{2,\alpha}(S_n) = \alpha.\alpha + (2-n)\lambda(\alpha.F)^2,$$

where F is the class of a generic fibre.

The remaining task is thus to compute the number λ, an invariant of the manifold-with-boundary $W = D^2 \times T^2$, and this brings us to our last piece of geometric input. We can get back into the setting of closed manifolds by considering the "double" $S^2 \times T^2$ of W. The gluing formula gives

$$2\lambda = \Psi_{2,H}(S^2 \times T^2)$$

where now we write H for the homology class of a slice S^2 in $S^2 \times T^2$. So the problem comes down to the calculation of an invariant defined by a 4-dimensional moduli space of instantons on a bundle E over $S^2 \times T^2$ and the relevant Chern classes are

$$c_1(E) = [S^2] + [T^2] \ , \ c_2(E) = 1.$$

Here we must be careful however because $b^+(S^2 \times T^2) = 1$ so the invariants of this manifold are only defined on a system of chambers, and the remarks in the previous section apply: we consider invariants defined by a metric (for example a product metric) in which the T^2 factor is very *small.* The 4-manifold $S^2 \times T^2$ is of course a complex surface, so we can interpret our moduli spaces as moduli spaces of stable bundles. In particular they have a natural complex orientation, which turns out to be the appropriate one for the gluing theory above. Let us write L_H for the Hopf line bundle over $S^2 = \mathbf{CP}^1$, and regard it also as a bundle over the product $S^2 \times T^2$. Similarly fix a line bundle L_T over T^2 with degree 1 (i.e. fix a base point in the torus). The moduli space corresponding to the $U(2)$ instantons is that of stable bundles with fixed determinant $L_H \otimes L_T$.

Proposition. *Let M be the moduli space of stable rank-2 holomorphic bundles E over the complex surface $S^2 \times T^2$ with $c_2(E) = 1$ and with $\Lambda^2 E = L_H \otimes L_T$ and with respect to a polarisation $\omega = F + \epsilon H \in H^2(S^2 \times T^2)$ where $\epsilon \ll 1$. The invariant $\Psi_{2,H}(S^2 \times T^2) =< \mu(H)^2, [M] >$ is -2.*

Given this Proposition we see that $\lambda = -1$ and we get:

$$\Psi_{2,\alpha}(S_n) = \alpha.\alpha + (n-2)(\alpha.F)^2,$$

in agreement with [F], for the case of no multiple fibres. (V. Munoz has recently extended this approach to obtain the results of [F] for elliptic surfaces with multiple fibres.)

The proof of this Proposition is a straightforward piece of algebraic geometry. The critical values for the polarisation parameter ϵ arise when there are decomposable semi-stable bundles and it is easy to see that the only reduction with our Chern classes is $L_H \oplus L_V$, which is semi-stable only if $\epsilon = 1$. We claim that if $\epsilon < 1$ the stable bundles are precisely the non-trivial extensions

$$0 \to L_H \otimes L_\xi \to E \to L_V \otimes L_\xi^{-1} \to 0,$$

where L_ξ is the pull-back to the product surface of a line bundle of degree 0 over the torus. To see this we note first that if E is a stable bundle of the given topological type then for any L_ξ the Euler characteristic of $E \otimes L_H^{-1} \otimes L_\xi^{-1}$ is zero. The second cohomology group of this bundle is dual to $H^0(E^* \otimes L_H \otimes K) = H^0(E \otimes L_H^{-2} \otimes L_T^{-1})$ and this must vanish for a stable bundle, with any polarisation, since:

$$\deg(L_H^2 \otimes L_T) > \tfrac{1}{2} \deg(E) = \tfrac{1}{2} \deg(L_H \otimes L_T).$$

So either there is some ξ for which there is a non-trivial section of $E \otimes L_H^{-1} \otimes L_\xi^{-1}$ or all the cohomology groups of these bundles vanish, for all ξ. To rule out the latter alternative we apply the Grothendieck-Riemann-Roch formula to the projection map π_2 from $S \otimes T$ to T, where T is identified with the Picard variety of degree-zero line bundles L_ξ over S. We take the direct images $R^i(\pi_2)_*$ of the bundle $\pi_1^*(E \otimes L_H^{-1}) \otimes \mathbf{P}$ where P is the Poincare bundle over $S \times T$. If the second alternative holds then the alternating sum of the Chern characters of these direct image sheaves is 0, but on the other hand this alternating sum can be computed in terms of the characteristic classes of E using the Grothendieck-Riemann-Roch formula. Actually, we can avoid making this calculation directly by just using the fact that it depends only on the topological type of E, and then finding the direct images for any other bundle of this topological type, such as $L_H \oplus L_T$. In any event this argument shows that the alternating sum of the Chern characters is non-zero, so ruling out the second alternative. Thus there is a ξ for which there is a non-trivial section of $E \otimes L_H^{-1} \otimes L_\xi^{-1}$. If this section vanishes on a divisor D then the stability condition implies that $D = L_H^a$ for some integer $a \geq 0$. But then $E \otimes L_H^{-(1+a)} \otimes L_\xi^{-1}$ has a section with isolated zeros, and hence has non-negative second Chern class. On the other hand $c_2(E \otimes L_H^{-(1+a)} \otimes L_\xi^{--1}) = -a$, so we must have $a = 0$ and there is a nowhere-vanishing holomorphic map from $L_H \otimes L_\xi$ to E, which gives the desired extension. Conversely it is easy to check that all extensions of this form are stable when $\epsilon < 1$. (When $\epsilon > 1$ this argument shows that the moduli space is *empty* and the behaviour here is an example of the general discussion of the change of moduli spaces with the polarisation given by Mong [Mo] and Qin [Q]. In fact another way of getting this description of the moduli space is to see that the space is empty when ϵ is large, then use these general results to show that the "new" stable bundles created when ϵ moves across 1 are just the extensions above.)

Now the extensions for fixed ξ are parametrised by

$$H^1(S;(L_T\otimes L_\xi)^{-1}\otimes L_H)=H^1(T;(L_T\otimes L_\xi)^{-1})\otimes H^0(S^2;L_H)=\mathbf{C}^2.$$

The isomorphism class of the middle term depends only on the corresponding projective space, which is canonically identified with $S^2=\mathbf{P}(H^0(L_H))$, since the other factor $H^1((L_T\otimes L_\xi)^{-1}$ is 1-dimensional. On the other hand the line bundles L_ξ are parametrised by $\xi\in T$, which is canonically identified with $Pic_0(T)$ by the choice of base point. So we conclude that the moduli space M can be identified with the surface $S^2\times T$ itself. Thus our remaining task is to identify the cohomology class $\mu(H)$ in the $H^2(M)$ in terms of this description.

Consider first the restriction of $\mu(H)$ to a slice $S^2=\mathbf{CP}^1$ in the moduli space. Thus we are considering the projective space in the moduli space obtained by fixing L_ξ and varying the extension. This is a familar problem in the subject: more generally one can consider a general base space X, line bundles L_1, L_2 over X and the family of bundles given by extensions $0\to L_1\to E\to L_2\to 0$, parametrised by the projective space $\mathbf{P}=\mathbf{P}(H^1(L_1\otimes L_2^*))$. If α is any class in $H_2(X)$ the corresponding class $\mu(\alpha)\in H^2(\mathbf{P})$ is

$$\tfrac{1}{2}<c_1(L_2)-c_1(L_1),\alpha>$$

times the standard generator of $H^2(\mathbf{P})$. (The calculation is explained, in the language of gauge theory in [DK], Chapter 5, and it is easy to translate this to the holomorphic setting.) So in our case we conclude that $\mu(H)$ restricts to $-1/2$ on the S^2 slice. (The sign here is that given by the complex orientation, which turns out to be the right one for the gluing discussion.)

The calculation of the restriction of $\mu(H)$ to a "vertical" slice $T\subset S^2\times T$ in the moduli space is less standard. We will exploit the action of the translations of the torus on itself, and hence on the product $S^2\times T$, but this involves a fiddly point. The translations do not act directly on the moduli space M because they do not preserve the fixed-determinant condition det $E=L_T$. They do act naturally on the moduli space, M' say, of projectivised bundles $\mathbf{P}(E)$ but this is not quite the same as M. In fact M is a 4-fold cover of M', since if η is a point of order 2 in T, so L_η^2 is trivial, a bundle given by an extension

$$0\to L_H\otimes L_\xi\to E\to L_T\otimes L_\xi^{-1}\to 0$$

is projectively equivalent to one given by a similar extension

$$0\to L_H\otimes L_{\xi+\eta}\to E\otimes L_\eta\to L_T\otimes L_{\xi+\eta}^{-1}\to 0.$$

Thus the moduli space M' is canonically identified with the original surface $S^2\times T$, and its 4-fold cover is only identified with this surface when a base point is fixed. In turn the translations in the T factor of the base space act in the obvious way on M'—simply transitively on each slice. Now let $\mathcal{P}$ be the universal projective bundle over $X \times M'$. For clarity we will write the original base space X as $S_1^2 \times T_1$ and the moduli space M' as $S_2^2 \times T_2$. The above property of the translations means that the restriction of $\mathcal{P}$ to the mixed product $S_1^2 \times T_2$ is isomorphic to the projectivisation of the original bundle E. Thus the restriction of $\mu(S_1^2)$ to the T_2 factor in M' is just

$$c_2(E) - \tfrac{1}{4}c_1^2(E) = 1 - \tfrac{1}{4}.2 = 1/2.$$

Passing to the cover M we introduce a factor of 4, so $\mu(H)$ is 2 on the vertical slice T in M. Finally, then, we see that in terms of the standard generators H, T for the cohomology of the moduli space M,

$$\mu(H) = 2H - \tfrac{1}{2}T$$

so $< \mu(H)^2, M >= -2.2.\frac{1}{2} = -2$, as desired.

§4. Curves of higher genus and quantum cohomology

In the previous section we have illustrated how the theory can be applied in a very simple case. We will now discuss some topics that arise when one considers the analogous programme for curves of higher genus. One would like to study the following set-up. Let Σ be a complex curve of genus g which embeds in two surfaces Z_1, Z_2 with trivial normal bundle. Let Z_0 be the singular surface with a normal crossing $Z_1 \cup_\Sigma Z_2$ and $Z_t, t \in \mathbf{C}$ be a standard smoothing of this surface, so for non-zero t the surface Z_t is smooth with underlying 4-manifold obtained by cutting out neighbourhoods of the curve and gluing the resulting boundaries; copies of $Y = \Sigma \times S^1$. Algebro-geometrically, Z_t is a degenerating family of surfaces, with Z_0 as the singular fibre. One would expect that the analogue of the differential-geometric gluing theory should be found in an algebro-geometric discussion of degeneration of moduli spaces, and the latter theory has been developed by Gieseker and Li [GL] (This extends extends work of Gieseker [G] in the case of curves, which has been applied by Thaddeus to the mathematical formulation of the "Verlinde fusion rules", which are intimately related to the gluing problems discussed in this article, compare [D]). Of course the notion of a "bundle" (locally free sheaf) over the singular space Z_0 makes sense, but in the Gieseker and Li theory the right moduli space to attach to this surface includes more general objects. One considers also bundles over repeated blow-ups $\tilde{Z}_0^{(k)}$ of Z_0. Here $\tilde{Z}_0^{(1)}$ is the blow-up of Z_0 along the double curve Σ, which is a surface with three components $Z_1, Z_2, \Sigma \times \mathbf{CP}^1$ glued along copies of Σ: $\tilde{Z}_0^{(2)}$ is obtained by blowing up one of the double curves in $\tilde{Z}_0^{(1)}$ and so on. So $\tilde{Z}_0^{(k)}$ contains a chain of k copies of $\Sigma \times \mathbf{CP}^1$ linking copies of Z_1, Z_2. A bundle over $\tilde{Z}_0^{(k)}$ is given by bundles over

Z_1, Z_2 and k bundles over $\Sigma \times \mathbf{CP}^1$ whose restrictions to the various double curves are isomorphic, and with a choice of isomorphisms along these curves. Plainly this should correspond to the differential-geometric description of instantons over the 4-manifold with a long neck in terms of a chain of instantons over the tube $Y \times \mathbf{R} = \Sigma \times S^1 \times \mathbf{R}$. To understand this picture more clearly one would like to have an algebro-geometric description of these instantons over tubes, in the same spirit as the description of instantons over a compact surface in terms of stable bundles. As in the rest of this paper, we will restrict attention to the situation when we have bundles whose restriction to Σ has odd degree, so we avoid complications from semi-stability and reducible connections. In this case we make the following conjecture:

Conjecture. *There is a natural correspondence between*

(1) *finite-energy $U(2)$-instantons over $\Sigma \times S^1 \times R$, with c_1 odd over Σ,*
(2) *rank-2 holomorphic bundles over $\Sigma \times \mathbf{CP}^1$ with odd degree over Σ, and whose restrictions to $\Sigma \times \{0\}, \Sigma \times \{\infty\}$ are stable.*

Note that the finite-energy instantons have limits which are projectively flat connections over $\Sigma \times S^1$. Up to a choice of holonomy ± 1 around the circle these correspond to projectively flat connections over Σ, and so in turn to stable holomorphic bundles over Σ by the Narasimhan-Seshadri theorem: in the conjectural correspondence above the restrictions of the holomorphic bundle should of course match up with the limits of the instanton. (The choice of ± 1 holonomy around the circle goes over to the parity of the first Chern class over the $\mathbf{CP}^1$ factor.)

One can also discuss the case of normal crossings along a curve with non-trivial normal bundle, when the relevant 3-manifold is a non-trivial circle bundle over Σ. In this case results in the same general direction as the above conjecture, relating the instantons over the tube to holomorphic bundles over a ruled surface, have been obtained by G-Y. Guo in his Oxford thesis: it seems very likely that similar techniques can be applied to prove the conjecture. In any event, if we assume something like this to be true we see how to translate much of the gluing theory into algebro-geometric language. First, the Floer homology groups $HF_*(\Sigma \times S^1)$ are just the ordinary homology groups of the moduli space $N(\Sigma)$ of stable bundles with fixed odd determinant over Σ (with the grading reduced modulo 8); which have been extensively studied from many points of view. To see this one has to discuss the "Morse-Bott" version of the Floer theory in which the Chern-Simons function has a critical manifold, extending the case of isolated critical points as discussed in §2. If γ is a class in $H_1(\Sigma \times S^1)$ there is a Morse-Bott description of the Fukaya-Floer groups $HFF_*(\Sigma \times S^1, \gamma)$ using a complex made up of an infinite number of copies of the ordinary singular chain groups of $N(\Sigma)$, with a differential defined by the instanton moduli spaces on the tube and the maps to $N(\Sigma)$ given by the limits at $\pm\infty$: that is, assuming our conjecture, by certain moduli spaces of bundles over $\Sigma \times \mathbf{CP}^1$ and the maps induced by restriction to Σ_0, Σ_∞. In turn one can describe the differentials in the spectral sequence $HF_*(\Sigma \times S^1) \otimes H_*(\mathbf{CP}^\infty) \Rightarrow HFF_*(\Sigma \times S^1, \gamma)$ in terms of these moduli spaces of holomorphic bundles. In fact a variant of an argument due to Furuta [DFK], shows that if γ is the class of the circle factor in $\Sigma \times S^1$ then the spectral sequence degenerates, so that $HFF_*(\Sigma \times S^1)$ is just an infinite sum of copies of the homology of $N(\Sigma)$ in this case.

We will now bring another theme into this discussion, involving the *symplectic geometry* of the moduli space $N(\Sigma)$ of bundles over the curve Σ, amplifying some remarks in [BD2]. In general, if (V, ω) is a "positive" symplectic manifold, with first Chern class a positive multiple of $[\omega]$—in particular for a complex projective Fano manifold— there is a "quantum product" structure on the cohomology of V defined by the holomorphic maps from the the Riemann sphere into V, with some compatible almost-complex structure. This is a very active area of research at the moment [RT],[KMa],[W]. Roughly speaking, if $\mathcal{M}$ is a moduli space of pseudo-holomorphic maps from S^2 to V, then evaluation at the points $0, 1, \infty$ gives a map $r : \mathcal{M} \to V \times V \times V$. Ignoring problems of compactification, the image $r_*(\mathcal{M})$ defines a homology class in $V \times V \times V$ which can be regarded as a map from $H^*(V) \otimes H^*(V)$ to $H^*(V)$. The quantum product is given by the sum, over all moduli spaces $\mathcal{M}$ of dimension less than $3 \dim V$, of these maps. It is an associative and graded-commutative product on the ordinary cohomology. The constant maps give the ordinary cup product and the other moduli spaces give correction terms, which are not compatible with the integer grading on the homology. This deformed product is an invariant of the symplectic structure on V, encoding data from the parametrised holomorphic curves. There are further invariants [RT] which encode more data of the same kind, for example from curves of higher genus.

Now the moduli space $N(\Sigma)$ is a Fano manifold, so this pseudo-holomorphic curve theory can be applied, and is intimately related to the questions in gauge theory and holomorphic bundle theory discussed in this article. This is a manifestation of a relation between the Yang-Mills instantons over $\Sigma \times S^1 \times \mathbf{R}$ and holomorphic maps from S^2 to $N(\Sigma)$, which can be understood in two ways. The first way, due to Dostoglou and Salamon [DS], involves studying the "adiabatic limit" of the instanton equations over $\Sigma \times S^1 \times \mathbf{R}$, for Riemannian metrics in which the Σ factor is very small. The second way goes through the description of these instantons in terms of holomorphic bundles, assuming our conjecture. If a holomorphic bundle E over $\Sigma \times \mathbf{CP}^1$ is stable on Σ_0, Σ_∞ it must be stable on the *generic* fibre $\Sigma_\lambda = \Sigma \times \{\lambda\}$ in the product. On the other hand it is clear that the bundles over $\Sigma \times \mathbf{CP}^1$ which are stable on *all* the fibres Σ_λ correspond precisely to the holomorphic maps from $\mathbf{CP}^1$ to $N(\Sigma)$. For a family of bundles U over Σ one expects that the set of unstable bundles forms a subvariety of codimension g, so if $g > 1$ one expects that the bundles over $\Sigma \times \mathbf{CP}^1$ which are stable on all the fibres to be an open subset whose complement has codimension $g-1$: i.e. one expects a natural isomorphism between the moduli spaces of instantons and holomorphic maps outside a subvariety of this codimension. (One can probably make this more precise using the same kind of techniques as in [F]: if E is a bundle over $\Sigma \times \mathbf{CP}^1$ which is stable over the generic slice Σ_λ but unstable over a finite number of exceptional slices then there is another bundle E' canonically associated to E which is stable on all the slices, and E should be obtained from E' by "elementary modifications" over these slices.)

In the above circle of ideas, Salamon has recently shown that the quantum product on the cohomology of $N(\Sigma)$ agrees with the cup product operation on Floer homology in the gauge theory setting. That is, if γ is a class in $H_1(\Sigma)$ it defines a 3-dimensional cohomology class $\mu(\gamma)$ in the space of connections over $S^1 \times \Sigma$ which gives a degree 3 multiplication on the Floer cohomology. (We will switch to cohomology here, to fit in with the discussion of the quantum product.) On the other hand $\mu(\gamma)$ can also be regarded, by restriction, as a class in $H^3(N(\Sigma))$, with $N(\Sigma)$ viewed as moduli space of flat connections. Salamon's result is that the quantum product with $\mu(\Sigma)$ in $H^*(N(\Sigma))$ agrees with the product with $\mu(\Sigma)$ in the Floer homology, under the isomorphism between $HF^*(\Sigma \times S^1)$ and $H^*(N(\Sigma))$. One can also see this, assuming the conjecture, by directly identifying the moduli spaces of instantons involved in the definition of one product with the moduli spaces of holomorphic maps involved in the definition of the other.

In summary, we see that if γ comes from $H_1(\Sigma)$ the first differential d_3 in the spectral sequence computing the Fukaya-Floer homology of $\Sigma \times S^1$ is given by the quantum product on $H^*(N_\Sigma)$. One should be able to go further and view the higher differentials as some kind of quantum Massey products. Indeed it should be possible to define these higher operations, and groups like the Fukaya-Floer groups for arbitrary positive symplectic manifolds (although it seems quite likely that the higher operations are trivial for Kähler manifolds).

Returning to our overall programme of studying the invariants for smoothings Z_t of $Z_1 \cup_\Sigma Z_2$; our general theory would give a procedure for doing this if we understand both the Fukaya-Floer groups for $\Sigma \times S^1$, which we have discussed above, and the invariants for various basic cases. In particular, as in the calculation for elliptic surfaces in §3, one would need to know invariants for the manifold $\Sigma \times S^2$, which can be described in terms of stable holomorphic bundles over this surface. Now there is again a close connection between these bundles and holomorphic maps from S^2 into N_Σ. Indeed one would expect that the moduli spaces of bundles and maps are birationally equivalent once $g > 1$, since the generic bundle on the product should be stable on all slices. However it is not clear if there is any useful relation between the instanton invariants of the product $S^2 \times \Sigma$ and the various invariants defined by holomorphic curves in $N(\Sigma)$, since questions of compactification become very important. In fact the case of genus 1, when the instanton invariants are non-trivial even though the moduli space $N(\Sigma)$ is just a point, suggests that the relation will not be very close, in contrast to the case of the product structures and Fukaya-Floer homology discussed above.

Finally, to illustrate these theoretical ideas, and perhaps make a small step towards the goal of calculating invariants for manifolds like the smoothings Z_t, we will calculate the quantum product on the cohomology of $N(\Sigma)$ in the case when Σ has genus 2, so $N = N(\Sigma)$ has complex dimension 3. This moduli space is very well understood. As a complex manifold it can be identified with the intersection of two quadrics in $\mathbf{CP}^5$ [N], and the cohomology $H^*(N, \mathbf{Z})$ is made up of:

(1) copies of $\mathbf{Z}$ in H^0, H^2, H^4, H^6;
(2) the image of an isomorphism $\mu : H_1(\Sigma, \mathbf{Z}) \to H^3(N, \mathbf{Z})$.

Let us write $1 = h_0, h = h_2, h_4, h_6$ for the integral generators in the even dimensions; the class h is just the restriction of the hyperplane class in $\mathbf{CP}^5$. The ordinary cup product structure is well-known to be:

$$h \cup h = 4h_4, \quad h \cup h_4 = h_6, \quad \mu(\gamma_1) \cup \mu(\gamma_2) = (\gamma_1 . \gamma_2) h_6,$$

where (.) is the intersection product on $H_1(\Sigma)$. The quantum cohomology is $\mathbf{Z}/4$ graded, so there are just 3 non-trivial groups which are $H^0(N) \oplus H^4(N)$, $H^2(N) \oplus H^6(N)$ and $H^3(N)$.

The first new term in the quantum product comes from the moduli space of degree 1 maps from S^2 to N, i.e. to the space of *lines* in N. This space of lines has been studied a good deal in algebraic geometry (compare the last chapter of [GH]) and is a copy of the Jacobian $J(\Sigma)$ of the curve. In terms of bundles the lines in N correspond to families of extensions

$$0 \to L_0 \to E \to L_1 \to 0$$

where L_0 has degree 0 and L_1 has degree 1. For any choice of line bundles $H^1(L_1^* \otimes L_0)$ has dimension 2, so we get a family of bundles parametrised by $\mathbf{CP}^1$, which gives a line in the moduli space. Then varying L_0, L_1 with fixed determinant $L_0 \otimes L_1$ we get a family of lines parametrised by the Jacobian of Σ. The calculations we need for the quantum product can be expressed in terms of the geometry of lines in the intersection of quadrics or in terms of the Chern classes of the Poincaré bundle for this family of extensions, a vector bundle over a S^2 bundle over the Jacobian. In any event the geometric information we need is the following.

(1) The number of lines in N passing through a generic point is 4.
(2) If l_1, l_2 are generic lines in N then the number of transversals in N to this pair (i.e. the number of lines that meet both l_1 and l_2) is 2.

To see the first of these: the lines in a single quadric through a given point P form a copy of a quadric surface in a $\mathbf{CP}^3$ inside the projectivised tangent space of $\mathbf{CP}^5$ at P. So the lines in two quadrics correspond to the intersection of two conics in $\mathbf{CP}^2$, which gives 4 points. To see the second we observe that l_1, l_2 span a $\mathbf{CP}^3$ in $\mathbf{CP}^5$, and the whole discussion takes place within this $\mathbf{CP}^3$. So we are reduced to counting lines in the intersection of quadrics in $\mathbf{CP}^3$, and we need to know the elementary fact that if two such quadrics meet in a pair of lines from one ruling then they also meet in a pair from the other ruling.

We will now explain how to compute the whole quantum cohomology structure from this geometric data. We first consider the component of the product which maps $H^2(N) \otimes H^2(N)$ to $H^0(N)$. This comes from the component of $r(\mathcal{M})$ in $H_2 \otimes H_2 \otimes H_6$. If we represent h by generic hyperplane sections $H_1, H_2 \subset N$ and the fundamental class by a point p in N then this component is given by the number of degree 1 holomorphic maps f with $f(0) = P, f(1) \in H_1, f(\infty) \in H_2$. But since any line meets H_1, H_2 in just one point this is the same as the number of lines through P computed above. So we see that the quantum product is

$$h^2 = 4(h_4 + 1).$$

In the same way, since a line is Poincaré dual to h_4, the second calculation gives the component of the quantum product from $H^2 \otimes H^4 \to H^2$—depending on the component of $r(\mathcal{M})$ in $H_2 \otimes H_4 \otimes H_4$—and we have

$$hh_4 = h_6 + 2h.$$

Now let μ_1, μ_2 be classes in $H^3(N)$ with cup product h_6. We know that

$$h\mu_1 = 0$$

since it lives in the group $H^1(N) \oplus H^5(N)$, which vanishes. Thus $h\mu_1\mu_2 = 0$. Now we can write $\mu_1\mu_2 = h_6 + \alpha h$ for some integer α, so we must have

$$hh_6 + \alpha h^2 = 0,$$

that is

$$hh_6 = -4\alpha(h_4 + 1).$$

Now the component of hh_6 in H^4 depends upon the same component in $H_2 \otimes H_2 \otimes H_6$ in the triple product that we used before (by the symmetry between the three points $0, 1, \infty$). So

$$hh_6 = 4h_4 + \beta,$$

for some β, and combining the information we see that $\alpha = -1$ and $\beta = 4$. So

$$hh_6 = 4(h_4 + 1) = h^2.$$

Now eliminating all the generators except h, we write

$$h_4 = \tfrac{1}{4}h^2 - 1,$$

$$h_6 = hh_4 - 2h = \tfrac{1}{4}h^3 - 3h;$$

and we get the relation

$$hh_6 = \tfrac{1}{4}h^4 - 3h^2 = h^2,$$

that is

$$h^4 = 16h^2.$$

We see then that the quantum cohomology ring is generated by h and classes $\mu(\gamma)$ with defining relations

$$h^4 = 16h^2, h\mu(\gamma) = 0, \mu(\gamma_1)\mu(\gamma_2) = (\gamma_1.\gamma_2)(\tfrac{1}{4}h^3 - 4h).$$

References

[A] V.I. Arnold, Volume dedicated to A. Floer, Birkhauser, 1994.

[AB] D. Austin and P. Braam, *Equivariant Floer cohomology and gluing 4-manifold invariants*, Preprint.

[BD1] P. Braam and S. Donaldson, *Floer's work on instanton homology, knots and surgery*, Volume dedicated to A. Floer, Birkhauser, 1994.

[BD2] P. Braam and S. Donaldson, *Fukaya-Floer homology and gluing formulae for polynomial invariants*, As [BD1].

[D] S. Donaldson, *Gluing techniques in the cohomology of moduli spaces*, Topological methods in modern mathematics (Ed. Goldberg, Phillips), Publish or Perish, 1993.

[DFK] S. Donaldson, M. Furuta and D. Kotschick, *Floer homology groups in Yang-Mills theory*, To appear.

[DK] S. Donaldson and P. Kronheimer, *The geometry of four-manifolds*, Oxford Univ. Press, 1990.

[DS] S. Dostoglou and D. Salamon, *Instanton homology and symplectic fixed points*, Symplectic Geometry (ed. Salamon), Cambridge U.P., 1993.

[F] R. Friedman, *Vector bundles and SO(3)-invariants for elliptic surfaces III: the case of odd fibre degree*, American Jour. Math. (To appear).

[FM] R. Friedman and J. Morgan, *Four manifolds and complex surfaces*, Springer, 1994.

[FS] R. Fintushel and R. Stern.

[G] D. Gieseker, *A degeneration of moduli spaces of stable bundles*, J. Diff. Geom. **19** (1984), 173-206.

[GH] P. Griffiths and J. Harris, *Principles of algebraic geometry*, John Wiley, 1978.

[GL] D. Gieseker and J. Li, J. Diff. Geom. (To appear).

[KM] D. Kotschick and J. Morgan, *SO(3) invariants for manifolds with* $b^+ = 1$, *II*, J. Diff. Geom. **39** (1994), 433–456.

[KMa] M. Kontsevich and Y. Manin, *Gromov-Witten classes, quantum cohomology and enumerative geometry*, Max Planck Inst. Preprint.

[L] J. Li, *Algebro-geometric interpretation of Donaldson polynomial invariants*, J. Diff. Geom. **37** (1993), 417–466.

[M] J. Morgan, *Comparison of the Donaldson polynomial invariants with their algebro-geometric analogues*, Topology **32** (1993), 449–489.

[MM] J. Morgan and T. Mrowka, International Mathematics Res. Notices (1993).

[Mo] K-C. Mong, *Some polynomials on the blow-up of P^2*, J. für die Reine Angew. Math. **419** (1991), 67–78.

[Mu] S. Mukai, *On the symplectic structures of moduli spaces of stable sheaves over abelian varieties and K3 surfaces*, Inv. Math. **7** (1984), 101–116.

[N] P. Newstead, *Topological properties of some spaces of stable bundles*, Topology **6** (1967), 241–262.

[O] P. Orzvath, *Some blow-up formulae for Donaldson invariants*, J. Diff. Geom. (To appear).

[OG] K. O'Grady, *Donaldson polynomials for K3 surfaces*, J. Diff. Geom. **35** (1992), 415–429.

[Q] Z. Qin, *Equivalence classes of polarisations and moduli spaces of stable sheaves*, J. Diff. Geom. (1993), 397–417.

[RT] Y. Ruan and G. Tian, *A mathematical theory of quantum cohomology*, Preprint.

[W] E. Witten, *The Verlinde algebra and the cohomology of the Grassmannian*, Preprint.

The Horrocks-Mumford Bundle

Klaus Hulek

This is a survey article on the Horrocks-Mumford bundle and its geometry. Since its discovery in 1972 this bundle has attracted the interest of several algebraic geometers (see e.g.[BHM1], [BHM2], [BM], [DS1], [DS2], [HL2], [HV], [HKW2], [K], [Sa1], [Sch], [Su2]). It is my point of view that we now understand the geometry of this bundle quite well (of course this does not mean that other interesting and surprising facts might not be discovered in the future). In any case the knowledge of the geometry of the Horrocks-Mumford bundle has already proved useful in other contexts, see e.g. the article by Decker and Popescu in this volume.

I would like to thank the organizers of the Durham symposium on Vector Bundles in Algebraic Geometry for organizing this meeting and for giving me the opportunity to speak about the Horrocks-Mumford bundle. I also lectured on the Horrocks-Mumford bundle during a stay at Tokyo Metropolitain University in March 1993. I would like to take this opportunity to thank Professor N. Sasakura for inviting me to Tokyo. This article is based on both my lectures in Tokyo and my talks in Durham.

I am grateful to the Deutsche Forschungsgemeinschaft for partial support over the last few years in the framework of the Schwerpunktprogramm "Komplexe Mannigfaltigkeiten".

Throughout this paper I shall work over the complex numbers $\mathbb{C}$.

I Vector bundles on $\mathbb{P}_n$

In order to put the Horrocks-Mumford bundle into perspective, I first want to recall some known results about the existence of vector bundles on $\mathbb{P}_n$.

1 The projective line $\mathbb{P}_1$

Here the situation is completely described by Grothendieck's theorem.

Theorem 1.1 *Every algebraic vector bundle E on $\mathbb{P}_1$ is a sum of line bundles:*

$$E = \mathcal{O}_{\mathbb{P}_1}(a_1) \oplus \ldots \oplus \mathcal{O}_{\mathbb{P}_1}(a_r).$$

The integers a_i are uniquely determined up to permutation.

Proof. [OSS, theorem 2.1.1] □

2 The projective plane $\mathbb{P}_2$

Topological $\mathbb{C}^2$-bundles over $\mathbb{P}_2$ were classified by Wu:

Theorem 2.1 *There is a bijection between the isomorphism classes of topological $\mathbb{C}^2$-bundles on $\mathbb{P}_2$ and $\mathbb{Z}^2$ given by associating to each vector bundle E its Chern classes $(c_1(E), c_2(E))$.*

Proof. For a proof see [Th]. □

It is natural to ask whether these bundles carry a holomorphic, or an algebraic structure. (By GAGA these two questions are equivalent). A positive answer was given by Schwarzenberger.

Theorem 2.2 ([**Sw**]) *Every topological $\mathbb{C}^2$-bundle on $\mathbb{P}_2$, and hence every complex topological vector bundle on $\mathbb{P}_2$ admits an algebraic structure.*

Proof. [Sw] □

An immediate consequence of this result is the existence of many indecomposable rank 2 bundles on $\mathbb{P}_2$.

3 Projective space $\mathbb{P}_3$

Here the situation is similar to $\mathbb{P}_2$. The topological $\mathbb{C}^2$-bundles were classified by Atiyah and Rees who also proved the following

Theorem 3.1 ([**AR**]) *Every topological $\mathbb{C}^2$-bundle on $\mathbb{P}_3$ admits an algebraic structure.*

Proof. [AR] □

This was generalized by Vogelaar to higher rank.

Theorem 3.2 ([**Vo**]) *Every topological $\mathbb{C}^3$-bundle on $\mathbb{P}_3$, and hence every complex topological bundle on $\mathbb{P}_3$, carries at least one algebraic structure.*

Proof. [Vo] □

4 Higher dimensional projective spaces $\mathbb{P}_n$, $n \geq 5$

If the characteristic of the base field is different from 2, then no indecomposable rank 2 bundles on $\mathbb{P}_n$, $n \geq 5$ are known. (Horrocks constructed one indecomposable rank 2 bundle on $\mathbb{P}_5$ in characteristic 2).

There are two conjectures concerning the non-existence of rank 2 bundles on $\mathbb{P}_n$, $n \geq 4$:

Conjecture 1 (Grauert-Schneider) *Every unstable rank 2 bundle on* $\mathbb{P}_n$*,* $n \geq 4$*, is the sum of two line bundles.*

Conjecture 2 (Hartshorne) *Every rank 2 vector bundle on* $\mathbb{P}_n$*,* $n \geq 6$*, splits.*

The latter conjecture was originally formulated in terms of complete intersections. We shall return to this later.

5 Projective 4-space $\mathbb{P}_4$

Here one knows essentially one indecomposable rank 2 bundle, namely the Horrocks-Mumford bundle F constructed by Horrocks and Mumford in 1972 [HM]. We shall always normalize F such that

$$c_1(F) = 5, \quad c_2(F) = 10.$$

Clearly F is indecomposable, since

$$c(F) = 1 + 5h + 10h^2 \in \mathbb{Z}[h]$$

is irreducible over $\mathbb{Z}$. In fact more is true, namely F is *stable*, since the twisted bundle $F(-1)$ has no sections.

Given F one can form its "satellites" $F(k) = F \otimes \mathcal{O}_{\mathbb{P}_4}(k)$, and $\pi^* F(k)$ where $\pi : \mathbb{P}_4 \to \mathbb{P}_4$ is a finite branched covering. All these bundles are stable, and in particular also indecomposable. No other indecomposable rank 2 bundles on $\mathbb{P}_4$ are known.

II Construction methods for the Horrocks-Mumford bundle

Meanwhile one knows several methods to construct the Horrocks-Mumford bundle.

1 The monad construction

Monads were first introduced by Horrocks. A *monad* is a complex

$$\text{(M)} \qquad A \xrightarrow{p} B \xrightarrow{q} C$$

of vector bundles, such that
(i) p is an injective bundle map
(ii) q is surjective.

Then the cohomology of (M)

$$E = \ker q / \operatorname{im} p$$

is again a vector bundle. The idea is to use simple bundles A, B and C and to define suitable maps p and q in order to construct a new vector bundle E. To construct the Horrocks-Mumford bundle using a monad we fix a vector space

$$V = \mathbb{C}^5$$

and denote its standard basis by e_i, $i \in \mathbb{Z}_5$ where we use cyclic notation for the indices. The Koszul complex on $\mathbb{P}_4 = \mathbb{P}(V)$ reads

$$0 \longrightarrow \mathcal{O}(-1) \xrightarrow{s} V \otimes \mathcal{O} \xrightarrow{\wedge s} \Lambda^2 V \otimes \mathcal{O}(1) \xrightarrow{\wedge s} \Lambda^3 V \otimes \mathcal{O}(2) \xrightarrow{\wedge s}$$
$$\xrightarrow{\wedge s} \Lambda^4 V \otimes \mathcal{O}(3) \longrightarrow \mathcal{O}(4) \longrightarrow 0,$$

where

$$\begin{array}{ccccc} \Lambda^2 V \otimes \mathcal{O}(1) & & \longrightarrow & & \Lambda^3 V \otimes \mathcal{O}(2) \\ & \searrow_{p_0} & & {}_{q_0}\nearrow & \\ & & (\Lambda^2 T)(-1) & & \\ & \nearrow & & \searrow & \\ 0 & & & & 0 \end{array}$$

The crucial ingredient is the following pair of maps found by Horrocks and Mumford

$$f^+ : V \longrightarrow \Lambda^2 V, \quad f^+\left(\sum v_i e_i\right) = \sum v_i e_{i+2} \wedge e_{i+3}$$
$$f^- : V \longrightarrow \Lambda^2 V, \quad f^-\left(\sum v_i e_i\right) = \sum v_i e_{i+1} \wedge e_{i+4}.$$

Using these maps one can define

$$p : V \otimes \mathcal{O}(2) \xrightarrow{(f_+, f_-)} 2\Lambda^2 V \otimes \mathcal{O}(2) \xrightarrow{2p_0(1)} 2\Lambda^2 T$$
$$q : 2\Lambda^2 T \xrightarrow{2q_0(1)} 2\Lambda^3 V \otimes \mathcal{O}(3) \xrightarrow{(-f_-, f_+)} V^* \otimes \mathcal{O}(3).$$

It is elementary to check that $q \circ p = 0$, and hence one obtains a monad

$$V \otimes \mathcal{O}(2) \xrightarrow{p} 2\Lambda^2 T \xrightarrow{q} V^* \otimes \mathcal{O}(3).$$

Its cohomology

$$F = \ker q / \operatorname{im} p$$

is the Horrocks-Mumford bundle. Clearly F is a rank 2 bundle and its Chern polynomial is

$$c(F) = 1 + 5h + 10h^2.$$

Since this polynomial is irreducible over $\mathbb{Z}$, the bundle F is *indecomposable*. Moreover one can deduce from the monad that $F(-1)$ has no sections, and this shows that F is in fact *stable*.

This is how Horrocks and Mumford constructed the bundle in [HM].

2 Serre construction

The idea of the Serre construction is to associate to a 2-codimensional subvariety Y of a variety X a rank 2 bundle E on X together with a section s which vanishes along Y. This construction goes back to Serre in the local case [Se] and was rediscovered by several authors [Ho], [BV], [Ha1], [GM]. When X is a projective space this result can be formulated as follows.

Theorem 2.1 *Let Y be a 2-codimensional subvariety of $\mathbb{P}_n$, $n \geq 3$. If Y is a locally complete intersection and $\omega_Y = \mathcal{O}_Y(l)$ for some $l \in \mathbb{Z}$, then there exists a rank 2 vector bundle E on $\mathbb{P}_n$ together with a section s such that $Y = \{s = 0\}$. The Chern classes of E are given by $c_1(E) = l + n + 1$ and $c_2(E) = \deg Y$.*

Proof. E.g. [OSS, theorem 5.1.1]. □

A well known result of Barth's says that the canonical bundle of any 2-codimensional submanifold $Y \subset \mathbb{P}_n$, $n \geq 6$ is induced, i.e. of the form $\omega_Y = \mathcal{O}_Y(l)$. Hence via the Serre correspondence conjecture 1 is equivalent to

Conjecture 3 (Hartshorne) *Every codimension-2 submanifold $Y \subset \mathbb{P}_n$, $n \geq 6$ is a complete intersection.*

In order to construct a rank 2 bundle on $\mathbb{P}_4$ one has to look for a subcanonical surface in $\mathbb{P}_4$. The natural idea is to look for smooth surfaces Y in $\mathbb{P}_4$ with trivial canonical bundle. The double point formula says that

$$d^2 = 10d + 5HK + K^2 - e(Y)$$

where d is the degree, H the hyperplane class, K the canonical class and $e(Y)$ the Euler number of Y. Hence the only possibilities are $K3$-surfaces of degree 4 or 6 and abelian surfaces of degree 10. The $K3$-surfaces are complete intersections and hence lead to decomposable bundles. Therefore, one is led to ask whether smooth abelian surfaces of degree 10 exist in $\mathbb{P}_4$. Indeed, Horrocks and Mumford first convinced themselves of the existence of such surfaces and then found the monad using the symmetries which such a surface would necessarily possess.

In fact the first to prove the existence of abelian surfaces in $\mathbb{P}_4$ was Comessatti in 1916 [Com]. He considered a 2-dimensional family of abelian surfaces which are special in the sense that they have real multiplication in $\mathbb{Q}(\sqrt{5})$. In 1985 Ramanan [R] gave a criterion for most abelian surfaces with a $(1,5)$-polarization determining whether they could be embedded in $\mathbb{P}_4$. The remaining cases were treated in [HL1]. Now it is an easy consequence of Reider's theorem to conclude the existence of abelian surfaces in $\mathbb{P}_4$.

Theorem 2.2 ([**Re**]) *Let S be a smooth projective surface. Assume D is a divisor on S which is big and nef. Moreover assume $D^2 \geq 9$. If $|D + K_S|$ is not very ample, then there exists an effective divisor E which fulfills one of the following properties:*
(i) $DE = 0$, $E^2 = -1$ *or* -2,
(ii) $DE = 1$, $E^2 = 0$ *or* -1,
(iii) $DE = 2$, $E^2 = 0$,
(iv) $E^2 = 1$, $D \equiv 3E$ *(and hence* $D^2 = 9$*).*

Proof. [Re] □

From this one easily obtains

Corollary 2.3 *There exist smooth abelian surfaces of degree* 10 *in* $\mathbb{P}_4$.

Proof. Let Y be an abelian surface which is not isogenous to a product of elliptic curves and which has a polarization H of type $(1, 5)$. Then $H^2 = 10$ and H is ample. If H were not very ample, it follows from Reider's theorem that there exists an effective divisor $E \subset X$ with $E^2 \leq 0$. Since there are no curves on an abelian surface with negative self-intersection number, one finds $E^2 = 0$. But this implies that every component of E is an elliptic curve, and hence that Y is isogenous to a product of elliptic curves. □

Remarks 2.4 *(i) One can use Reider's method to characterize all polarized abelian surfaces which can be embedded in* $\mathbb{P}_4$*, and this gives another proof of the results of [R] and [HL1]. For details see [LB, chapter 10].*
(ii) It was already shown by Horrocks and Mumford [HM] that all abelian surfaces in $\mathbb{P}_4$ *lead to the same bundle* F *(up to a change of coordinates).*
(iii) One can also use degenerations of abelian surfaces to construct the Horrocks-Mumford bundle. This was done in [HV] for multiplicity-2 structures on elliptic ruled surfaces. See also [Hul2].

3 Sasakura's construction

Sasakura gave a stratification theoretic construction of the Horrocks-Mumford bundle which also leads to further interesting reflexive rank 2 sheaves on higher dimensional projective spaces (see [Sa1], [Sa2], [Sa3], [SEK1], [SEK2]). In order to describe his construction consider the following sets in $\mathbb{P}_4$:

$$X_i^1 = \{x_i = 0\},\ i \in \mathbb{Z}_5; \qquad X^1 = \bigcup X_i^1$$
$$X_{ij}^2 = X_i \cap X_j\ (i \neq j); \qquad X^2 = \bigcup X_{ij}^2.$$

Moreover let

$$\dot{\mathbb{P}} = \mathbb{P}_4 \setminus X^2, \qquad \dot{X}^1 = X^1 \setminus X^2.$$

Then

$$\dot{\mathbb{P}} = (\mathbb{P}_4 \setminus X^1) \cup \dot{X}^1.$$

Choose open neighbourhoods N_i of $\dot{X}_i^1 = X_i^1 \setminus X^2$ in $\mathbb{P}_4$ (in the complex topology) such that $N_i \cap N_j = \emptyset$ for $i \neq j$ and let $N = \cup N_i$. Then

$$\dot{\mathbb{P}} = (\mathbb{P}^4 \setminus X^1) \cup N.$$

Every matrix

$$A \in \mathrm{GL}(2, \mathcal{O}_{N \setminus \dot{X}^1})$$

defines a rank 2 vector bundle $\dot{E}$ on $\dot{\mathbb{P}} = \mathbb{P}_4 \setminus X^2$ together with frames e_0 of $\dot{E}|_{\mathbb{P}_4 \setminus X^1}$ and e_1 of $\dot{E}|_N$ such that

$$e_0 = e_1 A \text{ on } N \setminus \dot{X}^1.$$

In order to obtain the Horrocks-Mumford bundle Sasakura considers the invertible matrix A defined by

$$A|_{N_i} = \begin{pmatrix} 1 & -x_{i+2}x_{i+3}/x_{i+1}x_{i+4} \\ 0 & x_i/x_{i+1} \end{pmatrix}.$$

Proposition 3.1 *Let $i : \dot{\mathbb{P}} \to \mathbb{P}_4$ be the inclusion. Then $i_*\dot{E}$ is a rank 2 vector bundle on $\mathbb{P}_4$. It is isomorphic to the Horrocks-Mumford bundle.*

Proof. [Sa1], [Sa2]. □

Remark 3.2 *(i) The idea of Sasakura's construction is to generalize the classical notion of matrix divisor on a Riemann surface (cf [Tj], [We]) to higher dimensions.*
(ii) One can also use this method to obtain explicit transition matrices for the Horrocks-Mumford bundle [Sa2], [Sa3].
(iii) The matrix A is closely related to the maps f^+ and f^- which were the crucial ingredient in the Horrocks-Mumford monad. Note that f^+ is obtained by starting with the pair of indices $(1, 4)$ which are the quadratic residues mod 5, whereas f^- arises from the pair $(2, 3)$ which are the non quadratic residues mod 5.
(iv) The above construction can be generalized to give reflexive rank 2 sheaves E_p on $\mathbb{P}_{p-1}$, where p is a prime congruent to 1 mod 4. For $p > 5$ the sheaves E_p are singular in codimension 4. Their Chern classes have very interesting arithmetic properties and are related to the number of rational points on certain K3 surfaces [SEK1], [SEK2].

4 Sumihiro's construction

For a rank 2 vector bundle E on a curve C let $\pi : \mathbb{P}(E) \to C$ be the associated $\mathbb{P}_1$-bundle. If we blow up $\mathbb{P}(E)$ in a point x, then the fibre through x becomes a (-1)-curve. Blowing this down one obtains another $\mathbb{P}_1$-bundle $\mathbb{P}(E')$ over C. In terms of vector bundles this process can be described as follows: let

$E_{\pi(x)}$ be the geometric fibre of the vector bundle E over x. Then the point x corresponds to a projection $\varphi_x : E_{\pi(x)} \to \mathbb{C}$. We obtain an exact sequence

$$0 \longrightarrow E' \longrightarrow E \longrightarrow \mathbb{C}_{\pi(x)} \longrightarrow 0$$

where the last map is restriction to $E_{\pi(x)}$ followed by the projection φ_x. The kernel E' is locally free of rank 2. This is the simplest example of an elementary transformation of a vector bundle. Such elementary transformations were studied by Maruyama [Ma1], [Ma2] who showed that every vector bundle can be obtained from the trivial bundle (up to tensoring with line bundles) by his elementary transformations, provided the dimension of the base space is at most 3. In order to remove this restriction Sumihiro generalized Maruyama's concept of elementary transformations of vector bundles in [Su1], [Su2]. As a consequence of his theory he obtains the following result: let X be a noetherian scheme, Z a normal Cartier divisor on X, and let $|W|$ be a linear system of dimension $r-1$ of effective Weil divisors on Z. Then there exists a rank r vector bundle E on X together with sections $s_1, \ldots, s_r$ of E such that the divisors $W_i = Z(s_1 \wedge \ldots \wedge \hat{s}_i \wedge \ldots \wedge s_r)$, $1 \le i \le r$, on Z span $|W|$. This can also be viewed as a generalization of the Serre construction (cf. also [Vo]).

In order to reconstruct the Horrocks-Mumford bundle, Sumihiro considers $X = \mathbb{P}_4$, and takes Z to be the quintic hypersurface

$$Z = \{x_0^5 + x_1^5 + x_2^5 + x_3^5 + x_4^5 - 5x_0x_1x_2x_3x_4 = 0\}.$$

This quintic hypersurface has 125 nodes. It was first studied by Schoen [Sch], who also noticed a connection with the Horrocks-Mumford bundle (for this see also [Sch]). The quintic Z contains the smooth quadric surface

$$S = \{\sigma_1 = \sigma_2 = 0\}$$

where σ_i is the i-th elementary symmetric polynomial in the coordinates x_k. Let τ be the transformation of $\mathbb{P}_4$ given by $e_k \mapsto \varepsilon^k e_k$ where $\varepsilon = e^{2\pi i/5}$ (see section III.1 below). Then

$$W = \bigcup_{i \in \mathbb{Z}_5} \tau^i(S)$$

is a union of five quadrics contained in Z. In this way one obtains a 1-dimensional linear system $|W|$ on Z whose general element is an abelian surface with real multiplication in $\mathbb{Q}(\sqrt{5})$ (see [Sch]). Applying his construction to the pair $(Z, |W|)$ Sumihiro was able to reconstruct the Horrocks-Mumford bundle. This also shows that the Horrocks-Mumford bundle has a section vanishing on a union of five quadrics. This was first noticed in [HM]. We shall return to this later, see e.g. theorem IV.5.1 below.

III First geometric properties of F

The original paper by Horrocks and Mumford [HM] has the title "A vector bundle with 15,000 symmetries". We shall first describe the symmetry goup of

F, and then discuss "jumping phenomena". This leads to results concerning the uniqueness and the non-extendability of F.

1 The symmetry group

Consider the following linear transformations on V:

(1.1)
$$\sigma : e_i \longmapsto e_{i-1}$$
$$\tau : e_i \longmapsto \varepsilon^i e_i \qquad (\varepsilon = e^{2\pi i/5}).$$

Then σ and τ are of order 5. Their commutator is

$$[\sigma, \tau] = \varepsilon \, \mathrm{id}_V .$$

The subgroup

$$H_5 = \langle \sigma, \tau \rangle \subset \mathrm{SL}(5, \mathbb{C})$$

of $\mathrm{SL}(5,\mathbb{C})$ generated by σ and τ is the *Heisenberg group* of *level* 5. The representation given by the inclusion $H_5 \subset \mathrm{SL}(5,\mathbb{C})$ is called the *Schrödinger representation.* The group H_5 is a central extension

$$\begin{array}{ccccccc} 1 \longrightarrow \mu_5 = \{\varepsilon^i \,;\, i \in \mathbb{Z}_5\} & \longrightarrow & H_5 & \longrightarrow & \mathbb{Z}_5 \times \mathbb{Z}_5 \longrightarrow 1 \\ \varepsilon & \longmapsto & \varepsilon \, \mathrm{id}_V & & \\ & & \sigma & \longmapsto & (1,0) \\ & & \tau & \longmapsto & (0,1). \end{array}$$

The image of H_5 in $\mathrm{PSL}(5,\mathbb{C})$ is isomorphic to $\mathbb{Z}_5 \times \mathbb{Z}_5$. In fact H_5 is the extension of $\mathbb{Z}_5 \times \mathbb{Z}_5$ which lifts the projective representation of $\mathbb{Z}_5 \times \mathbb{Z}_5$ in $\mathrm{PSL}(5,\mathbb{C})$ defined by (1.1) to a linear representation.

Let N_5 be the normalizer of H_5 in $\mathrm{SL}(5,\mathbb{C})$. It was shown in [HM] that

$$N_5/H_5 \cong \mathrm{SL}(2, \mathbb{Z}_5).$$

In fact N_5 is a semi-direct product

$$N_5 = H_5 \rtimes \mathrm{SL}(2, \mathbb{Z}_5)$$

and hence the order of N_5 is

$$|N_5| = |H_5| \cdot |\,\mathrm{SL}(2, \mathbb{Z}_5)| = 125 \cdot 120 = 15,000.$$

Proposition 1.1 *The action of N_5 on $\mathbb{P}_4$ lifts to an action on the bundle F.*

Proof. [HM] □

Remarks 1.2 *(i) Decker [De1] has shown that N_5 is in fact the full symmetry group of F.*
(ii) Both the Heisenberg group H_5 and the group N_5 are closely related to the symmetries of abelian surfaces in $\mathbb{P}_4$. We shall return to this in section V.2 below.

We finally note the element $\iota \in N_5$ given by

$$\iota : e_i \longmapsto e_{-i}. \tag{1.2}$$

The involution ι decomposes V into eigenspaces

$$V = V^+ \oplus V^-$$

where

$$V^+ = \langle e_0, e_1 + e_4, e_2 + e_3 \rangle, \qquad V^- = \langle e_1 - e_4, e_2 - e_3 \rangle.$$

We set

$$\mathbb{P}_1^- = \mathbb{P}(V^-), \qquad \mathbb{P}_2^+ = \mathbb{P}(V^+).$$

2 Jumping phenomena

The method of studying vector bundles on projective spaces by considering their restriction to all the lines in this projective space was first introduced by Van de Ven [V]. Since F is a stable bundle the theorem of Grauert and Mülich implies that

$$F|_L = \mathcal{O}_L(2) \oplus \mathcal{O}_L(3)$$

for a *general* line L in $\mathbb{P}_4$. This leads naturally to the

Definition *A line L in $\mathbb{P}_4$ is called a* jumping line *of F of* order $a \geq 1$ *if*

$$F|_L = \mathcal{O}_L(2-a) \oplus \mathcal{O}_L(3+a).$$

We define

$$J^i(F) = \{L \in \mathrm{Gr}(1,4) \ ; \ L \text{ is a jumping line of order } \geq i\}.$$

Clearly

$$J^3(F) \subset J^2(F) \subset J^1(F)$$

and by semi-continuity these are closed subvarieties of the Grassmannian $\mathrm{Gr}(1,4)$. The varieties of jumping lines and jumping planes were studied by Barth, Moore and the author in [BHM1]. Here we became aware of the close connection between the Shioda modular surface of level 5 and the Horrocks-Mumford bundle. I shall return to the Shioda modular surface S(5) in more detail in section IV.3.

Theorem 2.1 ([BHM1]) *F has jumping lines of order a = 1, 2 and 3. Moreover the following holds:*
(i) $J^1(F)$ is a rational 4-fold. Its singular locus is the set $J^2(F)$.
(ii) $J^2(F)$ is a surface. It is the Shioda modular surface S(5) with the 25 sections contracted to A_4-singularities. Its singular locus is the set $J^3(F)$.
(iii) $J^3(F)$ consists of 25 points. The 25 jumping lines of order 3 are the lines $L_{ij} = \sigma^i\tau^j(\mathbb{P}_1^-)$, $(i,j) \in \mathbb{Z}_5 \times \mathbb{Z}_5$.

Proof. [BHM1, theorem 2] □

It has also proved useful to study *jumping planes* as well as jumping lines.

Definition *A plane E in $\mathbb{P}_4$ is called a* jumping plane *of F if $F|_E$ is unstable.*

We set

$$\mathrm{S}(F) := \{E \in \mathrm{Gr}(2,4)\ ;\ E \text{ is a jumping plane of } F\}.$$

Proposition 2.2 ([BHM1]) *The jumping planes of F are parametrized by a smooth surface $\mathrm{S}(F)$ which is isomorphic to Shioda's modular surface $\mathrm{S}(5)$.*

Proof. Using the monad one can show that

$$\mathrm{S}(F) = \mathrm{Gr}(2,4) \cap (\mathbb{P}_1 \times \mathbb{P}_4) \subset \mathbb{P}_9$$

where the Grassmannian is embedded in $\mathbb{P}_9$ by the Plücker embedding, and where $\mathbb{P}_1 \times \mathbb{P}_4$ is embedded by a suitably normalized Segre map. It is then not difficult to identify this intersection with $\mathrm{S}(5)$. For details see [Hul2]. □

At this point it is appropriate to mention the following uniqueness result due to Decker and Schreyer.

Theorem 2.3 ([DS1]) *Every stable rank 2 bundle $\tilde{F}$ on $\mathbb{P}_4$ with Chern classes $c_1(\tilde{F}) = 5$ and $c_2(\tilde{F}) = 10$ is, up to possibly a change of coordinates, isomorphic to the Horrocks-Mumford bundle F.*

Proof. The idea of Decker and Schreyer is to analyse the variety $\mathrm{S}(\tilde{F})$ of jumping planes of $\tilde{F}$, and to show that $\mathrm{S}(\tilde{F})$ is isomorphic to $\mathrm{S}(5)$. It is then not too difficult to show that $\tilde{F}$ is isomorphic to F. Details can be found in [DS1] and [De2]. □

Remark 2.4 *The restriction of F to any projective 3-space in $\mathbb{P}_4$ is stable, i.e. F has no jumping 3-spaces.*

Let $M_{\mathbb{P}_4}(5,10)$ be the moduli space of stable rank 2 vector bundles on $\mathbb{P}_4$ with $c_1 = 5$ and $c_2 = 10$. Then the above theorem implies the following description of this moduli space:

Corollary 2.5 $M_{\mathbb{P}_4}(5,10) \cong \mathrm{PGL}(5,\mathbb{C})/((\mathbb{Z}_5 \times \mathbb{Z}_5) \rtimes \mathrm{SL}(2,\mathbb{Z}_5))$.

Remark 2.6 *In [DS2] Decker and Schreyer studied pullbacks φ^*F of F via finite maps $\mathbb{P}_4 \to \mathbb{P}_4$. They showed that every small deformation of φ^*F arises from a deformation of φ.*

The above ideas can also be used to prove the following theorem originally due to W. Decker

Theorem 2.7 ([**De2**]) *There exists no stable rank 2 bundle on $\mathbb{P}_5$ with Chern classes $c_1 = 5$ and $c_2 = 10$. In particular, the Horrocks-Mumford bundle cannot be extended to $\mathbb{P}_5$.*

Proof. Originally Decker [De2] proved this result by studying possible monads of such a bundle on $\mathbb{P}_5$. Ellingsrud and Strømme gave the following beautiful proof: consider a finite morphism of bidegree $(1,1)$

$$\pi : \mathbb{P}_4 \times \mathbb{P}_1 \longrightarrow \mathbb{P}_5$$

and set

$$\hat{F} = \pi^* \tilde{F}$$

where $\tilde{F}$ is a stable rank 2 bundle on $\mathbb{P}_5$ with $c_1 = 5$, $c_2 = 10$. Using the uniqueness result of Decker and Schreyer and remark 2.4 one shows that $\tilde{F}|_{\mathbb{P}_4}$ is stable for all $\mathbb{P}_4 \subset \mathbb{P}_5$. Hence we obtain a morphism

$$\varphi : \begin{cases} \mathbb{P}_1 & \longrightarrow & M_{\mathbb{P}_4}(5,10) \\ t & \longmapsto & \tilde{F}|_{\mathbb{P}_4 \times \{t\}} \end{cases} .$$

A simple Chern class calculation proves that φ cannot be constant. But then using corollary 2.5 one can lift this map to a non-constant morphism

$$\tilde{\varphi} : \mathbb{P}_1 \longrightarrow \mathrm{PGL}(5, \mathbb{C}).$$

Since $\mathrm{PGL}(5, \mathbb{C})$ is affine, this is a contradiction. □

Remark 2.8 *There are three different topological $\mathbb{C}^2$-bundles on $\mathbb{P}_5$ with Chern classes $c_1 = 5$ and $c_2 = 10$.*

IV Classification of HM-surfaces

In this section I want to describe the classification of Horrocks-Mumford surfaces.

1 HM-surfaces

For every non-zero section $s \in \Gamma(F)$ we can consider the zero set

$$X_s = \{s = 0\}.$$

Since $h^0(F(-1)) = 0$ it follows that X_s is a *surface* of degree

$$\deg X_s = c_2(F) = 10.$$

Definition *The surfaces X_s are called* HM (Horrocks-Mumford) -surfaces.

Two sections define the same HM-surface if and only if they differ by a non-zero scalar constant. Since $h^0(F) = 4$ the HM-surfaces are parametrized by the 3-dimensional projective space

$$\mathbb{P}\Gamma = \mathbb{P}(\Gamma(F)).$$

Proposition 1.1 ([HM]) (i) *The general* HM-*surface is smooth.*
(ii) *Every smooth* HM-*surface is abelian.*

Proof. (i) is clear if one constructs F via the Serre construction from an abelian surface. If one constructs F via a monad, then one can show that F is globally generated outside the 25 jumping lines L_{ij} of order 3. A calculation in local coordinates then shows that a general HM-surface is also smooth near these lines [HM].
(ii) It follows from the exact sequence

$$0 \longrightarrow \mathcal{O} \xrightarrow{s} F \longrightarrow \mathcal{I}_{X_s}(5) \longrightarrow 0$$

that

$$\omega_{X_s} = \mathcal{O}_{X_s}.$$

Hence X_s is abelian or $K3$. One can now either use the double point formula to show that X_s is abelian or use the fact that

$$c_2(X_s) = c_2(T_{X_s}) = 0.$$

□

2 Elliptic curves

Before we can describe the classification of singular HM-surfaces it is necessary to recall some facts about elliptic curves. Here we shall restrict ourselves to elliptic normal curves of degree 5 in $\mathbb{P}_4$, but this discussion goes through essentially unchanged for arbitrary degree $n \geq 3$.

For an elliptic curve E we can consider the group $E^{(5)}$ of 5-torsion points. As an abstract group $E^{(5)}$ is isomorphic to $\mathbb{Z}_5 \times \mathbb{Z}_5$. There is, however, no canonical isomorphism between these groups. On the other hand $E^{(5)}$ carries an intrinsically defined, non-degenerate, alternating form

$$\mu^{(5)} : E^{(5)} \times E^{(5)} \longrightarrow \mathbb{Z}_5$$

the so-called *Weil pairing.*

Definition *A* level-5 structure *on E is a symplectic basis of $E^{(5)}$, i.e. a basis a, b with $\mu^{(5)}(a,b) = 1$.*

Remark 2.1 *Alternatively one can define a level-5 structure as a symplectic isomorphism*

$$\alpha : E^{(5)} \longrightarrow \mathbb{Z}_5 \times \mathbb{Z}_5$$

where $E^{(5)}$ carries the Weil pairing and $\mathbb{Z}_5 \times \mathbb{Z}_5$ carries the standard symplectic form.

Let O be the origin of E. We consider the line bundle

$$\mathcal{L} = \mathcal{O}_E(5O).$$

Let

$$\lambda : \begin{cases} E & \longrightarrow & \operatorname{Pic}^0 E \\ x & \longmapsto & \mathcal{L}^{-1} \otimes T_x^* \mathcal{L} \end{cases}$$

where T_x denotes translation by x. Then

$$E^{(5)} = \ker \lambda.$$

I.e.

$$T_x^* \mathcal{L} = \mathcal{L}$$

if and only if $x \in E^{(5)}$. Moreover $\mathcal{L}$ is *symmetric*, i.e. if

$$\iota : \begin{cases} E & \longrightarrow & E \\ x & \longmapsto & -x \end{cases}$$

is the standard involution on E, then

$$\iota^* \mathcal{L} = \mathcal{L}.$$

Now fix a level-5 structure $\alpha : E^{(5)} \to \mathbb{Z}_5 \times \mathbb{Z}_5$. This defines an isomorphism

$$E^{(5)} \rtimes \langle \iota \rangle \xrightarrow{\cong} (\mathbb{Z}_5 \times \mathbb{Z}_5) \rtimes \mathbb{Z}_2. \tag{2.3}$$

Recall that we have an exact sequence

$$1 \longrightarrow \mu_5 = \{\varepsilon^i \operatorname{id}_V\} \longrightarrow H_5 \rtimes \mathbb{Z}_2 \longrightarrow (\mathbb{Z}_5 \times \mathbb{Z}_5) \rtimes \mathbb{Z}_2 \longrightarrow 0.$$

Proposition 2.2 *Let α be a level 5-structure on E. Then the action of $E^{(5)} \rtimes \langle \iota \rangle$ lifts via the isomorphism (2.3) to an action of $H_5 \rtimes \mathbb{Z}_2$ on the bundle $\mathcal{L}$. This lifting is unique, provided ι acts on the fibre of $\mathcal{L}$ over the origin by $+1$.*

Proof. This is a special case of a much more general fact which is true for polarized abelian varieties, see [LB, theorem 6.9.5]. □

Recall that by III.(1.1) the group $H_5 \rtimes \mathbb{Z}_2$ also acts on V, and hence on $\mathbb{P}_4$ and $\mathcal{O}(1)$.

Proposition 2.3 *Given a level-5 structure on E there exists an (up to a common scalar) unique basis $s_0, \dots, s_4$ of $\Gamma(\mathcal{L})$ such that the embedding*

$$\varphi_{|\mathcal{L}|} : \begin{cases} E & \longrightarrow & \mathbb{P}_4 = \mathbb{P}(V) \\ x & \longmapsto & (s_0(x) : \dots : s_4(x)) \end{cases}$$

is equivariant with respect to the action of $H_5 \rtimes \mathbb{Z}_2$ on $(E, \mathcal{L})$ and on $(\mathbb{P}_4, \mathcal{O}(1))$.

Proof. [BHM1, proposition 4] □

Remark 2.4 *The above proposition describes in fact a bijection between elliptic curves with a level-5 structure and $H_5 \rtimes \mathbb{Z}_2$-invariant elliptic normal curves in $\mathbb{P}_4$.*

3 Modular curves and Shioda modular surfaces

Again I shall restrict myself to the level-5 case, although most of what follows carries over immediately to the level-n case.

Recall the *upper half plane*

$$S_1 = \{\tau \in \mathbb{C}\ ;\ \operatorname{Im} \tau > 0\}.$$

The *modular group*

$$\Gamma = \mathrm{SL}(2, \mathbb{Z})$$

acts on S_1 by

$$\begin{pmatrix} a & b \\ c & d \end{pmatrix} : \tau \longmapsto \frac{a\tau + b}{c\tau + d}. \tag{3.4}$$

By associating to each point $\tau \in S_1$ the elliptic curve

$$E_\tau = \mathbb{C}/(\mathbb{Z} + \mathbb{Z}\tau)$$

one obtains a bijection between the set S_1/Γ and the set of isomorphism classes of elliptic curves. The j-function defines an isomorphism of Riemann surfaces

$$j : S_1/\Gamma \xrightarrow{\sim} \mathbb{C}.$$

The *principal congruence subgroup* of *level* 5 is defined as

$$\Gamma(5) := \{\gamma \in \mathrm{SL}(2, \mathbb{Z})\ ;\ \gamma \equiv \mathbf{1} \bmod 5\}.$$

As a subgroup of Γ it also acts on S_1 and the quotient $S_1/\Gamma(5)$ parametrizes elliptic curves with a level-5 structure: to each point $\tau \in S_1$ one associates the elliptic curve E_τ together with the symplectic basis $[\tau/5], [1/5] \in E_\tau$.

Clearly $X_0(1) = S_1/\Gamma$ can be compactified to $\mathbb{P}_1$. Intrinsically this can be done by setting

$$X(1) = (S_1 \cup \mathbb{Q} \cup \{i\infty\})/\Gamma$$

and defining a suitable complex structure on this quotient. The points in $\mathbb{Q} \cup \{i\infty\}$ form one orbit under Γ, and this means that one has to add one *cusp* ∞ to S_1/Γ to obtain the *modular curve* $X(1)$ of level 1. Similarly one can compactify $X^0(5) = S_1/\Gamma(5)$ by

$$X(5) = (S_1 \cup \mathbb{Q} \cup \{i\infty\})/\Gamma(5).$$

In this case one has to add 12 cusps. The modular curve $X(5)$ of level 5 is rational. (Note that the modular curves $X(n)$ of level n are no longer rational if $n \geq 6$).

The group

$$\Gamma/(\pm\Gamma(5)) = \mathrm{PSL}(2, \mathbb{Z}_5) = A_5$$

acts in a natural way on $X(5)$. If one identifies $X(5)$ with $\mathbb{P}_1$, resp. via stereographic projection with S^2 this becomes the action of the *icosahedral group* as a subgroup of SO(3) on S^2. The 12 cusps can then be interpreted as the vertices of an icosahedron inscribed in S^2. The group A_5 acts transitively on the cusps which, under the quotient map, are mapped to the unique cusp in $X(1)$.

There exists a universal elliptic curve $\pi_0 : S^0(5) \to X^0(5)$ (this is true for all levels $n \geq 3$). *Shioda's modular surface* $S(5)$ of *level* 5 compactifies this universal family, i.e. there is a commutative diagram

$$\begin{array}{ccc} S^0(5) & \subset & S(5) \\ {\scriptstyle \pi_0}\downarrow & & \downarrow{\scriptstyle \pi} \\ X^0(5) & \subset & X(5). \end{array}$$

$S(5)$ is a smooth projective surface. The fibres over the 12 cusps are pentagons of rational (-2)-curves:

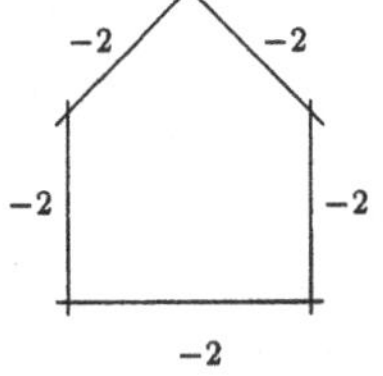

$S(5)$ has 25 sections. In the smooth fibres these are exactly the 5-torsion points. The sections have self-intersection number -5 (cf. theorem III.2.1). Also note that the involution which acts by $x \mapsto -x$ on the fibres of $S^0(5)$ extends to an involution ι on $S(5)$. We call the quotient

$$K(5) = S(5)/\langle\iota\rangle$$

the *Kummer modular surface* of level 5. The singular fibres over the cusps look as follows:

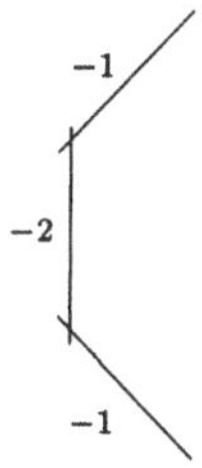

4 Elliptic normal curves in $\mathbb{P}_4$

Let $E \subset \mathbb{P}_4$ be an elliptic normal quintic curve which is $(H_5 \rtimes \mathbb{Z}_2)$-equivariantly embedded. It is well known that E lies on 5 independent quadrics, and that these quadrics describe E scheme-theoretically. As an H_5-module

$$\Gamma(\mathcal{O}_{\mathbb{P}_4}(2)) = V_0 \oplus V_1 \oplus V_2$$

where

$$\begin{aligned} V_0 &= \langle x_0^2, x_1^2, x_2^2, x_3^2, x_4^2 \rangle \\ V_1 &= \langle x_1x_4, x_2x_0, x_3x_1, x_4x_2, x_0x_3 \rangle \\ V_2 &= \langle x_2x_3, x_3x_4, x_4x_0, x_0x_1, x_1x_2 \rangle. \end{aligned}$$

Hence any H_5-invariant 5-dimensional space of quadrics is spanned by elements

$$Q_i(a,b,c) = ax_{i+2}x_{i+3} + bx_i^2 + cx_{i+1}x_{i+4}, \qquad i \in \mathbb{Z}_5.$$

The intersection of 5 such quadrics is non-empty if and only if

$$(a:b:c) = (\lambda^2 : \lambda\mu : -\mu^2)$$

for some $(\lambda : \mu) \in \mathbb{P}_1$. Hence we are led to systems of quadrics

$$Q_i(\lambda,\mu) = \lambda^2 x_{i+2}x_{i+3} + \lambda\mu x_i^2 - \mu^2 x_{i+1}x_{i+4}, \qquad i \in \mathbb{Z}_5.$$

We define

$$E(\lambda:\mu) = \bigcap_{i\in\mathbb{Z}_5} \{Q_i(\lambda,\mu) = 0\}.$$

The set

$$\Lambda = \{0, \infty, \varepsilon^k(\varepsilon^2+\varepsilon^3), \varepsilon^k(\varepsilon+\varepsilon^4)\ ;\ k \in \mathbb{Z}_5\}$$

consists of 12 points. As a subset of $\mathbb{P}^1 = S^2$ it can be identified with the 12 vertices of an icosahedron inscribed in the sphere S^2.

Proposition 4.1 (i) *If* $(\lambda : \mu) \notin \Lambda$, *then* $E(\lambda : \mu)$ *is a smooth elliptic quintic curve.*
(ii) *If* $(\lambda : \mu) \in \Lambda$, *then* $E(\lambda : \mu)$ *is a pentagon of lines.*

Proof. [BHM1, propositions 3 and 6] □

Next we define the surface

$$S_{15} = \bigcup_{(\lambda:\mu)\in\mathbb{P}_1} E(\lambda : \mu).$$

It contains all Heisenberg-invariant elliptic curves as well as 12 singular curves.

Proposition 4.2 ([BHM1]) (i) S_{15} *is a surface of degree* 15. *It is smooth outside* 30 *points where two smooth branches meet transversally.*
(ii) *There is a surjective morphism* $S(5) \to S_{15}$, *which is an immersion into* $\mathbb{P}_4$. *The only identifications which occur arise from singular fibres over opposite vertices of the icosahedron* Λ *which together form a complete pentagon.*

Proof. [BHM1, proposition 10] □

Remark 4.3 *(i) The map* $S(5) \to S_{15}$ *was also discussed in [BH].*
(ii) A pair of opposite vertices is given by $0, \infty \in \Lambda$. *There the situation is as follows:*

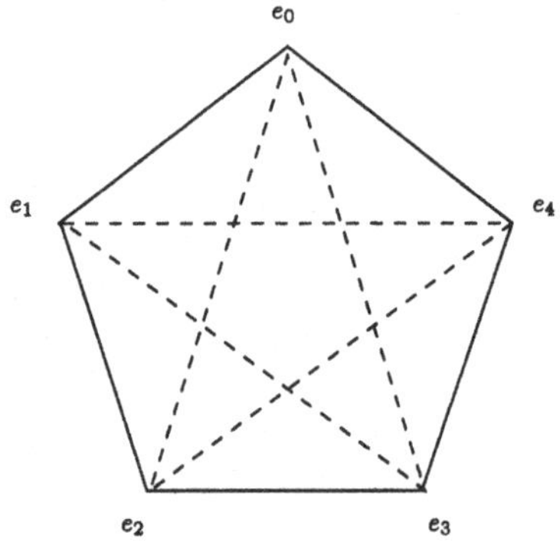

The dotted pentagon is the curve $E(0 : 1)$, *and the fully drawn pentagon is the curve* $E(1 : 0)$.

5 Description of HM-surfaces

The classification of HM-surfaces is given by

Theorem 5.1 ([BHM2]) *Every* HM-*surface is one of the following:*
(i) *a smooth abelian surface*
(ii) *a translation scroll of an elliptic normal quintic curve*

(iii) *a tangent scroll of an elliptic normal quintic curve*
(iv) *an elliptic quintic scroll with a double structure*
(v) *a union of five quadrics*
(vi) *a union of five planes with a double structure.*

Before giving an outline of the proof of this theorem, I would like to comment on these surfaces. Let $E \subset \mathbb{P}_4$ be an elliptic quintic curve. Consider a point $P_0 \in E$ which is not a 2-torion point, i.e. $2P_0 \neq 0$. For every point P we consider the secant line $L(P, P + P_0)$ joining P and P_0. Then the union of these secant lines

$$X(E, P_0) = \bigcup_{P \in E} L(P, P + P_0)$$

is called the *translation scroll* defined by the pair (E, P_0).

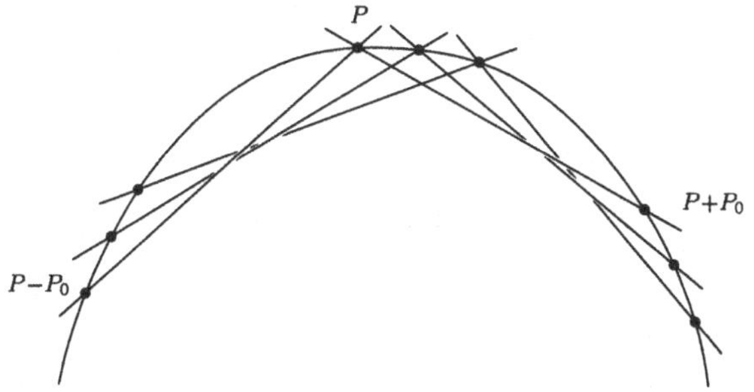

$X(E, P_0)$ is a singular surface of degree 10. Its singular locus is exactly the curve E. The translation scrolls which occur in (ii) are precisely those which are obtained from the smooth elliptic curves $E(\lambda : \mu)$, $(\lambda : \mu) \notin \Lambda$.

If P_0 goes to O, then the translation scroll $X(E, P_0)$ becomes the tangent scroll of E. Again this is a degree 10 surface, singular along E, where $\mathrm{Tan}\, E$ has a cuspidal edge. If P_0 becomes a non-zero 2-torsion point, then set-theoretically $X(E, P_0)$ is a quintic elliptic scroll. As an HM-surface it carries a multiplicity-2 structure. For a discussion of this multiplicity-2 structure see the article [HV].

The other degenerations arise from degenerations of the quintic elliptic curve $E(\lambda : \mu)$ to a pentagon. Let us consider the situation where the two pentagons belonging to the vertices $0, \infty \in \Lambda$ come together to form the complete pentagon given by the coordinate points $e_0, \ldots, e_4 \in \mathbb{P}_4$ (see remark 4.3 above). There is a 1-dimensional family of HM-surfaces containing this complete pentagon. This family of HM-surfaces was first written down in [HM]:

$$(5.5) \qquad X_\alpha = \bigcup_{i \in \mathbb{Z}_5} \{x_i = x_{i+1}x_{i+4} + \alpha x_{i+2}x_{i+3} = 0\}, \qquad \alpha \in \mathbb{P}_1.$$

For $\alpha \neq 0, \infty$ this is a union of 5 quadrics whose singular locus is the complete pentagon defined by $e_0, \ldots, e_4$. For $\alpha = 0$ or ∞ one obtains 5 planes with a multiplicity-2 structure. A union of five quadrics or five planes can also be interpreted as a translation scroll if one interprets the smooth part of a pentagon as $\mathbb{C}^* \times \mathbb{Z}_5$.

The hierarchy of degenerations is as follows:

$$\text{translation scrolls} \;\begin{array}{c}\swarrow\\ \longleftarrow\\ \nwarrow\end{array}\; \begin{array}{c}\text{tangent scrolls}\\ \text{double quintic scrolls}\\ \text{unions of 5 quadrics}\end{array} \;\begin{array}{c}\nwarrow\\ \longleftarrow\\ \swarrow\end{array}\; \text{unions of 5 planes}$$

Outline of the proof of theorem 5.1: This theorem was proved in [BHM2].
Step 1: We consider the following matrix which was found by R. Moore:

$$M(x) = \begin{pmatrix} x_0^3 & x_0x_1x_4 & x_0x_2x_3 & x_1x_2^2 + x_3^2x_4 & x_1^2x_3 + x_2x_4^2 \\ x_1^3 & \cdots & \cdots & \cdots & \cdots \\ \cdots & \cdots & \cdots & \cdots & \cdots \\ \cdots & \cdots & \cdots & \cdots & \cdots \\ x_4^3 & x_4x_0x_3 & x_4x_1x_2 & x_0x_1^2 + x_2^2x_3 & x_0^2x_2 + x_1x_3^2 \end{pmatrix}.$$

Its determinant is a non-zero form of degree 15. In fact it is the unique N_5-invariant form of minimal degree. Consider the degree 15 hypersurface

$$\mathbb{M} = \{\det M(x) = 0\}$$

in $\mathbb{P}_4$. Using the given form of $M(x)$ it is not hard to see that $\mathbb{M}$ is the union of the quintic elliptic scrolls and the unions of 5 planes which are obtained from the curves $E(\lambda : \mu)$, $(\lambda : \mu) \in \mathbb{P}_1$.

Step 2: The natural map

$$\Lambda^2\Gamma(F) \longrightarrow \Gamma(\Lambda^2 F) = \Gamma(\mathcal{O}_{\mathbb{P}_4}(5))$$

defines an isomorphism

$$\Lambda^2\Gamma(F) \xrightarrow{\sim} \Gamma_H(\mathcal{O}_{\mathbb{P}_4}(5))$$

where $\Gamma_H(\mathcal{O}_{\mathbb{P}_4}(5))$ denotes the 6-dimensional space of Heisenberg invariant quintic forms. Let

$$q : \left\{ \begin{array}{ccl} \mathbb{P}_4 & \dashrightarrow & \mathbb{P}_5 = \mathbb{P}(\Lambda^2\Gamma(F)) = \mathbb{P}(\Gamma_H(\mathcal{O}_{\mathbb{P}_4}(5))) \\ x & \longmapsto & (s_i \wedge s_j)(x)_{i<j} \end{array} \right.$$

(here $s_0, \ldots, s_3$ is a basis of $\Gamma(F)$) be the rational map associated to the linear system of Heisenberg invariant quintics. Since the scheme-theoretic intersection of these quintics is the union of the 25 lines $L_{ij} = \sigma^i\tau^j\mathbb{P}_1^-$, we obtain a commutative diagram

$$\begin{array}{ccc} \tilde{\mathbb{P}}_4 & & \\ {\scriptstyle\pi}\downarrow & \overset{\tilde{q}}{\searrow} & \\ \mathbb{P}_4 & \overset{q}{\dashrightarrow} & \mathbb{P}(\Lambda^2\Gamma(F)) \end{array}$$

where $\pi : \tilde{\mathbb{P}}_4 \to \mathbb{P}_4$ is the blow-up of $\mathbb{P}_4$ along the lines L_{ij}. Let $\mathbb{D}$ be the ramification locus of $\tilde{q}$. Since two general HM-surfaces intersect in 100 points, the map $\tilde{q}$ is generically $100 : 1$. One can also show that

$$\pi(\mathbb{D}) = \mathbb{M}.$$

Using the construction of the map q, resp. $\tilde{q}$, one can then prove that $\mathbb{M}$ is the union of the singular loci of all HM-surfaces.
Step 3: Let x be a singular point of an HM-surface X_s. Then $x \in \mathbb{M}$ and there are two cases:
(i) $x \in S_{15}$. In this case there exists a $\mathbb{P}_1$ of HM-surfaces which are all singular at x. If x is on a smooth elliptic curve $E(\lambda : \mu)$, $(\lambda : \mu) \notin \Lambda$, then this $\mathbb{P}_1$ is the family of HM-surfaces which is obtained by varying the point P_0 in the construction of the tangent scroll $X(E, P_0)$. If x is on a singular curve, then $\mathbb{P}_1$ is, up to the action of N_5, the pencil of unions of 5 quadrics given by (5.5).
(ii) $x \notin S_{15}$. In this case there are at most two HM-surfaces which are singular at x. It turns out that they are either a quintic elliptic scroll or an union of 5 planes with a double structure.
Step 3 requires a careful geometric analysis. □

V The Horrocks-Mumford bundle and abelian surfaces

In this section I want to describe the relation between the Horrocks-Mumford bundle and moduli of abelian surfaces.

1 Moduli of abelian surfaces

Every compact 2-dimensional complex torus is of the form

$$A = \mathbb{C}^2/L$$

where L is a rank 4 lattice in $\mathbb{C}^2$. A *Riemann form* with respect to L is a semi-positive definite Hermitian form H whose imaginary part is integer-valued on L. Then

$$H' = \mathrm{Im}(H) : L \otimes L \longrightarrow \mathbb{Z}$$

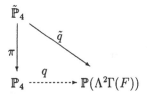

is an alternating form. The complex torus A is an *abelian surface* (i.e. a projective variety) if and only if there exists a positive-definite Riemann form with respect to L. In this case H' is, with respect to a suitable basis of L, of the form

$$\Lambda = \left(\begin{array}{cc|cc} & 0 & d_1 & \\ & & & d_2 \\ \hline -d_1 & & & 0 \\ & -d_2 & & \end{array}\right)$$

where d_1 and d_2 are positive integers with $d_1|d_2$. The form H is then called a *polarization* of *type* (d_1, d_2). If $(d_1, d_2) = (1, 1)$ one also calls H a *principal polarization.* Principally polarized abelian surfaces have been studied the most. They are either Jacobians of genus 2 curves or products of elliptic curves.

In order to construct moduli spaces of abelian surfaces we consider *Siegel space of degree* 2:

$$S_2 = \{\tau \in \mathrm{Mat}(2 \times 2, \mathbb{C})\ ;\ \tau = {}^t\tau, \mathrm{Im}\,\tau > 0\}.$$

To every $\tau \in S_2$ one can associate a period matrix

$$\Omega_\tau = \begin{pmatrix} \tau \\ E \end{pmatrix}$$

where

$$E = \begin{pmatrix} d_1 & 0 \\ 0 & d_2 \end{pmatrix}.$$

Let L_τ be the lattice spanned by the rows of the period matrix Ω_τ. Then

$$A_\tau = \mathbb{C}^2/L_\tau$$

carries a (d_1, d_2)-polarization given by

$$H(x, y) = x(\mathrm{Im}\,\tau)^{-1}\,{}^t\bar{y}\,.$$

Every (d_1, d_2)-polarized abelian surface arises in this way. In order to construct the moduli space of (d_1, d_2)-polarized abelian surfaces we have to consider the group

$$\mathrm{Sp}(\Lambda, \mathbb{Z}) = \{g \in \mathrm{GL}(4, \mathbb{Z})\ ;\ g\Lambda\,{}^tg = \Lambda\}.$$

This group acts on S_2 by

$$\begin{pmatrix} A & B \\ C & D \end{pmatrix} : \tau \longmapsto (A\tau + BE)(C\tau + DE)^{-1}E.$$

Here A, B, C and D are 2×2-blocks. (Note that this generalizes the action of $\mathrm{Sp}(2, \mathbb{Z}) = \mathrm{SL}(2, \mathbb{Z})$ on the usual upper half plane.) Then

$$\mathcal{A}(d_1, d_2) = S_2/\,\mathrm{Sp}(\Lambda, \mathbb{Z})$$

is the moduli space of (d_1, d_2)-polarized abelian surfaces.

In connection with the Horrocks-Mumford bundle we are interested in abelian surfaces of degree 10 in $\mathbb{P}_4$. The hyperplane section is then a polarization of type $(1,5)$. I shall therefore now concentrate on abelian surfaces with a $(1,p)$-polarization where p is a prime greater than or equal to 5. Let H be a $(1,p)$-polarization on an abelian surface A, and let $\mathcal{L}$ be a line bundle on A representing H. The map

$$\lambda : \left\{ \begin{array}{ccl} A & \longrightarrow & \hat{A} = \operatorname{Pic}^0 A \\ x & \longmapsto & T_x^* \mathcal{L} \otimes \mathcal{L}^{-1} \end{array} \right.$$

is independent of the choice of $\mathcal{L}$. Since H is a polarization of type $(1,p)$ the kernel of λ is (non-canonically) isomorphic to $\mathbb{Z}_p \times \mathbb{Z}_p$. The group $\ker \lambda$ carries an intrinsically defined alternating form given by the *Weil pairing.*

Definition *A* level structure *(*of canonical type*) is a symplectic isomorphism* $\alpha : \ker \lambda \to \mathbb{Z}_p \times \mathbb{Z}_p$*, where* $\mathbb{Z}_p \times \mathbb{Z}_p$ *carries the standard form.*

We can now consider the moduli problem of abelian surfaces with a $(1,p)$-polarization and a (canonical) level structure. For the lattice L we define the *dual lattice* with respect to H by

$$L^\vee = \{y \in L \otimes_\mathbb{Z} \mathbb{R} \ ; \ \operatorname{Im} H(x,y) \in \mathbb{Z} \text{ for all } x \in L\}.$$

If $L = \mathbb{Z}^4$ and $\operatorname{Im} H$ is given by Λ with $d_1 = 1$ and $d_2 = p$, then

$$L^\vee = \mathbb{Z} \oplus \frac{1}{p}\mathbb{Z} \oplus \mathbb{Z} \oplus \frac{1}{p}\mathbb{Z}.$$

It is easy to see [HKW2, I.1] that the above moduli problem leads to the following subgroup of $\operatorname{Sp}(\Lambda, \mathbb{Z})$:

$$\Gamma_{1,p} = \{g \in \operatorname{Sp}(\Lambda, \mathbb{Z}) \ ; \ vg \equiv v \mod L \text{ for all } v \in L^\vee\}.$$

In other words the space

$$\mathcal{A}(1,p) = S_2 / \Gamma_{1,p}$$

is the moduli space of $(1,p)$-polarized abelian surfaces with a (canonical) level structure.

Since we are particulary interested in abelian surfaces which are embedded in $\mathbb{P}_4$, we want to consider the following open part of the moduli space $\mathcal{A}(1,p)$:

$$\mathcal{A}^0(1,p) = \{(A, H, \alpha) \in \mathcal{A}(1,p) \ ; \ H \text{ is very ample}\}.$$

One can describe this variety more precisely. For this let

$$\begin{aligned} \mathcal{H}_1 &= \left\{ \tau = \begin{pmatrix} \tau_1 & \tau_2 \\ \tau_2 & \tau_3 \end{pmatrix} \in S_2 \ ; \ \tau_2 = 0 \right\} \\ \mathcal{H}_2 &= \left\{ \tau = \begin{pmatrix} \tau_1 & \tau_2 \\ \tau_2 & \tau_3 \end{pmatrix} \in S_2 \ ; \ \tau_2 = p/2 \right\}. \end{aligned}$$

The images

$$H_i = \pi(\mathcal{H}_i), \qquad i = 1, 2$$

under the quotient map $\pi : S_2 \to \mathcal{A}(1,p)$ are closed surfaces. They are examples of Humbert surfaces [HKW2].

Proposition 1.1 $\mathcal{A}^0(1,p) = \mathcal{A}(1,p) \setminus (H_1 \cup H_2)$.

Proof. See [HW1]. □

Remarks 1.2 *(i) The polarized abelian surfaces parametrized by points in H_1 are products of elliptic curves with a product polarization.*
(ii) The surfaces parametrized by points in H_2 are bielliptic *abelian surfaces, i.e. they are certain covers of Jacobians of bielliptic curves of genus 2.*

Finally note that the spaces $\mathcal{A}(1,p)$ are 3-dimensional quasi-projective varieties which are singular along 2 curves where we have transversal A_2, resp. transversal $C_{3,1}$-singularities [HKW1].

2 The Horrocks-Mumford map

Recall the 3-dimensional projective space

$$\mathbb{P}\Gamma = \mathbb{P}(\Gamma(F))$$

which parametrizes HM-surfaces X_s, $s \in \Gamma(F)$. Every HM-surface is invariant under the group $H_5 \rtimes \langle\iota\rangle$. Hence the action of N_5 on F induces an action of

$$A_5 = \mathrm{PSL}(2, \mathbb{Z}_5) = N_5/H_5 \rtimes \langle\iota\rangle$$

on $\mathbb{P}\Gamma$. Let

$$\mathbb{P}\Gamma_{\mathrm{smooth}} = \{X_s \ ; \ X_s \text{ is a smooth HM-surface}\}$$

be the space of smooth HM-surfaces, and

$$\mathbb{P}\Gamma_{\mathrm{sing}} = \{X_s \ ; \ X_s \text{ is a singular HM-surface}\}$$

be the space of singular HM-surfaces. Clearly $\mathbb{P}\Gamma_{\mathrm{smooth}}$ is A_5-invariant.

We have already remarked that the hyperplane section defines a (very ample) polarization of type $(1,5)$ on smooth HM-surfaces $X_s \subset \mathbb{P}_4$. Let $\mathcal{A}(1,5)$ be the moduli space of $(1,5)$-polarized abelian surfaces with a level-structure, and let $\mathcal{A}^0(1,5)$ be the open part where the polarization is very ample. The group

$$A_5 = \mathrm{PSL}(2, \mathbb{Z}_5) = \mathrm{Sp}(\Lambda, \mathbb{Z})/\Gamma_{1,5}$$

acts on both $\mathcal{A}(1,5)$ and $\mathcal{A}^0(1,5)$. This action leaves the polarization fixed, but acts transitively on the possible level-structures of a polarized abelian surface.

The following theorem is due to Horrocks and Mumford.

Theorem 2.1 ([HM]) *There exists a natural A_5-equivariant isomorphism*

$$\varphi_0 : \mathcal{A}^0(1,5) \xrightarrow{\sim} \mathbb{P}\Gamma_{\text{smooth}}.$$

Outline of proof. Consider a point $(A, H, \alpha) \in \mathcal{A}^0(1,5)$ and let $\mathcal{L}$ be a symmetric line bundle which represents H. The group $\ker\lambda \rtimes \langle\iota\rangle$ acts on A. Here $\ker\lambda$ acts by translation and ι is the involution $x \mapsto -x$. The level-structure α identifies $\ker\lambda$ with $\mathbb{Z}_5 \times \mathbb{Z}_5$. The action of the group $(\mathbb{Z}_5 \times \mathbb{Z}_5) \rtimes \langle\iota\rangle$ on A lifts to an action of $H_5 \rtimes \langle\iota\rangle$ on $\mathcal{L}$. Such a lifting is unique if one requires that ι acts on the fibre of $\mathcal{L}$ over the origin by $+1$. There exists a unique basis $s_0, \ldots, s_4 \in \Gamma(A, \mathcal{L})$ such that the map

$$\varphi_{|\mathcal{L}|} : \left\{ \begin{array}{ccc} A & \longrightarrow & \mathbb{P}_4 \\ x & \longmapsto & (s_0(x) : \ldots : s_4(x)) \end{array} \right.$$

is equivariant with respect to the action of $H_5 \rtimes \langle\iota\rangle$ on $(A, \mathcal{L})$ and $(\mathbb{P}_4, \mathcal{O}(1))$. The image $\varphi_{|\mathcal{L}|}(A)$ is independent of the choice of the symmetric line bundle $\mathcal{L}$. Moreover there exists a section $0 \neq s \in \Gamma(F)$, which is unique up to a scalar, such that $\varphi_{|\mathcal{L}|}(A) = X_s$. The map φ_0 maps (A, H, α) to X_s. Conversely let X_s be a smooth HM-surface. Then $\mathcal{O}_{X_s}(1)$ defines a $(1,5)$-polarization H on X_s. Choose any point $O \in X_s$ as origin. Then the pair $(\sigma(O), \tau(O))$ defines a level-structure on (X_s, H). This defines an inverse map for φ_0. Using the universal property of the moduli space $\mathcal{A}(1,5)$ one can show that φ_0^{-1} is a morphism. Then φ_0 and φ_0^{-1} are isomorphisms by Zariski's main theorem. For details see [HM, theorem 5.2] and [HKW2, III.2]. □

Definition φ_0 *is called the* Horrocks-Mumford map.

It is now natural to ask whether φ_0 extends to the whole of the moduli space $\mathcal{A}(1,5)$, or to a suitable compactification of $\mathcal{A}(1,5)$, and what the relation between such an extension of φ_0 and the degeneration of HM-surfaces is.

3 The Igusa compactification of $\mathcal{A}(1,p)$

In subsection (IV.3) we discussed the compact modular curves

$$X(n) = (S_1 \cup \mathbb{Q} \cup \{i\infty\})/\Gamma(n)$$

which arise from the open Riemann surfaces $X^0(n)$ by adding finitely many cusps. In general, the moduli spaces of abelian varieties of dimension g have dimension $g(g+1)/2$. Whenever $g \geq 2$ there are several ways to compactify these spaces. In particular in the case of $\mathcal{A}_g$, the moduli space of principally polarized abelian varieties of dimension g, this problem has a long history: The first compactification is due to Satake. The Satake compactification $\bar{\mathcal{A}}_g$ is in some sense the minimal compactification of $\mathcal{A}_g$. The boundary of $\bar{\mathcal{A}}_g$ has dimension $g(g-1)/2$, and although $\bar{\mathcal{A}}_g$ is normal, it is highly singular along the boundary. The Satake compactification was later generalized by Baily and Borel to quotients of other bounded symmetric domains. Igusa constructed a compactification $\mathcal{A}_g^*$ by blowing up the Satake compactification along the boundary. The Igusa compactification $\mathcal{A}_g^*$ has at worst finite quotient singularities. The boundary of $\mathcal{A}_g^*$, however, has codimension 1. The ideas of Igusa together with work of Hirzebruch on Hilbert modular surfaces led to Mumford's very general theory of toroidal compactifications of quotients of bounded symmetric domains. Namikawa then showed that Igusa's compactification can be interpreted as a toroidal compactification. Finally Chai and Faltings constructed compactifications of $\mathcal{A}_g$ over the integers.

In a joint book with Kahn and Weintraub [HKW2] we constructed and described a toroidal compactification $\mathcal{A}^*(1,p)$ of $\mathcal{A}(1,p)$. Since in the case $p = 1$ this compactification is the Igusa compactification $\mathcal{A}_2^*$, we also called $\mathcal{A}^*(1,p)$ the Igusa compactification of $\mathcal{A}(1,p)$. One has to compactify $\mathcal{A}(1,p)$ at various "ends"; this corresponds to adding the cusps in the case of modular curves. These ends are enumerated by the vertices of the *Tits-building* which belongs to the group $\Gamma_{1,p}$. In our case the Tits-building is a graph. There are two types of vertices. These are given by
(i) (Λ-isotropic) lines l in $\mathbb{Q}^4$ (modulo $\Gamma_{1,p}$)
(ii) Λ-isotropic planes h in $\mathbb{Q}^4$ (modulo $\Gamma_{1,p}$).
The edges are given by joining l and h whenever $l \subset h$. To each vertex l (resp. h) one can associate the stabilizer subgroup $P(l)$ (resp. $P(h)$) in $\Gamma_{1,p}$. In this way one obtains the usual description of the Tits-building in terms of parabolic subgroups.

Proposition 3.1 ([HKW2]) *The Tits-building of $\Gamma_{1,p}$ contains $1+(p^2-1)/2$ vertices corresponding to lines l and $p+1$ vertices corresponding to isotropic planes h. It looks as follows:*

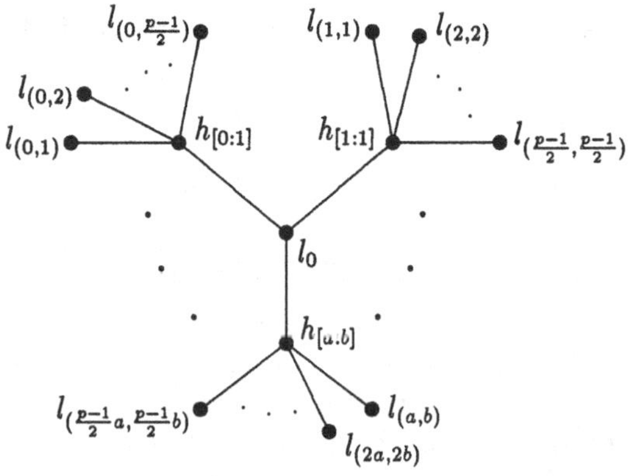

(Here $(a,b) \in (\mathbb{Z}_p \times \mathbb{Z}_p \setminus \{(0,0)\})/\pm 1$ and $[a:b] \in \mathbb{P}_1(\mathbb{Z}_p)$.)

Proof. [HKW2, theorem I.3.40] □

Set-theoretically one obtains the Igusa compactification $\mathcal{A}^*(1,p)$ as a disjoint union

$$\mathcal{A}^*(1,p) = \mathcal{A}(1,p) \coprod D(l_0) \coprod \left(\coprod_{(a,b)} D(l_{(a,b)}) \right) \coprod \left(\coprod_{[a:b]} E(h_{[a:b]}) \right).$$

Here $D(l)$ are (open) surfaces and $E(h)$ are unions of projective lines. Of course the crucial point is that $\mathcal{A}^*(1,p)$ can be given the structure of a projective variety. Two vertices l and h are joined by an edge in the Tits-building if and only if $D^*(l) \cap D(h) \neq \emptyset$ where $D^*(l)$ denotes the closure of $D(l)$ in $\mathcal{A}^*(1,p)$. We call $D(l)$ the *corank* 1 *boundary components* and $E(h)$ the *corank* 2 *boundary components*. We refer to $D(l_0)$ also as the *central boundary surface* and to the $D(l_{(a,b)})$ as *peripheral boundary surfaces.*

It would go far beyond the limitation of a survey article to give all technical details which are necessary to construct $\mathcal{A}^*(1,p)$. For details see [HKW2]. I do, however, briefly want to comment on the toroidal nature of this compactification. For this purpose I shall concentrate on the vertex

$$h = h_{[0:1]} = (0,0,1,0) \wedge (0,0,0,1) \subset \mathbb{Q}^4.$$

The stabilizer $P(h)$ of the Λ-isotropic plane h in $\Gamma_{1,p}$ is an extension

$$1 \longrightarrow P'(h) \longrightarrow P(h) \longrightarrow P''(h) \longrightarrow 1$$

where

$$P'(h) = \left\{ \left(\begin{array}{c|cc} \mathbf{1} & \begin{matrix} m & n \\ pn & pk \end{matrix} \\ \hline 0 & \mathbf{1} \end{array} \right) \; ; \; m,n,k \in \mathbb{Z} \right\}$$

is a lattice of rank 3. The quotient $P''(h)$ is isomorphic to the following subgroup of $\mathrm{GL}(2,\mathbb{Z})$:

$$G = \left\{ Q \in \mathrm{SL}(2,\mathbb{Z}) \; ; \; Q \equiv \begin{pmatrix} * & 0 \\ * & 1 \end{pmatrix} \mod p \right\}.$$

The group $P'(h)$ acts on S_2 by

$$\begin{pmatrix} \tau_1 & \tau_2 \\ \tau_2 & \tau_3 \end{pmatrix} \longmapsto \begin{pmatrix} \tau_1 + m & \tau_2 + pn \\ \tau_2 + pn & \tau_3 + p^2 k \end{pmatrix}.$$

Dividing out by $P'(h)$ gives rise to a partial quotient

$$\begin{array}{ccc} S_2 & \longrightarrow & S_2/P'(h) \subset (\mathbb{C}^*)^3 \\ \begin{pmatrix} \tau_1 & \tau_2 \\ \tau_2 & \tau_3 \end{pmatrix} & \longmapsto & (e^{2\pi i \tau_1}, e^{2\pi i \tau_2/p}, e^{2\pi i \tau_3/p^2}). \end{array}$$

Let $\mathrm{Sym}^+(2,\mathbb{R})$ be the case of positive-definite symmetric real matrices in $\mathrm{Sym}(2,\mathbb{R}) \cong \mathbb{R}^3$. Let $\sigma_0 \subset \mathrm{Sym}(2,\mathbb{R})$ be the simplicial cone

$$\sigma_0 = \mathbb{R}_{\geq 0}\begin{pmatrix}1 & 0\\ 0 & 0\end{pmatrix} + \mathbb{R}_{\geq 0}\begin{pmatrix}1 & 1\\ 1 & 1\end{pmatrix} + \mathbb{R}_{\geq 0}\begin{pmatrix}0 & 0\\ 0 & 1\end{pmatrix}$$

in $\overline{\mathrm{Sym}^+(2,\mathbb{R})}$. The group $\mathrm{GL}(2,\mathbb{Z})$ acts naturally on $\mathrm{Sym}(2,\mathbb{R})$, leaving $\overline{\mathrm{Sym}^+(2,\mathbb{R})}$ invariant. The *Legendre decomposition* Σ_L is the collection of cones in $\overline{\mathrm{Sym}^+(2,\mathbb{R})}$ consisting of all $\mathrm{GL}(2,\mathbb{Z})$-translates of σ_0 together with their respective faces. It is a fan and as such defines a toroidal variety T_Σ of dimension 3. Consider

$$S_2 \longrightarrow S_2/P'(h) \subset (\mathbb{C}^*)^3 \subset T_\Sigma.$$

The next step is to take the interior points of the closure of $S_2/P'(h)$ in T_Σ. Call this set $X_{\Sigma(h)}(h)$. The situation can be envisaged as follows: Recall that

$$T_\Sigma = \bigcup_{\sigma \in \Sigma_L} T_\sigma.$$

If σ is a 3-dimensional cone then $T_\sigma \cong \mathbb{C}^3$ and the image of S_2 in T_σ is a neighbourhood of the coordinate axes whose intersection with the coordinate planes looks roughly as follows:

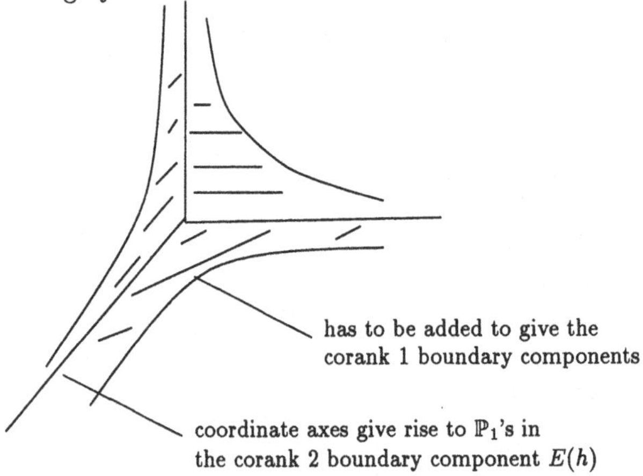

Finally the action of $P''(h)$ on $S_2/P'(h)$ extends to an action of $P''(h)$ on T_Σ. The quotient $X_{\Sigma(h)}(h)/P''(h)$ is a neighbourhood of the corank 2 boundary component $E(h)$ in $\mathcal{A}^*(1,p)$.

It remains to describe the boundary components geometrically.

Proposition 3.2 (i) *The central boundary surface $D(l_0)$ is isomorphic to the open Kummer modular surface $K^0(p)$.*
(ii) *The peripheral boundary surfaces $D(l_{(a,b)})$ are isomorphic to the open Kummer modular surface $K^0(1)$.*

Proof. [HKW2, I.3] □

Proposition 3.3 *The corank 2 boundary components $E(h_{[a:b]})$ are configurations of $(p-1)(p+5)/8$ copies on $\mathbb{P}_1$.*

Proof. [HKW2, I.4] □

In general these configurations are very complicated. Pictures for $p \leq 37$ can be found in [HKW2, I.4]. For $p = 5$ the situation is still very simple: we obtain 5 lines which intersect as follows:

(3.6)

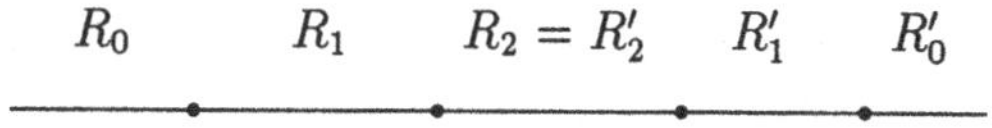

In the terminology of [HKW2] the lines R_0 and R_0' are the *cp*-lines, R_1 and R_1' are the adjacent *cc*-lines and $R_2 = R_2'$ is the non-adjacent *cc*-line (here *c* stands for central and *p* for peripheral).

Finally we consider the closures $D^*(l)$ of the boundary surfaces $D(l)$ in the compactified moduli space $\mathcal{A}^*(1,p)$.

Proposition 3.4 (i) *The closure $D^*(l_{(a,b)})$ of the peripheral boundary surface $D(l_{(a,b)})$ is isomorphic to the Kummer modular surface $K(1)$.*
(ii) *There is a map $K(p) \to D^*(l_0)$ which is an isomorphism locally around each singular fibre, but not a global isomorphism for $p \geq 5$.*

Proof. [HKW2, theorem I.4.8] □

I want to describe briefly the map $K(p) \to D^*(l_0)$ in case $p = 5$. Recall that the 12 cusps of $X(5)$ correspond to the vertices of an icosahedron. On the other hand these cusps can be labelled by pairs $(a,b) \in (\mathbb{Z}_5 \times \mathbb{Z}_5 \backslash \{(0,0)\}/\pm 1$. Opposite vertices of the icosahedron correspond to pairs (a,b), (a',b') with $[a:b] = [a':b']$. Let us consider $(0,1)$ and $(0,2)$. Then the fibres of $K(5)$ over these cusps consist of 3 rational curves each. The map $K(5) \to D^*(l_0)$ identifies two rational curves in different fibres as follows:

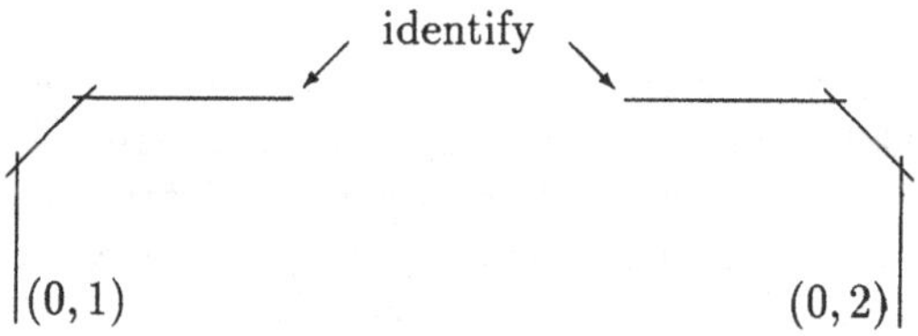

The result is a configuration of 5 lines as depicted in (3.6), namely the corank 2 boundary component $E(h_{[0:1]})$.

4 Geometry of the space of HM-surfaces

The geometry of the space $\mathbb{P}\Gamma$ of HM-surfaces was studied in detail by Barth and Moore [BM]. Some of their results can be summarized as follows.

Theorem 4.1 ([BM]) (i) *The variety* $\mathbb{P}\Gamma_{\text{sing}}$ *of singular HM-surfaces is an irreducible singular rational surface of degree* 10 *in* $\mathbb{P}\Gamma$.
(ii) *The tangent scrolls together with the* 12 *singular* HM*-surfaces which consist of five planes with a double structure form a smooth rational curve* C_{12} *of degree* 12.
(iii) *The double elliptic scrolls together with the* 12 *singular* HM*-surfaces consisting of five planes with a double structure form a smooth rational curve* C_6 *of degree* 6.

Proof. [BM, section 1] □

Remark 4.2 *Barth and Moore also proved the following:*
(i) C_6 *and* C_{12} *are the unique* A_5*-invariant rational curves of degree 6, resp. 12 in* $\mathbb{P}\Gamma$.
(ii) $\mathbb{P}\Gamma_{\text{sing}}$ *is the trisecant surface of* C_6.
(iii) The six lines which parametrize HM-surfaces which consist of five quadrics or five double planes are the six double tangents of C_6.

It is easy to describe an explicit rational parametrization of $\mathbb{P}\Gamma_{\text{sing}}$: consider the map

$$S^0(5) \longrightarrow \mathbb{P}\Gamma_{\text{sing}}$$

which maps each point P in $S^0(5)$ to the translation scroll $X(E,P)$ where E is the fibre of $S^0(5)$ containing P. (Recall that E is naturally embedded in $\mathbb{P}_4$.) Since $X(E,P) = X(E,-P)$ this factors through a map

$$K^0(5) \longrightarrow \mathbb{P}\Gamma_{\text{sing}}.$$

It is not difficult to see that this can be extended to a morphism

$$\varphi : K(5) \longrightarrow \mathbb{P}\Gamma_{\text{sing}}$$

and it is also possible to interprete the map φ geometrically (see e.g. [BHM2]). The map φ has interesting geometric properties. Let O resp. B be the image of the zero-section, resp. the 3-section of non-zero 2-torsion points in $S(5)$. The fibres of $K(5)$ are mapped to lines in $\mathbb{P}\Gamma_{\text{sing}}$. These are all HM-surfaces which are singular along a fixed quintic elliptic curve $E(\lambda : \mu)$. The O-section is mapped to the curve C_{12}, and the curve B is mapped $3:1$ onto C_6. (This reflects the fact that every quintic elliptic scroll contains three ellipt normal quintic curves). Finally consider the situation over a cusp where we have the

following picture

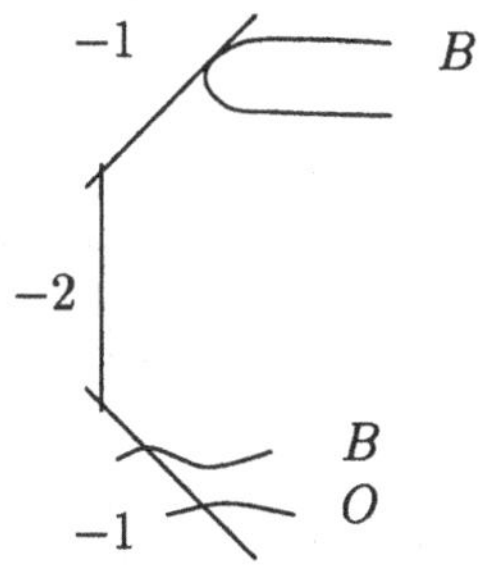

The map φ contracts the two rational curves where B intersects transversally, resp. which B does not intersect. The remaining rational curve is mapped to a double tangent of C_6. Singular fibres over cusps belonging to opposite vertices of the icosahedron are mapped to the same double tangent of C_6.

5 The extension theorem

The Horrocks-Mumford map $\varphi^0 : \mathcal{A}^0(1,5) \to \mathbb{P}\Gamma$ can be extended to the Igusa compactification $\mathcal{A}^*(1,5)$. First, however, recall that the line

$$\mathbb{P}_1^- = \mathbb{P}(V^-), \qquad V^- = \langle e_1 - e_4, e_2 - e_3 \rangle$$

is a jumping line of order 3 of F, i.e.

$$F|_{\mathbb{P}_1^-} = \mathcal{O}_{\mathbb{P}_1^-}(6) \oplus \mathcal{O}_{\mathbb{P}_1^-}(-1).$$

There exists a subgroup $G \subset N_5$, isomorphic to $\mathrm{SL}(2, \mathbb{Z}_5)$ which fixes $\mathbb{P}_1^-$ (as a line). Hence we can view $\Gamma(\mathcal{O}_{\mathbb{P}_1^-}(6))$ as an $\mathrm{SL}(2,\mathbb{Z}_5)$-module. As such it decomposes into two irreducible modules of rank 4 and 3 respectively:

$$\Gamma(\mathcal{O}_{\mathbb{P}_1^-}(6)) = \chi_4 \oplus \chi_3$$

(For a character table of $\mathrm{SL}(2,\mathbb{Z}_5)$ see [HM, appendix]). Recall that $\Gamma(F)$ is also an $\mathrm{SL}(2,\mathbb{Z}_5)$-module with

$$\Gamma(F) = \chi_4.$$

Lemma 5.1 *The restriction map rest:* $\Gamma(F) \to \Gamma(F|_{\mathbb{P}_1^-}) = \chi_4 \oplus \chi_3$ *defines an isomorphism of* $\Gamma(F)$ *with the* χ_4*-part of* $\Gamma(\mathcal{O}_{\mathbb{P}_1^-}(6))$.

Proof. Since an abelian surface contains no line the restriction map is non-zero. The claim follows from Schur's lemma. □

Remark 5.2 *(i) This shows that the HM-surfaces* X_s *are determined by the (unordered) 6-tuple* $X_s \cap \mathbb{P}_1^-$. *The multiplicities of this 6-tuple also describe the type of* X_s *(see [BM, p. 233]):*

multiplicities						*type of X_s*
1	*1*	*1*	*1*	*1*	*1*	*abelian surface*
2	*1*	*1*	*1*	*1*		*translation scroll*
3	*1*	*1*	*1*			*tangent scroll*
2	*2*	*2*				*double elliptic quintic scroll*
2	*2*	*1*	*1*			*union of five quadrics*
4	*2*					*five planes with a double structure*

(ii) If X_s is smooth abelian, then the 6 points $X_s \cap \mathbb{P}_1^-$ are the 6 odd 2-torsion points of the symmetric line bundle $\mathcal{O}_{X_s}(1)$.

We can now state the extension theorem.

Theorem 5.3 *The Horrocks-Mumford map $\varphi^0 : \mathcal{A}^0(1,5) \to \mathbb{P}\Gamma$ can be extended to a morphism $\varphi^* : \mathcal{A}^*(1,5) \to \mathbb{P}\Gamma$. The map φ^* has the following properties:*
(i) *It maps the closure H_1^* of the Humbert surface H_1 to the rational curve C_{12} parametrizing tangent scrolls and unions of five planes with a double structure.*
(ii) *It maps H_2^* to the rational curve C_6 parametrizing quintic elliptic scrolls and unions of five planes with a double structure.*
(iii) *The closure $D^*(l_0)$ of the central boundary surface is mapped onto $\mathbb{P}\Gamma_{\text{sing}}$.*
(iv) *The 12 surfaces $D^*(l_{(a,b)})$ are contracted to the 12 points corresponding to unions of five planes with a double structure.*
(v) *All lines in $E(h_{[a:b]})$ with the exception of the non-adjacent cc-line $R_2 = R_2'$ are contracted to points parametrizing unions of five planes with a double structure. The line $R_2 = R_2'$ is mapped to a line in $\mathbb{P}\Gamma$ parametrizing unions of five quadrics, resp. five planes with a double structure, i.e. a double tangent of C_6.*

Outline of proof. The problem consists of two parts: One has to extend φ^0 to the Humbert surfaces H_1 and H_2, and to the boundary of $\mathcal{A}^*(1,5)$. In both cases the same method can be applied. Consider a point $(A, H, \alpha) \in \mathcal{A}^0(1,5)$ and assume that

$$A = \mathbb{C}^2 / L_\tau$$

where L_τ is spanned by the rows of the period matrix

$$\Omega_\tau = \begin{pmatrix} \tau_1 & \tau_2 \\ \tau_2 & \tau_3 \\ 1 & 0 \\ 0 & p \end{pmatrix}$$

with

$$\tau = \begin{pmatrix} \tau_1 & \tau_2 \\ \tau_2 & \tau_3 \end{pmatrix} \in S_2.$$

For $k \in \mathbb{Z}$, let

$$\underline{k} = \left(0, \frac{k}{5}\right).$$

We consider the theta functions

$$\Theta_k(\tau, z) = \sum_{q \in \mathbb{Z}^2} e^{2\pi i\left[\frac{1}{2}(q+\underline{k})\tau^t(q+\underline{k})+(q+\underline{k})^t z\right]}, \qquad k \in \mathbb{Z}_5.$$

(Note that this notation differs slightly from that of [HKW2, part III].) These theta functions are all sections of the same line bundle $\mathcal{L}_\tau$ on A, and the map

$$\varphi_{|\mathcal{L}_\tau|} : \begin{cases} A & \longrightarrow & \mathbb{P}_4 \\ x & \longmapsto & (\Theta_0(x) : \ldots : \Theta_4(x)) \end{cases}$$

is $H_5 \rtimes \mathbb{Z}_2$-equivariant. In particular, if $(A, H, \alpha) \in \mathcal{A}^0(1,5)$ then

$$\varphi^0([\tau]) = \varphi^0(A, H, \alpha) = \varphi_{|\mathcal{L}_\tau|}(A) \in \mathbb{P}\Gamma.$$

Let $p_1, \ldots, p_6$ be the 6 odd 2-torsion points of A with respect to $\mathcal{L}_\tau$. In view of our above discussion the point $\varphi^0([\tau])$ is determined by the 6-tuple

$$(\Theta_1(p_i) : \Theta_2(p_i)) \in \mathbb{P}_1^-, \qquad i = 1, \ldots, 6.$$

Explicit calculation then shows what happens when τ is a point on either $\mathcal{H}_1$ or $\mathcal{H}_2$, or when τ goes towards the boundary. For details see [HKW2, III.3 and III.4]. □

In section II.2 we mentioned Comessati's result who embedded abelian surfaces with real multiplication in $\mathbb{Q}(\sqrt{5})$ in $\mathbb{P}_4$. It is interesting to identify these surfaces in the space $\mathbb{P}\Gamma$. Consider the Humbert surface

$$\mathcal{H}_3 = \left\{\tau = \begin{pmatrix} \tau_1 & \tau_2 \\ \tau_2 & \tau_3 \end{pmatrix} \in S_2 \; ; \; 5\tau_1 - 5\tau_2 + \tau_3 = 0\right\}$$

in S_2, resp. its image H_3 in $\mathcal{A}(1,5)$. This parametrizes abelian surfaces with real multiplication in $\mathbb{Q}(\sqrt{5})$.

Proposition 5.4 ([HL2]) *The closure H_3^* of the Humbert surface H_3 in $\mathcal{A}^*(1,5)$ is mapped to the unique A_5-invariant cubic in $\mathbb{P}\Gamma$ which is the Clebsch diagonal cubic.*

Proof. This follows immediately from [HL2, theorem 5.1]. □

Remark 5.5 *It was first shown by Hirzebruch [Hi] that the symmetric Hilbert modular surface $Y_0(5, \sqrt{5})$ associated to the ideal generated by $\sqrt{5}$ in the ring of integers in $\mathbb{Q}(\sqrt{5})$ has a natural compactification which is isomorphic to the Clebsch diagonal cubic.*

6 Degenerations of abelian surfaces

In [HKW2, part II] the boundary points of $\mathcal{A}^*(1,p)$ were interpreted as degenerate polarized abelian surfaces. It is natural to ask, how in the case $p=5$, these abstract degenerations relate to degenerations of the embedded abelian surfaces, i.e. to degenerations of HM-surfaces. Here we shall discuss the two most important cases.

Again let

$$\tau = \begin{pmatrix} \tau_1 & \tau_2 \\ \tau_2 & \tau_3 \end{pmatrix} \in S_2$$

and consider the period matrix

$$\Omega_\tau = \begin{pmatrix} \tau_1 & \tau_2 \\ \tau_2 & \tau_3 \\ 1 & 0 \\ 0 & 5 \end{pmatrix}.$$

Denote the rows of Ω_τ by $e_1, \dots, e_4$. They define the rank 4 lattice

$$L_\tau = \mathbb{Z}e_1 + \mathbb{Z}e_2 + \mathbb{Z}e_3 + \mathbb{Z}e_4.$$

The point τ goes to the central boundary component if τ_1 goes to $i\infty$. In this case the lattice L_τ degenerates to the rank 3 lattice

$$L'_\tau = \mathbb{Z}e_2 + \mathbb{Z}e_3 + \mathbb{Z}e_4.$$

The quotient

$$A'_\tau = \mathbb{C}^2 / L'_\tau$$

is a semi-abelian surface of rank 1. More precisely it is an extension

$$1 \longrightarrow \mathbb{C}^* \longrightarrow A'_\tau \longrightarrow E_{\tau_{3,5}} \longrightarrow 1$$

where

$$E_{\tau_{3,5}} = \mathbb{C}/(\mathbb{Z}\tau_3 + \mathbb{Z}5).$$

Adding a 0-section and a section at infinity one can compactify A'_τ to a $\mathbb{P}_1$-bundle $\tilde{A}_\tau$. It is easy to compute that

$$\tilde{A}_\tau = \mathbb{P}(\mathcal{O}_{E_{\tau_{3,5}}} \oplus \mathcal{O}_{E_{\tau_{3,5}}}(5(e - O)))$$

where

$$e = [\tau_2] \in E_{\tau_{3,5}}$$

and O denotes the origin. Let

$$\bar{A}_\tau = \tilde{A}_\tau / \sim$$

be the singular surface which arises by glueing the section at infinity and the 0-section of $\tilde{A}_\tau$ with the shift e:

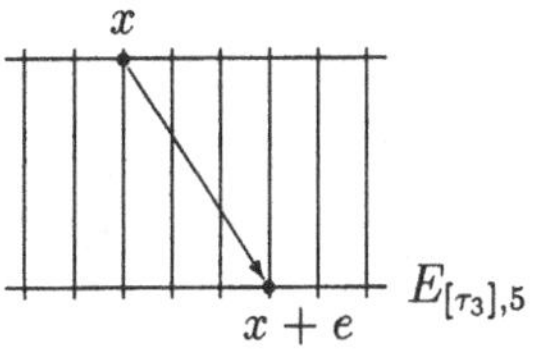

Considering limits of theta functions one can prove the existence of a polarization $\bar{\mathcal{L}}_\tau$ on $\bar{A}_\tau$ (and one can in fact also generalize the notion of level-structure to the singular case). The degenerate abelian surface $(\bar{A}_\tau, \bar{\mathcal{L}}_\tau)$ depends on the pair (τ_2, τ_3) or, more precisely on the elliptic curve $E_{\tau_{3,5}}$ and the point $e = [\tau_2]$ on $E_{\tau_{3,5}}$. The points $\pm e$ give the same degenerate abelian surface. If we identify $E_{\tau_{3,5}}$ with the appropriate fibre in the Shioda modular surface $S^0(5)$, then $\pm e$ is a point in $K^0(5)$. Recall that the central boundary surface $D^0(l_0)$ is isomorphic to $K^0(5)$. Under this identification $(\bar{A}_\tau, \bar{\mathcal{L}}_\tau)$ is the degenerate abelian surface associated to the point in the central boundary surface defined by the pair (τ_2, τ_3) (cf. [HKW2, proposition II.4.5]).

Proposition 6.1 ([HW2]) (i) *If* $2e \neq 0$ *then the line bundle* $\bar{\mathcal{L}}_\tau$ *is very ample and embeds* $\bar{A}_\tau$ *in* $\mathbb{P}_4$ *as the translation scroll* $X(E_{\tau_{3,5}}, e)$ *(cf. section III.5).*
(ii) *If* $e \neq 0$ *but* $2e = 0$ *then* $\bar{\mathcal{L}}_\tau$ *is base point free. It maps* $\bar{A}_\tau$ *generically* $2:1$ *onto a quintic elliptic scroll in* $\mathbb{P}_4$.

Proof. This can be shown by studying the degenerate theta functions which define $\bar{\mathcal{L}}_\tau$, cf. [HW2], [HKW2]. □

Remark 6.2 *If* $e = 0$ *then* $\bar{\mathcal{L}}_\tau$ *has base points. After a suitable modification this leads to tangent scrolls of elliptic quintic curves in* $\mathbb{P}_4$ *(cf. [HW2, 4.3]).*

If τ goes to a point on a peripheral boundary component $D^0(l_{(a,b)})$ then $\bar{A}_\tau$ is a cycle of five elliptic ruled surfaces. In this case the line bundle $\bar{\mathcal{L}}_\tau$ has always base points.

If τ goes to a corank 2 boundary component then L_τ degenerates further to a rank 2 lattice L''_τ. Since $\mathbb{C}^2/L''_\tau \cong (\mathbb{C}^*)^2$ the resulting degenerate abelian surfaces break up into rational parts. The surfaces constructed in [HKW2] are of two types: They are either a union of five quadrics or a union of five projective planes blown up in 3 points and 10 planes, resp. a union of 10 planes, depending on which degeneration one associates to the "deepest" points in $\mathcal{A}^*(1,5)$ (for details see [HKW2, II.4] resp. [Mu]). The most interesting case are the degenerate abelian surfaces associated to points on the non-adjacent cc-curves $R_2 = R'_2$. Here we find a union of five planes with the following identifications:

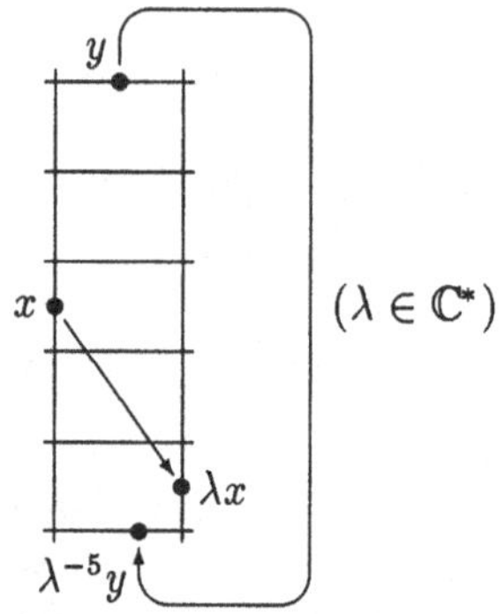

In this case $\bar{\mathcal{L}}_\tau$ is still very ample and embeds $\bar{A}_\tau$ as a union of five quadrics in $\mathbb{P}_4$. It was already remarked in [HM] that the above diagram interpreted as a real picture describes a real 2-torus.

References

[AR] M. F. Atiyah and M. Rees: Vector bundles on projective 3-space. Invent. math. **35** (1976), 131–153.

[BH] W. Barth and K. Hulek: Projective modules of Shioda modular surfaces. Manuscripta math. **50** (1985), 73–132.

[BHM1] W. Barth, K. Hulek, and R. Moore: Shioda's modular surface $S(5)$ and the Horrocks-Mumford bundle. In: Vector bundles on algebraic varieties. Papers presented at the Bombay colloquinum 1984, 35–106. Oxford University Press, Bombay 1987.

[BHM2] W. Barth, K. Hulek, and R. Moore: Degenerations of Horrocks-Mumford surfaces. Math. Ann. **277** (1987), 735–755

[BM] W. Barth and R. Moore: Geometry in the space of Horrocks-Mumford surfaces. Topology **28** (1989), 231–245

[BV] W. Barth and A. Van de Ven: A decomposability criterion for algebraic 2-bundles on projective spaces. Invent. math. **25** (1974), 91–106.

[Com] A. Comessatti: Sulle superficie di Jacobi. Tipografia della R. Accademia dei Lincei, Rome, 1919.

[Del] W. Decker: Das Horrocks-Mumford Bündel und das Modul-Schema für stabile 2-Vektorbündel über $\mathbb{P}_4$ mit $c_1 = -1$, $c_2 = 4$. Math. Z. **188** (1984), 101–110.

[De2] W. Decker: Stable rank 2 bundles with Chern classes $c_1 = -1$, $c_2 = 4$. Math. Ann. **275** (1986), 481–500.

[DS1] W. Decker and F.-O. Schreyer: On the uniqueness of the Horrocks-Mumford bundle. Math. Ann. **273** (1986), 415–443

[DS2] W. Decker and F.-O. Schreyer: Pullbacks of the Horrocks-Mumford bundle. J. reine angew. Math. **382** (1987), 215–220.

[GM] H. Grauert and G. Mülich: Vektorbündel vom Rang 2 über dem n-dimensionalen komplex projektiven Raum. Manuscripta math. **16** (1975), 75–100.

[Ha1] R. Hartshorne: Varieties of small codimension in projective space. BAMS **80** (1974), 1017–1032.

[Ha2] R. Hartshorne: *Algebraic Geometry.* Graduate Texts in Math. **52**. New York: Springer-Verlag 1977

[Ha3] R. Hartshorne: Stable vector bundles of rank 2 on $\mathbb{P}^3$. Math. Ann. **238** (1978), 229–280

[Hi] F. Hirzebruch: The ring of Hilbert modular forms for real quadratic fields of small discriminant. Springer Lecture Notes in Math. Vol. **627** (1977), 287–323.

[Ho] G. Horrocks: A construction for locally free sheaves. Topology **7** (1968), 117–120.

[HM] G. Horrocks and D. Mumford: A rank 2 vector bundle on $\mathbb{P}^4$ with 15,000 symmetries. Topology **12** (1973), 63–81

[Hul1] K. Hulek: *Projective Geometry of Elliptic Curves.* Astérisque **137** (1986)

[Hul2] K. Hulek: Geometry of the Horrocks-Mumford bundle. In: S. Bloch (ed.): *Algebraic Geometry, Bowdoin 1985. Proceedings.* Proc. Sympos. Pure Math. **46**(2) (1987), 69–85

[Hul3] K. Hulek: Elliptische Kurven, abelsche Flächen und das Ikosaeder. Jahresber. Deutsch. Math.-Verein. **91** (1989), 126–147

[HKW1] K. Hulek, C. Kahn, and S. H. Weintraub: Singularities of the moduli spaces of certain abelian surfaces. Compositio Math. **79** (1991), 231–253

[HKW2] K. Hulek, C. Kahn, and S. H. Weintraub: Moduli spaces of abelian surfaces: compactification, degenerations and theta functions. de Gruyter, Expositions in Mathematics **12**, Berlin, New York 1993.

[HL1] K. Hulek and H. Lange: Examples of abelian surfaces in $\mathbb{P}_4$. J. Reine Angew. Math. **363** (1985), 201–216

[HL2] K. Hulek and H. Lange: The Hilbert modular surface for the ideal $(\sqrt{5})$ and the Horrocks-Mumford bundle. Math. Z. **198** (1988), 95–116

[HV] K. Hulek and A. Van de Ven: The Horrocks-Mumford bundle and the Ferrand construction. Manuscripta Math. **50** (1985), 313–335.

[HW1] K. Hulek and S. H. Weintraub: Bielliptic abelian surfaces. Math. Ann. **283** (1989), 411–429

[HW2] K. Hulek and S. H. Weintraub: The principal degenerations of abelian surfaces and their polarisations. Math. Ann. **286** (1990), 281–307

[K] F. Klein: *Vorlesungen über das Ikosaeder und die Auflösung der Gleichungen vom fünften Grade. Kommentiert und herausgegeben von P. Slodowy.* Basel: Birkhäuser 1992

[LB] H. Lange and C. Birkenhake: *Complex Abelian Varieties.* Grundlehren Math. Wiss. **302**. Berlin: Springer-Verlag 1992

[Ma1] M. Maruyama: *On a family of algebraic vector bundles.* Number theory, algebraic geometry and commutative algebra. Kinokuniya, 1973, 95-149.

[Ma2] M. Maruyama: *Elementary transformations in the theory of algebraic vector bundles.* Lecture Notes in Math., **961**, Springer Verlag, 1983, 241-266.

[Mu] D. Mumford: An analytic construction of degenerating abelian varieties over complete rings. Compositio Math. **24** (1972), 239–272

[OSS] C. Okonek, M. Schneider, and H. Spindler: *Vector Bundles on Complex Projective Spaces.* Progress in Math. **3**. Boston: Birkhäuser 1980

[R] S. Ramanan: Ample divisors on abelian surfaces. Proc. London Math. Soc. **51** (1985), 231–245

[Re] I. Reider: Vector bundles of rank 2 and linear systems on algebraic surfaces. Ann. Math. **127** (1988), 309–316.

[Sa1] N. Sasakura: *A stratification theoretical method of construction of holomorphic vector bundles.* Advanced Studies in Pure Mathematics. **8** (1986), 527-581.

[Sa2] N. Sasakura: *Configuration of divisors and reflexive sheaves.* Report note R.I.M.S. Kyoto University **634** (1987), 407-513.

[Sa3] N. Sasakura: *Configuration of divisors and reflexive sheaves.* Proc. Japan Acad. **65A** (1989), 27-30.

[SEK1] N. Sasakura, Y. Enta, M. Kagesawa: *Construction of rank 2 reflexive sheaves with similar properties to the Horrocks-Mumford bundle.* Proc. Japan Acad. **69A** (1993), 144-148.

[SEK2] N. Sasakura, Y. Enta, M. Kagesawa: *Construction of rank 2 reflexive sheaves by means of quadratic residue graphs.* Preprint Tokyo 1993.

[Sch] C. Schoen: *On the geometry of a special determinantal hypersurface associated to the Horrocks-Mumford vector bundle.* J. reine angew. Math. **364** (1986), 85-111.

[Se] J. P. Serre: Sur les modules projectifs. Sèm. Dubrieul-Pisot 1960/61, exposé 2.

[Su1] H. Sumihiro: *Elementary transformations of algebraic vector bundles.* In: Algebraic and topological theories (volume in honour of T. Miyata), 1985, 305-327.

[Su2] H. Sumihiro: *Elementary transformations of algebraic vector bundles.* In: Algebraic geometry and commutative algebra (volume in honour of M. Nagata), 1987, 503-516.

[Sw] R. L. E. Schwarzenberger: Vector bundles on algebraic surfaces. Proc. London Math. Soc. **11** (1961), 601–622.

[Th] A. Thomas: Almost complex structures on complex projective spaces. TAMS **193** (1974), 123–132.

[Tj] A. N. Tjurin: *The classification of vector bundles over algebraic curves of arbitrary genus.* Izv. Nawk. U.S.S.R. **29** (1965), 657-688.

[V] A. Van den Ven: On uniform vector bundles. Math. Ann. **119** (1972), 245–248.

[Vo] J. A. Vogelaar: Constructing vector bundles from codimension 2 subvarieties. Thesis, Leiden 1978.

[We] A. Weil: *Generalisations des fonctions abeliennes.* J. Math. Pures et Appl. **17** (1939), 47-87.

Faisceaux semi-stables et systèmes cohérents

J. Le Potier

Université Paris 7, UFR de Mathématiques et URA 212, Case Postale 7012
2, Place Jussieu, 75251 Paris Cedex 05

Sommaire

Introduction

Soit X une variété projective lisse irréductible X de dimension n, munie d'un faisceau très ample $\mathcal{O}_X(1)$; considérons l'espace de modules $M = M_X(r, c_1, \ldots, c_n)$ des classes de S−équivalence de faisceaux semi-stables sur X, de rang r et classes de Chern $c_1, \ldots, c_n$ fixées dans l'anneau d'équivalence numérique Num(X). C'est une variété projective, dont on ne peut pas dire grand chose en général. Mais dans nombre de situations, on obtient une variété irréductible et normale dont on peut préciser la dimension : c'est le cas sur les courbes et sur le plan projectif ; c'est encore le cas sur toute surface pourvu que la classe de Chern c_2 soit assez grande [9] .

Notre principale préoccupation dans le travail présenté ici est l'étude du groupe de Picard de ces variétés de modules. Il existe une méthode efficace pour construire des fibrés inversibles sur ces variétés : c'est la notion de fibré déterminant, qui permet d'associer à certaines classes $u \in K(X)$ de l'algèbre de Grothendieck un fibré inversible $\lambda_M(u)$. Pour préciser quelles sont les classes qui conviennent, il convient de munir K(X) de sa forme quadratique standard q, décrite dans la section 1, et plutôt que d'imposer le rang et les classes de Chern, on fixe ici la classe c des faisceaux semi-stables étudiés dans l'algèbre $K_{num}(X) = K(X)/\ker q$, ce qui revient au même. Cette méthode, inspirée de la construction du fibré déterminant de Donaldson [3] pour les espaces de modules de fibrés vectoriels sur les surfaces, s'étend sans difficulté (*cf.* théorème 2.4) à tous les espaces de modules de faisceaux semi-stables, même aux espaces de modules de faisceaux cohérents de torsion considérés par Simpson.

Dans le cas du plan projectif, l'introduction de cet homomorphisme λ_M permet de donner une présentation unifiée pour le calcul du groupe de Picard des espaces de modules $M_{\mathbf{P}_2}(c)$ de classe $c \in K(\mathbf{P}_2)$. L'énoncé donné ici (*cf.* théorème 3.10) regroupe à la fois le théorème de Drézet, relatif aux faisceaux semi-stables sans torsion [5] , notre énoncé [15] , relatif aux faisceaux semi-stables de dimension 1, et l'énoncé classique de Fogarty relatif à la puissance symétrique $\mathrm{Sym}^{\ell}(\mathbf{P}_2)$, correspondant aux faisceaux de dimension 0 et de longueur ℓ. La démonstration est en fait la même en dimension 2 et 1, quand on dispose de la première étape dans le problème de Brill-Noether relatif à ces espaces de modules. Pour les faisceaux semi-stables sans torsion nous avons bénéficié de l'aide de L. Göttsche et A. Hirschowitz [10] qui ont établi pour nous le théorème de Brill-Noether dont nous avions besoin. Pour les faisceaux de dimension 1, cet énoncé ne pose pas de difficulté et s'obtient en fait directement en identifiant par projection sur $\mathbf{P}_1$ certains ouverts des espaces de modules considérés à des espaces de modules de paires stables (G, φ), où G est un fibré vectoriel sur la droite projective $\mathbf{P}_1$, et $\varphi : G \to G(1)$; ces paires sont analogues aux fibrés de Higgs.

La description du groupe de Picard des espaces de modules de fibrés semi-stables de déterminant fixé sur les courbes due à Drézet et Narasimhan [7] s'obtient aussi grâce à l'homomorphisme λ_M. Nous obtiendrons en fait leur énoncé comme conséquence de la description du groupe de Picard de certains espaces de modules de systèmes cohérents semi-stables. Ces systèmes cohérents, introduits dans la section 4, sont des paires (Γ, F) formées d'un faisceau algébrique cohérent F, et d'un sous-espace vectoriel $\Gamma \subset H^0(F)$ de l'espace vectoriel des sections de F ; ils généralisent les systèmes linéaires. Sur les courbes, cette notion n'est pas nouvelle : sous le nom de paires semi-stables, elles ont été abondamment utilisées par divers auteurs : S. Bradlow, A. Bertram, O. Garcia-Prada, N. Raghavendra et P.A. Vishwanath [22], A. King et P. Newstead [11], ..., et notamment pour les fibrés de rang 2 par Thaddeus [24]. Mais la notion de semi-stabilité qui nous a parue la plus adaptée pour les applications envisagées est différente de celle qu'ont introduite ces auteurs. Sur une variété X de dimension n, elle permet de ramener certains énoncés relatifs aux espaces de modules de faisceaux cohérents sans torsion à des énoncés sur des faisceaux cohérents dont le support est de dimension $n-1$. Ce dévissage ne pose aucune difficulté sur les courbes et sur les surfaces (*cf.* corollaires 4.21 et 4.22). Ainsi, sur les courbes, on ramène par cette méthode l'étude des espaces de modules de fibrés vectoriels de rang r, munis d'un espace vectoriel de sections de dimension r, à celle de faisceaux cohérents de dimension 0, munis d'un espace vectoriel de sections de dimension r. Parmi ces faisceaux cohérents de dimension 0, les faisceaux structuraux $\mathcal{O}_{\mathrm{D}}$ de diviseurs $\mathrm{D} \subset \mathrm{X}$ jouent un rôle fondamental dans le calcul.

Nous utilisons aussi les systèmes cohérents semi-stables sur le plan projectif dans la section 5.2 pour ramener l'étude de l'espace de modules M_4 des faisceaux semi-stables de rang 2, de classe de Chern $(0,4)$ à celle de faisceaux semi-stables de dimension 1 portés par des coniques, éventuellement singulières. Ce dévissage permet d'étudier l'application

$$\gamma : \mathrm{M}_4 \to |\mathcal{O}_{\mathbf{P}_2^*}(4)| = \mathbf{P}_{14}$$

qui associe à la classe d'un faisceau F la quartique de ses droites de saut. On sait classiquement [1] que ce morphisme est génériquement fini, et que l'image contient des courbes lisses. L'espace de modules M_4 est une variété irréductible de dimension 13, et l'image de cette application est donc une hypersurface $\mathscr{L}$ appelée hypersurface des quartiques de Lüroth : quand elles sont lisses, il s'agit des quartiques qui sont circonscrites à un pentagone. L'intersection de $\mathscr{L}$ avec le diviseur $\mathscr{S}$ des quartiques singulières a deux composantes irréductibles, dont l'une est l'image du diviseur correspondant aux classes de faisceaux semi-stables provenant des systèmes cohérents sur les coniques singulières dans le dévissage évoqué ci-dessus : ce diviseur est aussi le diviseur des classes de faisceaux semi-stables qui possèdent une droite de saut d'ordre ≥ 2. Les points de cette composante sont classiquement appelés quartiques de Lüroth singulière de type II. Une telle quartique de Lüroth singulière $\mathfrak{q}$ de type II, si elle est générique, possède une géométrie abondante, qui permet en fait de reconstruire le point $p \in \mathrm{M}_4$ dont c'est l'image : la fibre de γ au-dessus de $\mathfrak{q}$ est alors réduite au point p ; compte-tenu du fait que γ est non ramifié en p, c'est l'argument essentiel qui permet de montrer que l'application γ est de degré 1 sur son image.

Par variété algébrique, on entend schéma de type fini séparé sur $\mathbf{C}$. Les points considérés seront les points fermés.

1. L'espace de modules de Simpson

Soit X une variété algébrique projective lisse et irréductible, de dimension n, munie d'un faisceau très ample $\mathcal{O}_X(1)$. On désigne par K(X) l'algèbre de Grothendieck des classes de faisceaux algébriques cohérents, et par h la classe dans K(X) du faisceau structural $\mathcal{O}_Y$ d'une section hyperplane $Y \subset X$. Cette algèbre est équipée d'une forme quadratique $q : u \mapsto \chi(u^2)$, dont on désigne par $<,>$ la forme polaire. Cette forme quadratique $q(u)$ se calcule en termes du rang et des classes de Chern de u.

Exemples

(i) Si X est une courbe de genre g, et si u est de rang r et degré k on a $q(u) = 2rk + r^2(1-g)$.

(ii) Si X est une surface, et si $u \in K(X)$ est de rang r, de classe de Chern c_1 et de caractéristique d'Euler-Poincaré χ, on a

$$q(u) = 2r\chi + c_1^2 - r^2\chi(\mathcal{O}_X).$$

Le noyau $\ker q$ est constitué des classes dont le rang et les classes de Chern, vues dans l'anneau d'équivalence numérique sont nulles. On considère le quotient

$$K_{num}(X) = K(X)/\ker q$$

et on désigne encore par q la forme quadratique induite, et par h la classe d'une section hyperplane dans ce quotient.

Soit F un faisceau algébrique cohérent sur X. On appelle *dimension* de F la dimension d du support de F. Le polynôme de Hilbert de F s'écrit

$$\begin{aligned} P_F(m) &= < F, \mathcal{O}_X(m) > \\ &= < F, (1-h)^{-m} > \\ &= \sum_{0 \le i \le d} C^i_{m+i-1} < F, h^i > \end{aligned}$$

Définition 1.1. — *Soit* F *un faisceau algébrique cohérent de dimension* d *sur* X. *On appelle multiplicité de* F *le nombre* $r = < F, h^d >$, *et polynôme de Hilbert réduit de* F *le polynôme*

$$p_F = \frac{P_F}{r}$$

Exemples

(i) Si F est un faisceau sans torsion de rang rg(F) sur X, la multiplicité est $r = \mathrm{rg}(\mathrm{F}) \deg(\mathrm{X})$.

(ii) Si F est de dimension $n-1$, la multiplicité de F est le degré de la variété de Fitting de F. Cette variété de Fitting est une hypersurface en dehors du fermé, de codimension ≥ 2, des points x tels que F_x n'est pas de Cohen-Macaulay. En dehors de ce fermé, le faisceau F a une résolution localement libre

$$0 \to \mathrm{R}_1 \xrightarrow{f} \mathrm{R}_0 \to \mathrm{F} \to 0$$

et l'hypersurface définie par det f s'étend à la variété X. Cette hypersurface Z s'appelle le *support schématique* de F. Le faisceau F est muni d'une structure naturelle de $\mathcal{O}_\mathrm{Z}$–module.

Définition 1.2. — *Un faisceau algébrique cohérent de dimension d sur* X *est dit pur de dimension d s'il n'a pas de sous-faisceau cohérent non nul de dimension $< d$.*

Définition 1.3. — *Soit* F *un faisceau algébrique cohérent de dimension d et de multiplicité r. On dit que* F *est semi-stable (resp. stable) si*

(i) *il est pur de dimension d;*

(ii) *pour tout sous-faisceau cohérent* $\mathrm{F}' \subset \mathrm{F}$ *de multiplicité* $0 < r' < r$ *on a*

$$p_{\mathrm{F}'} \leq p_\mathrm{F} \quad (\text{resp. } <)$$

Dans la dernière inégalité, l'ordre considéré sur les polynômes est l'ordre lexicographique, en commençant par les termes de plus haut degré ; autrement dit, étant donnés deux polynômes P et Q à coefficients réels, $\mathrm{P} \leq \mathrm{Q}$ signifie que $\mathrm{P}(m) \leq \mathrm{Q}(m)$ pour m assez grand.

La catégorie additive des faisceaux algébriques semi-stables F de polynôme de Hilbert réduit fixé est une catégorie abélienne dans laquelle les objets simples correspondent aux faisceaux stables. Dans cette catégorie, tout faisceau semi-stable F a une filtration de Jordan-Hölder. On entend par là une suite croissante $\{0\} = \mathrm{F}_0 \subset \mathrm{F}_1 \subset \ldots \subset \mathrm{F}_k = \mathrm{F}$ telle que le gradué

$$gr_i = \mathrm{F}_i / \mathrm{F}_{i-1}$$

soit stable. Une telle filtration n'est pas unique, mais le théorème de Jordan-Hölder affirme que le gradué $gr(\mathrm{F}) = \oplus_i gr_i$ est bien défini à isomorphisme près. Suivant Seshadri, on dit que deux faisceaux semi-stables sont S−équivalents si leurs gradués de Jordan-Hölder sont isomorphes.

Théorème 1.4. — (C. Simpson) *La famille des faisceaux semi-stables de polynôme de Hilbert fixé* P *est limitée.*

Définition 1.5. — *Une classe* $c \in \mathrm{K}_{num}(\mathrm{X})$ *est dite effective si c'est la classe d'un faisceau cohérent non nul.*

Il revient au même de demander que le polynôme de Hilbert de c

$$\mathrm{P}_c(m) = < c, (1-h)^{-m} >$$

est lui-même > 0. Si c est une telle classe effective, la dimension de c est le degré de ce polynôme, c'est-à-dire le plus grand entier d tel que $< c, h^d > \neq 0$.

Soit $c \in \mathrm{K}_{num}(\mathrm{X})$ une classe effective. On considère le foncteur $\mathrm{S} \mapsto \underline{\mathrm{M}}_{\mathrm{X}}(c)(\mathrm{S})$ associant à la variété algébrique S l'ensemble des classes d'isomorphisme de faisceaux cohérents F sur $\mathrm{S} \times \mathrm{X}$, plats sur S, de dimension relative d, et tels que pour tout point fermé $s \in \mathrm{S}$ le faisceau induit F_s sur la fibre au-dessus de s soit de classe c.

Théorème 1.6. — (C. Simpson) *Soit* $c \in \mathrm{K}_{num}(\mathrm{X})$ *effective de dimension* d. *Il existe pour le foncteur* $\underline{\mathrm{M}}_{\mathrm{X}}(c)$ *un espace de modules grossier* $\mathrm{M}_{\mathrm{X}}(c)$. *C'est une variété projective, dont les points fermés sont les classes de* S−*équivalence de faisceaux semi-stables de classe* c.

Exemples

(i) Si la dimension d est 0, les faisceaux considérés sont de dimension 0 ; ils sont tous semi-stables ; se fixer le polynôme de Hilbert revient à fixer la longueur $\ell = \chi(\mathrm{F})$ de ces faisceaux. L'espace de modules $\mathrm{M}_{\mathrm{X}}(c)$ est isomorphe à la puissance symétrique de X :

$$\mathrm{M}_{\mathrm{X}}(c) \simeq \mathrm{Sym}^{\ell}(\mathrm{X})$$

(ii) Si la dimension d est $n = \dim \mathrm{X}$, la notion de stabilité ci-dessus coïncide avec celle qui avait été introduite par Gieseker dans le cas des surfaces [8] , et par Maruyama [19] en toute dimension. L'espace de modules ci-dessus est alors une réunion de composantes connexes de l'espace de modules de Gieseker et Maruyama.

(iii) Si c est de dimension $n-1$, désignons par c_1 la première classe de Chern de c dans $\mathrm{Num}^1(X)$. Considérons le schéma de Hilbert $\mathrm{Div}^{c_1}(X)$ des hypersurfaces de X de classe fondamentale c_1. On a alors un morphisme canonique

$$M_X(c) \to \mathrm{Div}^{c_1}(X)$$

qui associe à la classe d'un faisceau F son support schématique. Au-dessus d'un point représentant une hypersurface Y de classe fondamentale c_1 la fibre est l'espace de modules des $\mathcal{O}_Y$-modules cohérents semi-stables dont la classe dans $K_{num}(X)$ est c. Si l'hypersurface Y est intègre, ces faisceaux sont les faisceaux cohérents sans torsion et de rang 1 sur Y, dont la classe est c dans $K_{num}(X)$: la stabilité est dans ce cas automatique.

Dans le cas où X est de dimension 2, l'hypothèse de pureté signifie que les faisceaux F sont de Cohen-Macaulay de dimension 1 ; si Y est une courbe lisse, la fibre au-dessus de Y s'identifie alors à la composante du groupe de Picard $\mathrm{Pic}_\chi(Y)$ des faisceaux inversibles sur Y de caractéristique d'Euler-Poincaré χ.

2. Fibrés déterminants sur $M_X(c)$

Dans cette section, on montre comment associer à certaines classes $u \in K(X)$ un fibré inversible $\lambda_M(u) \in \mathrm{Pic}(M_X(c))$.

2.1. L'homomorphisme $\lambda_{\mathcal{F}}$

Soit G un groupe algébrique opérant sur une variété algébrique S. On désigne par $K^G(S)$ l'algèbre de Grothendieck des classes de fibrés vectoriels algébriques, munis d'une action de G au-dessus de l'action donnée. L'action de G induit une action naturelle sur le produit $S \times X$. On désigne par $K^G(S \times X)$ l'algèbre de Grothendieck des classes de faisceaux algébriques cohérents $\mathcal{G}$, plats sur S, et munis d'une action de G au-dessus de l'action de G sur S × X. Considérons les projections canoniques

$$\begin{array}{ccc} S \times X & \stackrel{pr_2}{\longrightarrow} & X \\ {\scriptstyle pr_1}\downarrow & & \\ S & & \end{array}$$

On a alors un morphisme

$$pr_{1!} : K^G(S \times X) \to K^G(S)$$

qui associe à la classe du faisceau $\mathscr{G}$ la classe de $\sum_i (-1)^i R^i pr_{1*}(\mathscr{G})$. Ces faisceaux de cohomologie $R^i pr_{1*}(\mathscr{G})$ sont les faisceaux de cohomologie d'un complexe fini de fibrés vectoriels $Rpr_{1*}(\mathscr{G})$ muni d'une action de G.

On se fixe maintenant sur S × X un faisceau algébrique cohérent $\mathscr{F}$, plat sur S, muni d'une action de G au-dessus de l'action donnée sur S × X. Sur la fibre au-dessus de $s \in S$ la classe dans $K_{num}(X)$ du faisceau induit $\mathscr{F}_s$ est indépendante de s et notée c ; on dira que $\mathscr{F}$ est une G–famille plate de faisceaux cohérents de classe $c \in K_{num}(X)$ paramétrée par S.

On désigne par $\mathrm{Pic}^G(S)$ le groupe des classes d'isomorphisme de G–fibrés inversibles sur S, et on pose, dans $K^G(S \times X)$, $\mathscr{F}(u) = \mathscr{F} \otimes pr_2^*(u)$ pour $u \in K(X)$, et dans $\mathrm{Pic}^G(S)$

$$\lambda_{\mathscr{F}}(u) = \det pr_{1!}(\mathscr{F}(u)).$$

Compte-tenu des remarques ci-dessus concernant le morphisme $pr_{1!}$, cette formule a bien un sens. L'application $\lambda_{\mathscr{F}} : K(X) \to \mathrm{Pic}^G(S)$ est un homomorphisme de groupes abéliens qui satisfait aux propriétés suivantes :

Lemme 2.1. —

(i) *La formation de $\lambda_{\mathscr{F}}(u)$ est compatible aux changements de base G–équivariants.*

(ii) *Etant donnée une suite exacte de G–faisceaux cohérents plats sur S*

$$0 \to \mathscr{F}' \to \mathscr{F} \to \mathscr{F}'' \to 0$$

on a $\lambda_{\mathscr{F}}(u) = \lambda_{\mathscr{F}'}(u) \otimes \lambda_{\mathscr{F}''}(u)$ dans $\mathrm{Pic}^G(S)$.

(iii) *Pour tout G–fibré inversible A sur S on a*

$$\lambda_{\mathscr{F} \otimes pr_1^*(A)}(u) = \lambda_{\mathscr{F}}(u) \otimes A^{\otimes(<c,u>)}$$

On peut en effet supposer que u est la classe d'un fibré vectoriel sur X. Le lemme est une conséquence facile de la construction du complexe image directe $\mathrm{R}pr_{1*}(\mathscr{F} \otimes pr_2^*(\mathrm{E}))$ (*cf.* par exemple [13]).

2.2. Application au schéma de Hilbert

Rappelons d'abord la notion de bon quotient, introduite par Seshadri.

Définition 2.2. — *Soit* G *un groupe algébrique affine, opérant sur une variété algébrique* R. *Un morphisme de variétés algébriques* $f : \mathrm{R} \to \mathrm{Y}$ *est appelé bon quotient de* R *par* G *si* f *est équivariant, et si les conditions suivantes sont satisfaites :*

(i) *le morphisme* f *est affine et surjectif;*

(ii) *l'image d'un fermé* G*−invariant est un fermé de* Y, *et* f *sépare les fermés* G*−invariants disjoints ;*

(iii) *le morphisme d'algèbres*

$$\mathscr{O}_{\mathrm{Y}} \to f_*(\mathscr{O}_{\mathrm{R}})^{\mathrm{G}}$$

où le membre de droite désigne le faisceau des sections G*−invariantes de l'image directe du faisceau structural, est un isomorphisme.*

Définition 2.3. — *Un bon quotient* $f : \mathrm{R} \to \mathrm{Y}$ *d'une variété algébrique* R *par l'action d'un groupe algébrique* G *est dit géométrique si les orbites de* G *dans* R *sont fermées.*

Soit $c \in \mathrm{K}_{num}(\mathrm{X})$ une classe effective de dimension d; considérons le polynôme de Hilbert $\mathrm{P}(m) =< c, (1-h)^{-m} >$. Soit m un entier assez grand, et H un espace vectoriel de dimension $\mathrm{P}(m)$. Considérons le fibré vectoriel $\mathrm{B} = \mathrm{H} \otimes \mathscr{O}_{\mathrm{X}}(-m)$, et le schéma de Hilbert-Grothendieck $\mathrm{R} = \mathbf{Hilb}(\mathrm{B}, c)$ des faisceaux cohérents quotients de B de classe c. Sur $\mathrm{R} \times \mathrm{X}$, on dispose d'un faisceau quotient universel $\mathscr{F}$. Sur le schéma R, le groupe $\mathrm{G} = \mathrm{GL}(\mathrm{H})$ des automorphismes de B opère de manière naturelle, et le faisceau universel $\mathscr{F}$ est lui-même muni d'une action de G au-dessus de l'action ci-dessus. Par suite, on obtient par la construction du paragraphe ci-dessus un homomorphisme de groupes

$$\lambda_{\mathscr{F}} : \mathrm{K}(\mathrm{X}) \to \mathrm{Pic}^{\mathrm{G}}(\mathrm{R})$$

Considérons l'ouvert R^{ss} des points $t \in R$ satisfaisant aux conditions suivantes

— *le faisceau* $\mathcal{F}_t$ *induit au-dessus de* t *est semi-stable ;*

— *l'application canonique* $H \to H^0(\mathcal{F}_t(m))$ *est un isomorphisme.*

Cet ouvert est invariant, et la propriété de module grossier fournit un morphisme $\pi : R^{ss} \to M_X(c)$ équivariant pour l'action de G. La construction de Simpson [23] montre qu'en fait, pourvu que m ait été choisi assez grand, ce morphisme est un bon quotient. L'action du groupe G sur R^{ss} n'est pas libre ; le stabilisateur d'un point $t \in R^{ss}$ s'identifie au groupe des automorphismes du faisceau $\mathcal{F}_t$. Ainsi, cette action se factorise à travers $\overline{G} = GL(H)/\mathbb{C}^*$.

L'ouvert R^s des points $t \in R^{ss}$ tels que $\mathcal{F}_t$ soit stable est invariant, et l'action du groupe $\overline{G} = GL(H)/\mathbb{C}^*$ est libre sur cet ouvert. Cet ouvert R^s est l'image réciproque d'un ouvert $M_X^s(c)$ et la projection $\pi : R^s \to M_X^s(c)$ est un quotient géométrique. Les points fermés de $M_X^s(c)$ correspondent aux classes d'isomorphisme de faisceaux stables de classe c.

Il n'existe pas en général de faisceau universel paramétré par la variété algébrique $M_X^s(c)$. Ceci arrive toutefois si les coefficients $< c, h^i >$ qui figurent dans le polynôme de Hilbert sont premiers entre eux. Dans de telles circonstances, tous les faisceaux semi-stables sont stables : l'espace de modules $M_X^s(c) = M_X(c)$ est alors un espace de modules fin ; mais un tel faisceau universel $\mathcal{G}$ n'est pas unique, puisque si A est un fibré inversible sur $M_X(c)$, le faisceau $\mathcal{G} \otimes pr_1^*(A)$ est encore un faisceau universel.

2.3. L'homomorphisme λ_M

En général, pour $u \in K(X)$ le faisceau $\lambda_{\mathcal{F}}(u)$ ne provient pas d'un fibré inversible sur $M_X(c)$. Désignons par $\mathbf{H}$ la sous-algèbre unitaire de K(X) engendrée par h, et pour $c \in K_{num}(X)$, par $c^{\perp}$ l'orthogonal de c dans K(X) ; enfin, soit $\mathbf{H}^{\perp}$ l'orthogonal de $\mathbf{H}$, et $\mathbf{H}^{\perp\perp}$ le biorthogonal.

Théorème 2.4. — (i) *L'homomorphisme canonique*

$$\lambda_{\mathcal{F}} : c^{\perp} \cap \mathbf{H}^{\perp\perp} \to \mathrm{Pic}^G(R^{ss})$$

se factorise de manière unique suivant le diagramme

$$\begin{array}{ccc} c^{\perp} \cap \mathbf{H}^{\perp\perp} & \xrightarrow{\lambda_{\mathscr{F}}} & \mathrm{Pic}^{\mathrm{G}}(\mathrm{R}^{ss}) \\ & \lambda_{\mathrm{M}}\searrow & \uparrow \pi^* \\ & & \mathrm{Pic}(\mathrm{M_X}(c)) \end{array}$$

(ii) *Pour* $u \in c^{\perp} \cap \mathbf{H}^{\perp\perp}$, *le fibré inversible* $\lambda_{\mathrm{M}}(u)$ *est caractérisé par la propriété universelle suivante : pour toute famille* $\mathscr{G}$ *de faisceaux semi-stables de classe* c *paramétrée par* S, *on a dans* Pic(S)

$$f_{\mathscr{G}}^*(\lambda_{\mathrm{M}}(u)) = \lambda_{\mathscr{G}}(u),$$

où $f_{\mathscr{G}} : \mathrm{S} \to \mathrm{M_X}(c)$ *désigne le morphisme modulaire.*

Le fait que $\pi : \mathrm{R}^{ss} \to \mathrm{M_X}(c)$ soit un bon quotient entraîne évidemment que le morphisme π^* est injectif, et donc l'unicité de la factorisation. La démonstration de l'existence de λ_{M} repose sur un lemme de descente dû à Kempf, Drézet et Narasimhan :

Lemme 2.5. — *Soit* G *un groupe algébrique opérant sur une variété algébrique* X, *et* $\pi : \mathrm{X} \to \mathrm{Y}$ *un bon quotient de* X *par l'action de* G. *Le morphisme image réciproque*

$$\pi^* : \mathrm{Pic(Y)} \to \mathrm{Pic}^{\mathrm{G}}(\mathrm{X})$$

est injectif et a pour image le groupe des classes de G*−fibrés inversibles* L *satisfaisant à la propriété suivante : pour tout point* $x \in \mathrm{X}$ *d'orbite fermée, le stabilisateur* G_x *de* x *agit trivialement sur la fibre* L_x *de* L *au point* x.

Démonstration du théorème 2.4

Ce théorème est démontré pour l'espace de modules des faisceaux semi-stables de dimension $d = n$ dans [13] . La démonstration s'étend sans difficulté au cas des faisceaux de dimension quelconque. Les points $t \in \mathrm{R}^{ss}$ d'orbite fermée correspondent aux faisceaux $\mathscr{F}_t$ qui sont polystables, c'est-à-dire somme directe de faisceaux stables de même polynôme de Hilbert réduit. Un tel faisceau s'écrit

$$\mathscr{F}_t = \mathrm{F}_1^{k_1} \oplus \ldots \oplus \mathrm{F}_\ell^{k_\ell}$$

où les F_i sont des faisceaux stables de même polynôme de Hilbert réduit, et deux à deux non isomorphes. En un tel point $t \in R$, le stabilisateur Stab(t) du point t s'identifie au groupe produit $GL(k_1) \times \ldots \times GL(k_\ell)$; compte-tenu du lemme 2.1 ci-dessus, on peut calculer l'action du stabilisateur sur la fibre en t, c'est-à-dire l'espace vectoriel $\lambda_{\mathcal{F}_t}(u)$: si on désigne par $c(F_i)$ la classe de F_i dans K(X) on constate qu'elle est donnée, pour $(g_1, \ldots, g_\ell) \in \mathrm{Stab}(t)$ par

$$(g_1, \ldots, g_\ell) \mapsto \prod_{i=1,\ldots,\ell} (\det g_i)^{<c(F_i),u>}.$$

Or, dans $K(X) \otimes \mathbf{Q}$, le fait que F et F_i aient même polynôme de Hilbert réduit signifie que la différence

$$\frac{c}{r} - \frac{c(F_i)}{r_i}$$

où r_i désigne la multiplicité de F_i, appartient à $\mathbf{H}^\perp$. Par suite, si $< c, u >= 0$, on a aussi $< c(F_i), u >= 0$; ainsi, l'hypothèse $u \in c^\perp \cap \mathbf{H}^{\perp\perp}$ entraîne que l'action du stabilisateur du point t est triviale sur $\lambda_{\mathcal{F}}(u)$. On peut donc appliquer le lemme de descente, ce qui donne la construction de $\lambda_M(u)$. La propriété universelle de ce fibré inversible résulte évidemment de la construction du morphisme modulaire $f_{\mathcal{G}}$. La démonstration est identique à celle qui est donnée dans [13] . □

Remarque 2.6. — Sur l'ouvert R^s des points stables, les conditions de descente sont moins restrictives ; il suffit, pour pouvoir descendre le fibré $\lambda_{\mathcal{F}}(u)$ de s'assurer que $u \in c^\perp$. On obtient donc une factorisation

$$\begin{array}{ccc} c^\perp & \xrightarrow{\lambda_{\mathcal{F}}} & \mathrm{Pic}^G(R^s) \\ & \lambda_{M^s}\searrow & \uparrow \pi^* \\ & & \mathrm{Pic}(M_X^s(c)) \end{array}$$

et le fibré $\lambda_{M^s}(u)$ ainsi obtenu pour $u \in c^\perp$ est caractérisé par une propriété universelle analogue, relative aux familles de faisceaux stables.

2.4. Exemples

Pour $u \in K(X)$, on a $< u, h^n >= \mathrm{rg}(u)\deg(X)$ et par conséquent les éléments de $\mathbf{H}^\perp$ sont de rang 0. Ceci implique que le groupe $A^n(X) \subset K(X)$

des classes d'équivalence linéaire de cycles de dimension 0 est contenu dans $\mathbf{H}^{\perp\perp}$. Pour $u \in \mathrm{A}^n(\mathrm{X})$, et $c \in \mathrm{K}_{num}(\mathrm{X})$, on a

$$< c, u >= \mathrm{rg}\,(c)\deg(u).$$

(1) *Faisceaux sans torsion.*

Supposons $d = n$; désignons par $\mathrm{Z}^n(\mathrm{X})$ le groupe des classes d'équivalence linéaire de 0-cycles de degré 0. On a alors $\mathrm{Z}^n(\mathrm{X}) \subset c^\perp \cap \mathbf{H}^{\perp\perp}$. Pour $u \in \mathrm{Z}^n(\mathrm{X})$ le fibré $\lambda_\mathrm{M}(u)$ a alors un sens.

Proposition 2.7. — [13] *Soit $c \in \mathrm{K}_{num}(\mathrm{X})$ de dimension n, et c_1 son image dans $\mathrm{Num}^1(\mathrm{X})$. Pour $u \in \mathrm{Z}^n(\mathrm{X})$ le fibré $\lambda_\mathrm{M}(u)$ est l'image réciproque du fibré $\lambda_{\mathrm{Pic}^{c_1}(\mathrm{X})}(u)$ par le morphisme*

$$\mathrm{M}_\mathrm{X}(c) \to \mathrm{Pic}^{c_1}(\mathrm{X})$$

qui associe à un faisceau F son déterminant.

Il résulte de cet énoncé que si on considère l'espace de modules des $\mathrm{M}_\mathrm{X}(c,\mathrm{L})$ des faisceaux semi-stables de classe c et de déterminant L fixé, pour $u \in c^\perp \cap \mathbf{H}^{\perp\perp}$, la restriction du fibré inversible $\lambda_\mathrm{M}(u)$ à cet espace de modules ne dépend que de la classe de u modulo $\mathrm{Z}^n(\mathrm{X})$.

En fait, cet énoncé s'étend en toute dimension $d \leq n$.

(2) *Faisceaux de codimension* 1.

Pour $d = n - 1$ nous avons $\mathrm{A}^n(\mathrm{X}) \subset c^\perp \cap \mathbf{H}^{\perp\perp}$. Pour $u \in \mathrm{A}^n(\mathrm{X})$, le fibré $\lambda_\mathrm{M}(u)$ s'interprète de la manière suivante. Soit c_1 l'image de c dans $\mathrm{Num}^1(\mathrm{X})$. Considérons l'hypersurface universelle Ξ dans $\mathrm{Div}^{c_1}(\mathrm{X}) \times \mathrm{X}$ paramétrée par $\mathrm{Div}^{c_1}(\mathrm{X})$, et le fibré inversible associé $\lambda_{\mathcal{O}_\Xi}(u)$ dans $\mathrm{Pic}(\mathrm{Div}^{c_1}(\mathrm{X}))$.

Proposition 2.8. — *Soit $c \in \mathrm{K}_{num}(\mathrm{X})$ de dimension $n-1$, et c_1 son image dans $\mathrm{Num}^1(\mathrm{X})$. Pour $u \in \mathrm{A}^n(\mathrm{X})$ le fibré $\lambda_\mathrm{M}(u)$ est l'image réciproque de $\lambda_{\mathcal{O}_\Xi}(u)$ par le morphisme*

$$\sigma : \mathrm{M}_\mathrm{X}(c) \to \mathrm{Div}^{c_1}(\mathrm{X})$$

qui associe à la classe d'un faisceau F son support schématique.

En effet, étant donnée une famille plate $\mathscr{F}$ de faisceaux cohérents purs de classe c paramétrée par une variété algébrique S, on peut trouver une résolution localement libre

$$0 \to \mathrm{R}_{n-1} \to \ldots \to \mathrm{R}_1 \xrightarrow{d_1} \mathrm{R}_0 \to \mathscr{F} \to 0$$

D'après le lemme 2.1 on a donc

$$\lambda_{\mathscr{F}}(u) = \otimes_i(\lambda_{\mathrm{R}_i}(u))^{(-1)^i}$$

Il suffit de vérifier l'énoncé lorsque u est la classe du faisceau structural d'un point a ; alors

$$\lambda_{\mathscr{F}}(a) = \otimes_i(\det \mathrm{R}_i(a))^{(-1)^i} = (\det \mathscr{F})(a).$$

Or, $\mathscr{F}$ définit une famille plate de faisceaux purs de codimension 1, donc relativement de Cohen-Macaulay en dehors d'un fermé A de $\mathrm{S} \times \mathrm{X}$ de codimension relative ≥ 3 au-dessus de S : ceci implique que $\mathrm{B}_0 = \mathrm{Im}\, d_1$ est un fibré vectoriel en dehors de A. En dehors de A, on peut considérer le fibré inversible $\det \mathscr{F} = \det \mathrm{R}_0 \otimes (\det \mathrm{B}_0)^{-1}$; il est muni d'une section canonique, dont le schéma des zéros est la trace sur le complémentaire de A du support schématique Σ de $\mathscr{F}$. Cette section s'étend de manière unique à $\mathrm{S} \times \mathrm{X}$: il en résulte une présentation

$$0 \to \mathcal{O}(-\det \mathscr{F}) \to \mathcal{O} \to \mathcal{O}_\Sigma \to 0.$$

Par suite, $\lambda_{\mathscr{F}}(a) = \lambda_{\mathcal{O}_\Sigma}(a)$. D'après la propriété universelle de Ξ, le morphisme canonique $\varphi : \mathrm{S} \to \mathrm{Div}^{c_1}(\mathrm{X})$ est tel que $(\varphi \times id_{\mathrm{X}})^*(\Xi) = \Sigma$, et par changement de base (*cf.* lemme 2.1), on obtient l'égalité $\lambda_{\mathscr{F}}(a) = \varphi^*(\lambda_{\mathcal{O}_\Xi}(a))$. La propriété universelle de $\lambda_{\mathrm{M}}(a)$ fournit alors l'égalité

$$\sigma^*(\lambda_{\mathcal{O}_\Xi}(a)) = \lambda_{\mathrm{M}}(a),$$

c'est-à-dire l'énoncé attendu. □

La démonstration a montré que si u est la classe d'un cycle effectif de dimension 0, le fibré inversible $\lambda_{\mathcal{O}_\Xi}(u)$ est muni d'une section dont le schéma des zéros correspond aux hypersurfaces qui rencontrent ce cycle.

(3) *Faisceaux de dimension* $d \leq n-2$.

On a encore comme ci-dessus $A^n(X) \subset c^\perp \cap \mathbf{H}^{\perp\perp}$, et donc pour $u \in A^n(X)$ le fibré inversible $\lambda_M(u)$ a un sens. En fait, si $\mathscr{F}$ est une famille plate de faisceaux relativement purs de dimension d, paramétrée par S, en dehors du support de $\mathscr{F}$, le fibré inversible $\det \mathscr{F}$ est trivial. Ce fermé est de codimension relative $n-d$ dessus de S, puisque $n-d \geq 2$ cette trivialisation s'étend à $S \times X$. Il en résulte que pour tout point $a \in X$ le fibré inversible $\lambda_{\mathscr{F}}(a) = (\det \mathscr{F})(a)$ est lui aussi trivial.

On a ainsi obtenu :

Proposition 2.9. — *Soit $c \in K_{num}(X)$, de dimension $d \leq n-2$. Alors*

$$A^n(X) \subset \ker \lambda_M.$$

Plus généralement, si on considère la filtration décroissante $F^pK(X)$ de $K(X)$ donnée par le sous-groupe des classes de faisceaux cohérents de codimension $\geq p$, une variante de l'argument ci-dessus montre que

$$F^{d+2}K(X) \cap \mathbf{H}^{\perp\perp} \subset \ker \lambda_M.$$

3. Faisceaux semi-stables sur le plan projectif

Le but de cette section est de donner une description du groupe de Picard des espaces de modules des faisceaux semi-stables sur le plan projectif. On commencera par rappeler d'abord sous quelles conditions ces espaces de modules sont de dimension ≥ 1.

3.1. Irréductibilité et conditions d'existence.

Quand X est le plan projectif $\mathbf{P}_2$, le groupe de Grothendieck $K(\mathbf{P}_2)$ est isomorphe à $\mathbf{Z}^3$, l'isomorphisme est donné par $c \mapsto (r, c_1, \chi)$, où r est le rang, c_1 la première classe de Chern (vue comme un entier), et χ la caractéristique d'Euler-Poincaré. La forme quadratique est alors donnée par

$$q(c) = 2r\chi + c_1^2 - r^2$$

et est donc non dégénérée. Par suite, $K_{num}(\mathbf{P}_2) = K(\mathbf{P}_2)$, et $K(\mathbf{P}_2)$ coïncide avec la sous-algèbre $\mathbf{H}$. Par suite, l'orthogonal $\mathbf{H}^\perp$ est réduit à $\{0\}$. On désigne par $u \mapsto u^*$ l'involution de $K(\mathbf{P}_2)$ qui associe à la classe d'un fibré, la classe de son dual ; on a en termes de rang et classes de Chern

$$\begin{aligned} < c^*, c > &= r^2 - (2rc_2 - (r-1)c_1^2) \\ &= 2r\chi - (r^2 + 3rc_1 + c_1^2) \end{aligned}$$

On sait déjà que si c est de dimension 0, et de longueur $\chi > 0$, l'espace de modules n'est pas vide et s'identifie alors à la puissance symétrique de $\mathrm{Sym}^{\chi}\mathbf{P}_2$. Dans ce cas, il n'existe pas de points stables, sauf si $\chi = 1$.

Théorème 3.1. — *Soit $c \in K(\mathbf{P}_2)$ une classe effective de dimension ≥ 1. Quand il n'est pas vide, l'ouvert $M^s_{\mathbf{P}_2}(c)$ des classes d'isomorphisme de faisceaux stables de classe c est une variété irréductible et lisse de dimension $1 - < c^*, c >$.*

En dimension 2, l'irréductibilité de l'ouvert des points stables est une conséquence d'un énoncé analogue de Ellingsrud et Hulek portant sur les fibrés vectoriels [6] . En dimension 1, ceci se voit en considérant le morphisme donné par le support schématique :

$$\sigma : M_{\mathbf{P}_2}(c) \to \mathrm{Div}^{c_1}(X)$$

On vérifie alors que l'image réciproque de l'ouvert des courbes lisses est dense dans $M_X(c)$: cet ouvert s'identifie à la variété de Picard relative des fibrés inversibles de caractéristique d'Euler-Poincaré χ [15]). La lissité résulte du fait que dans la description de $M_{\mathbf{P}_2}(c)$ comme quotient de l'ouvert R^{ss} du schéma de Hilbert $R = \mathbf{Hilb}(H \otimes \mathcal{O}_{\mathbf{P}_2}(-m), c)$ (*cf.* §2.2) l'ouvert R^s est déjà une variété lisse pourvu que m ait été choisi assez grand, variété sur laquelle le groupe $\overline{G} = GL(H)/\mathbf{C}^*$ opère librement.

En fait, l'ouvert R^{ss} est déjà une variété lisse : compte-tenu du fait que le quotient d'une variété affine normale par l'action d'un groupe réductif est une variété normale, on obtient :

Théorème 3.2. — *L'espace de modules $M_{\mathbf{P}_2}(c)$ est une variété normale.*

Reste à déterminer les conditions sous lesquelles ces espaces de modules sont non vides. Ceci fait appel à la notion de classe exceptionnelle.

Définition 3.3. — *Etant donnée une classe effective* $c \in \mathrm{K}(\mathbf{P}_2)$ *de dimension 2, de rang* r *et première classe de Chern* c_1*, on appelle pente de* c *le nombre rationnel*

$$\mu = \frac{c_1}{r}.$$

Proposition 3.4. — [6] *Soit* E *un faisceau stable sur* $\mathbf{P}_2$*, de classe* $a \in \mathrm{K}(\mathbf{P}_2)$*. Les conditions suivantes sont équivalentes :*

(i) *le faisceau* E *est rigide, c'est-à-dire tel que* $\mathrm{Ext}^1(\mathrm{E}, \mathrm{E}) = 0$;

(ii) *on a* $< a^*, a >= 1$.

Définition 3.5. — *Un faisceau stable* E *sur* $\mathbf{P}_2$ *est appelé exceptionnel s'il satisfait à l'une des conditions* (i), (ii) *ci-dessus. Une classe* $a \in \mathrm{K}(\mathbf{P}_2)$ *est dite exceptionnelle si c'est la classe d'un faisceau exceptionnel.*

Si a est la classe d'un faisceau exceptionnel E, l'espace de modules $\mathrm{M}_{\mathbf{P}_2}(a)$ est réduit au seul point défini par le faisceau E, et ce faisceau est en fait localement libre. Cet espace de modules est donc de dimension 0. Une telle classe exceptionnelle a est donc de rang > 0; elle est en fait déterminée par sa pente. Désignons $\mathfrak{C}$ l'ensemble des classes exceptionnelles. Cet ensemble est invariant par l'involution $a \mapsto a^*$ et par l'action du groupe des classes de fibrés inversibles $a \mapsto a(i)$ dite translation par $i \in \mathbf{Z}$. On peut en fait construire toutes les classes exceptionnelles à partir de l'élément unité par des opérations algébriques élémentaires. Ainsi, si $\mathfrak{C}$ contient deux classes a et b de pentes respectives α et β telles que $0 < \beta - \alpha < 3$, et telles que $< b^*, a >= 0$, il contient l'élément

$$< a^*, b > a - b.$$

On démontre [4] que $\mathfrak{C}$ est contruit à partir de 1 en faisant agir les translations, l'involution $a \mapsto a^*$, et cette opération. Par exemple, il contient les classes $e_i = (1-h)^{-i}$ des fibrés inversibles, la classe du fibré tangent $3e_1 - 1$, la classe $6 - e_2$ du noyau (de rang 5) du morphisme d'évaluation

$$\mathrm{H}^0(\mathcal{O}_{\mathbf{P}_2}(2)) \otimes \mathcal{O}_{\mathbf{P}_2} \to \mathcal{O}_{\mathbf{P}_2}(2).$$

On sait que s'il existe un faisceau stable et non rigide de classe c, on doit avoir $< c^*, c > \leq 0$. Réciproquement, étant donnée une classe effective

c, cette condition n'est pas suffisante pour assurer l'existence d'un tel fibré stable.

Théorème 3.6. — [6] *Soit $c \in \mathrm{K}(\mathbf{P}_2)$ une classe effective.*

(i) *L'espace de modules $\mathrm{M}_{\mathbf{P}_2}(c)$ est non vide de dimension 0 si et seulement si c est un multiple entier $c = ka$ d'une classe exceptionnelle a. Dans ce cas, cet espace de modules est réduit au point correspondant à la somme directe de k exemplaires du fibré exceptionnel défini par a.*

(ii) *L'espace de modules $\mathrm{M}_{\mathbf{P}_2}(c)$ est non vide de dimension ≥ 1 si et seulement si l'une des conditions suivantes est satisfaite*

— la classe c est de dimension ≤ 1 ;

— la classe c est de dimension 2, et pour toute classe exceptionnelle $a \in \mathfrak{E}$ de pente α telle que $-3 < \alpha + \mu \leq 0$, où μ désigne la pente de c, on a

$$< a, c > \leq 0.$$

Si l'une des conditions (ii) est satisfaite, l'espace de modules $\mathrm{M}_{\mathbf{P}_2}(c)$ est en fait de dimension ≥ 2, et si en outre c est de dimension ≥ 1, l'ouvert des points stables est dense.

Compte-tenu de l'énoncé 3.1 on obtient :

Corollaire 3.7. — *Soit $c \in \mathrm{K}(\mathbf{P}_2)$ une classe effective. Alors l'espace de modules $\mathrm{M}_{\mathbf{P}_2}(c)$ est une variété irréductible et normale.*

Etant donnée une classe effective c de dimension 2 de pente μ, on désigne par $\mathfrak{E}(c)$ l'ensemble des classes exceptionnelles a de pente α telles que $-3 < \alpha + \mu \leq 0$, et on pose

$$\delta(c) = \sup_{a \in \mathfrak{E}(c)} < a, c >$$

D'après l'énoncé ci-dessus, pour une telle classe c, l'espace de modules $\mathrm{M}_{\mathbf{P}_2}(c)$ est non vide et de dimension ≥ 1 si et seulement si $\delta(c) \leq 0$.

Proposition 3.8. — *Soit $c \in \mathrm{K}(\mathbf{P}_2)$ une classe effective.*

(i) *Si c est de dimension 2 et telle que $\delta(c) = 0$ il existe une classe exceptionnelle $a \in \mathfrak{E}(c)$ et une seule telle que $< a, c >= 0$.*

(ii) *Si c est de dimension 1, il existe au plus une classe exceptionnelle telle que $< a, c >= 0$.*

Cette classe est appelée la *classe exceptionnelle associée* à c. Cet énoncé résulte de la description de $\delta(c)$ dans le cas (i) [4] . L'assertion (ii) résulte trivialement de la définition de la forme quadratique. Un telle classe exceptionnelle n'existe dans le cas (ii) que si le rapport $\dfrac{\chi}{c_1}$, où χ est la caractéristique d'Euler-Poincaré et c_1 la première classe de Chern, est la pente d'un fibré exceptionnel. C'est toujours le cas si $c_1 = 1$ ou 2, seul cas qui nous sera utile dans l'énoncé qui suit.

3.2. Le groupe de Picard

Nous sommes maintenant en mesure de décrire le groupe de Picard de $\mathrm{M}_{\mathbf{P}_2}(c)$. La question du calcul du groupe de Picard est en fait très liée à la question de la factorialité locale de M. Rappelons d'abord de quoi il s'agit :

Définition 3.9. — *Soit* M *une variété normale. On dit que* M *est localement factorielle si ses anneaux locaux sont des anneaux factoriels.*

Si on considère le groupe Cl(M) des classes d'équivalence linéaire de diviseurs de Weil, le fait que la variété soit normale implique que le morphisme canonique Pic(M) $\to$ Cl(M) est injectif. Dire que X est localement factorielle signifie que ce morphisme est surjectif, ou encore que tout fibré inversible sur un ouvert de M se prolonge à M.

Théorème 3.10. — *Soit* $c \in \mathrm{K}(\mathbf{P}_2)$ *une classe effective satisfaisant à l'une des conditions suivantes, assurant que l'espace de modules* $\mathrm{M}_{\mathbf{P}_2}(c)$ *est non vide et non réduit à un point :*

— *la classe* c *est de dimension* ≤ 1;

— *la classe* c *est de dimension 2 et* $\delta(c) \leq 0$.

Considérons le morphisme canonique

$$\lambda_{\mathrm{M}} : c^{\perp} \to \mathrm{Pic}(\mathrm{M}_{\mathbf{P}_2}(c)).$$

(i) *Le morphisme* λ_{M} *est surjectif.*

(ii) *Il est injectif, sauf dans les cas suivants :*

— *la classe* c *est de dimension* 0 *et alors* $\ker \lambda_{\mathrm{M}} = \mathrm{A}^2(\mathbf{P}_2)$;

— *la classe* c *est de dimension* 1, *et* $c_1 = 1$ *ou* 2;

— *la classe* c *est de dimension 2, et* $\delta(c) = 0$.

Dans les deux derniers cas, le noyau $\ker \lambda_M$ *est le sous-groupe engendré par la classe exceptionnelle a associée à c.*

(iii) *La variété* $M_{P_2}(c)$ *est localement factorielle, excepté si c est la classe d'un cycle de dimension* 0 *et de longueur* ≥ 2.

Cet énoncé est dû à Fogarty dans le cas de la puissance symétrique, et à Drézet [5] dans le cas où c est de dimension 2. Le cas où c est de dimension 1 est traité dans [15] . En fait, la démonstration que nous allons esquisser traite à la fois ces deux derniers cas.

Remarquons que du fait que $< a^*, a >= 1$ pour une classe exceptionnelle, on a si $a \in c^\perp$ une décomposition en somme directe

$$c^\perp = (c, a^*)^\perp \oplus (a)$$

Il en résulte que le groupe quotient $c^\perp/(a)$ est isomorphe à $(c, a^*)^\perp$; sous l'hypothèse du théorème, c et a^* ne sont pas liés, par suite, ce sous-groupe est un groupe cyclique. On a ainsi obtenu :

Corollaire 3.11. — *Si l'espace de modules* $M_{P_2}(c)$ *est de dimension* > 0 *le groupe de Picard* $\mathrm{Pic}(M_{P_2}(c))$ *est un groupe abélien libre de rang 2, sauf dans les cas particuliers* (ii) *où c'est un groupe cyclique infini.*

3.3. L'outil fondamental

Considèrons un groupe algébrique G agissant sur une variété algébrique X (intègre pour simplifier). On désigne par $\mathcal{O}^*(X)$ le groupe multiplicatif des fonctions régulières inversibles. Un morphisme croisé $\varphi : G \to \mathcal{O}^*(X)$ est un morphisme de variétés $\varphi : G \times X \to \mathbf{C}^*$ tel que pour $g, g' \in G$ et $x \in X$ on ait $\varphi(gg', x) = \varphi(g, g'x)\varphi(g', x)$. Un tel morphisme croisé φ est dit principal s'il existe $\psi \in \mathcal{O}^*(X)$ tel que

$$\varphi(g, x) = \frac{\psi(gx)}{\psi(x)}.$$

On a une suite exacte

$$0 \to H^1(G, \mathcal{O}^*(X)) \to \mathrm{Pic}^G(X) \to \mathrm{Pic}(X)^G$$

où $H^1(G, \mathcal{O}^*(X))$ est le quotient du groupe des morphismes croisés $G \to \mathcal{O}^*(X)$ par le sous-groupe des morphismes croisés principaux et $\mathrm{Pic}(X)^G$ le sous-groupe du groupe de Picard des éléments invariants sous l'action de G. Si $\mathcal{O}^*(X) = \mathbb{C}^*$, le groupe des morphismes croisés s'identifie au groupe Char(G) des caractères $G \to \mathbb{C}^*$.

3.4. Le groupe de Picard de $\mathrm{Sym}^\ell(\mathbf{P}_2)$

On se propose de vérifier l'énoncé 3.10 dans le cas de la puissance symétrique. Soit ℓ un entier > 0. La puissance symétrique $\mathrm{Sym}^\ell(\mathbf{P}_2)$ est le quotient de ${\mathbf{P}_2}^\ell$ par l'action du groupe symétrique $\mathfrak{S}_\ell$ par permutation des facteurs. On peut donc appliquer l'énoncé ci-dessus pour déterminer le groupe $\mathrm{Pic}^{\mathfrak{S}_\ell}({\mathbf{P}_2}^\ell)$. Le groupe $\mathrm{H}^1(\mathfrak{S}_\ell, \mathbb{C}^*)$ s'identifie au groupe des caractères de $\mathfrak{S}_\ell$; le groupe des invariants $\mathrm{Pic}({\mathbf{P}_2}^\ell)^{\mathfrak{S}_\ell}$ est isomorphe à $\mathbf{Z}$, avec pour générateur le fibré inversible $\mathcal{O}(1, \ldots, 1)$. D'autre part par le lemme de descente 2.5 déjà évoqué, on a une suite exacte

$$0 \to \mathrm{Pic}(\mathrm{Sym}^\ell(\mathbf{P}_2)) \to \mathrm{Pic}^{\mathfrak{S}_\ell}({\mathbf{P}_2}^\ell) \xrightarrow{\rho} \prod_{x \in {\mathbf{P}_2}^\ell} \mathrm{Char}(\mathfrak{S}_x)$$

où $\mathfrak{S}_x$ est le stabilisateur du point x. Evidemment, l'intersection de $\ker \rho$ avec l'image de $\mathrm{Char}(\mathfrak{S}_\ell)$ est réduite à $\{0\}$. Par suite, on obtient un plongement

$$\mathrm{Pic}(\mathrm{Sym}^\ell(\mathbf{P}_2)) \to \mathrm{Pic}({\mathbf{P}_2}^\ell)^{\mathfrak{S}_\ell}$$

et par le lemme de Kempf, le générateur $\mathcal{O}(1, \ldots, 1)$ provient d'un fibré inversible sur $\mathrm{Sym}^\ell(\mathbf{P}_2)$: ce plongement est donc un isomorphisme.

Il reste à vérifier que l'homomorphisme $\lambda_M : c^\perp / A^2 \to \mathrm{Pic}(M)$, où $M = \mathrm{Sym}^\ell(\mathbf{P}_2)$ est un isomorphisme. Le membre de gauche est un groupe cyclique engendré par la classe de h. Pour calculer $\lambda_M(h)$, on considère la famille universelle de faisceaux semi-stables paramétrée par ${\mathbf{P}_2}^\ell$: si Δ est la diagonale de $\mathbf{P}_2 \times \mathbf{P}_2$ et $p_i : {\mathbf{P}_2}^\ell \to \mathbf{P}_2$ la i-ème projection, il s'agit du faisceau

$$\mathcal{F} = \oplus_{i=1}^{\ell} (p_i \times id)^*(\mathcal{O}_\Delta)$$

On est bien sûr ramené par fonctorialité à faire le calcul dans le cas $\ell = 1$. Dans ce cas, on obtient $\mathcal{O}_\Delta(h) = \mathcal{O}_\Delta - \mathcal{O}_\Delta(0, -1)$ dans $K(\mathbf{P}_2 \times \mathbf{P}_2)$ ce qui fournit, après projection, l'égalité $\lambda_{\mathcal{O}_\Delta}(h) = \mathcal{O}_{\mathbf{P}_2}(1)$ dans $\mathrm{Pic}(\mathbf{P}_2)$. Par suite $\lambda_{\mathcal{F}}(h) = \mathcal{O}_{{\mathbf{P}_2}^\ell}(1, \ldots, 1)$ dans $\mathrm{Pic}({\mathbf{P}_2}^\ell)$.

Le groupe symétrique $\mathfrak{S}_\ell$ opère librement sur le complémentaire de la diagonale : ainsi, tout caractère non trivial de $\mathfrak{S}_\ell$ définit un fibré sur l'ouvert U complémentaire de la diagonale dans $\mathrm{M} = \mathrm{Sym}^\ell(\mathbf{P}_2)$ et un tel élément est non trivial et de torsion dans Pic(U). Puisque le complémentaire de U est de codimension 2, le morphisme de restriction $\mathrm{Pic(M)} \to \mathrm{Pic(U)}$ est injectif. Ces éléments de torsion ne peuvent donc pas s'étendre à M. Un tel caractère non trivial existe si $\ell \geq 2$; ainsi, dans ce cas, la variété de modules M n'est pas localement factorielle.

3.5. Le problème de Brill-Noether

On se propose dans la suite de donner une esquisse de la démonstration du théorème 3.10. Les détails sont exposés dans le cas où c est de dimension 2 dans [16] ou [17]. Nous aurons d'abord besoin de renseignements concernant le problème de Brill-Noether.

Soit $c \in \mathrm{K}(\mathbf{P}_2)$ une classe effective de dimension ≥ 1, de rang r, première classe de Chern c_1, et de caractéristique d'Euler-Poincaré χ. On suppose que $c_1 + 3r > 0$, de sorte que pour tout faisceau semi-stable F de classe c on a $\mathrm{H}^2(\mathrm{F}) = 0$: ceci résulte du théorème de dualité de Serre si la dimension de c est 2 ; si la dimension est 1, c'est trivial.

Théorème 3.12. — *Soient $c \in \mathrm{K}(\mathbf{P}_2)$ une classe effective satisfaisant aux conditions ci-dessus, et $\mathrm{M} = \mathrm{M}_{\mathbf{P}_2}(c)$ l'espace de modules des faisceaux semi-stables de classe c.*

(i) *Si $\chi > 0$ (resp. < 0) le fermé de* M *défini par les classes des faisceaux semi-stables* F *satisfaisant à la condition déterminantielle* $\mathrm{H}^1(\mathrm{F}) \neq 0$ *(resp.* $\mathrm{H}^0(\mathrm{F}) \neq 0$) *est de codimension* ≥ 2.

(ii) *Si $\chi = 0$, le sous-schéma de* M *défini par la condition déterminantielle* $\mathrm{H}^1(\mathrm{F}) \neq 0$ *est, s'il n'est pas vide, une hypersurface intègre.*

Cette hypersurface est en fait un diviseur de Cartier : le fibré inversible $\lambda_{\mathrm{M}}(1)$ est muni d'une section canonique (appelée fonction thêta) et l'hypersurface en question est le schéma des zéros de cette section.

Cet énoncé est dû dans le cas où c est de dimension 2 à L. Göttsche et A. Hirschowitz [10] . Nous ne traitons ici que le cas où c est de dimension 1. Si $c_1 \leq 2$, les variétés déterminantielles qui figurent dans l'énoncé sont vides. On suppose donc $c_1 > 2$. Dans ce cas, le fermé des points non stables est de codimension ≥ 2 [15] , de sorte qu'il suffit de se placer sur l'ouvert des points stables. On fixe un point $p \in \mathbf{P}_2$ et on considère l'ouvert U_p de M des faisceaux stables dont le support ne passe pas par p. Ces ouverts U_p recouvrent l'ouvert des points stables. Par projection sur une droite ne passant pas par p

$$\pi : \mathbf{P}_2 - \{p\} \to \mathbf{P}_1$$

on voit que l'ouvert U_p est isomorphe à l'espace de modules $\mathrm{Higgs}_{\mathbf{P}_1}(c)$ des «paires de Higgs» (G, φ) où G est un fibré sur $\mathbf{P}_1$ de rang c_1 et de caractéristique d'Euler-Poincaré χ, et $\varphi : G \to G(1)$ un morphisme de fibrés ; ces paires sont soumises à la condition de stabilité suivante : pour tout sous-fibré $G' \subset G$ tel que $\varphi(G') \subset G'(1)$ on a $\mu(G') < \mu(G)$.

L'identification s'obtient en associant au faisceau F l'image directe $G = \pi_*(F)$; le morphisme φ provient de la structure de module sur U_p, contenu du fait que U_p s'identifie à l'espace total du fibré normal $\mathcal{O}_{\mathbf{P}_1}(1)$ de la droite $\mathbf{P}_1$. On a donc

$$H^i(F) = H^i(G)$$

de sorte qu'il suffit de vérifier le même énoncé pour l'espace des paires de Higgs stables. Considérons la sous-variété localement fermée V_G des classes d'isomorphisme de paires de Higgs stables (G, φ) où G est donné par

$$G = \oplus_{i \in \mathbf{Z}} \mathcal{O}(i)^{r_i}$$

où les r_i sont des entiers fixés ≥ 0. Ces sous-variétés, appelées strates, sont en nombre fini et recouvrent l'espace de modules $\mathrm{Higgs}_{\mathbf{P}_1}(c)$. Si $\mathrm{Hom}^s(G, G(1))$ est l'ouvert de $\mathrm{Hom}(G, G(1))$ des morphismes φ définissant une paire stable, alors l'action naturelle par conjugaison du groupe $\mathrm{PAut}(G) := \mathrm{Aut}(G)/\mathbf{C}^*$ est libre et le morphisme

$$\mathrm{Hom}^s(G, G(1)) \mapsto V_G$$

est un quotient géométrique de $\mathrm{Hom}^s(G, G(1)$. Il en résulte que

$$\begin{aligned} \dim V_G &= \dim \mathrm{Hom}^s(G, G(1)) - \dim \mathrm{Hom}(G, G) + 1 \\ &= \sum_{j \leq i+1} r_i r_j \end{aligned}$$

Par suite la codimension de V_G est

$$c_1^2 - \sum_{j\leq i+1} r_i r_j = \sum_{j>i+1} r_i r_j.$$

On vérifie facilement que la stabilité d'une telle paire de Higgs (G, φ) entraîne que le spectre de G, c'est-à-dire l'ensemble des $i \in \mathbb{Z}$ tels que $r_i \neq 0$, est connexe. On dit que G est rigide si ce spectre est connexe et réduit au plus à deux points ; il revient au même de demander que $\mathrm{Ext}^1(G, G) = 0$. Un tel fibré est déterminé par son rang et sa caractéristique d'Euler-Poincaré.

Supposons d'abord $\chi \neq 0$.

Les variétés déterminantielles sont des réunions de strates. Dans le cas où G est rigide, V_G est un ouvert qui ne rencontre aucune des variétés déterminantielles étudiées ; la seule strate V_G de codimension 1 correspond au cas où le spectre de G est formé de 3 points consécutifs $(i-1, i, i+1)$ et où l'on a $r_{i-1} = r_{i+1} = 1$. Une telle strate ne peut rencontrer les variétés déterminantielles étudiées que si $i = -1$; mais une telle strate n'existe que si $\chi = 0$. Ceci démontre (i).

Supposons maintenant $\chi = 0$.

Alors la variété déterminantielle Σ définie par la condition $H^1(G) \neq 0$ est le schéma des zéros d'une section du fibré inversible correspondant à $\lambda_M(1)$; c'est une réunion de strates de codimension ≥ 1 : par suite, cette section n'est pas identiquement nulle. Le schéma Σ contient comme ouvert partout dense la strate V_G où G est donné par $G = \mathcal{O} \oplus \mathcal{O}^{c_1-2}(-1) \oplus \mathcal{O}(-2)$. Pour voir que Σ est intègre, il suffit de constater que la différentielle de Petri en un point s de cette strate correspondant à la paire de Higgs $(G, \varphi) \in \Sigma$

$$T_s\mathrm{Higgs}(c) \to L(H^0(G), H^1(G))$$

est surjective.

Lemme 3.13. — *Soit* (G, φ) *une paire de Higgs stable. Considérons le complexe* K *à 2 termes*

$$0 \to \underline{\mathrm{Hom}}(G, G) \to \underline{\mathrm{Hom}}(G, G(1)) \to 0$$

où la différentielle est donnée par $u \mapsto \varphi u - u\varphi$. *L'espace tangent de Zariski* $T_s\mathrm{Higgs}(c)$ *au point* s *correspondant à* (G, φ) *s'identifie à l'hypercohomologie* $\mathbf{H}^1(\mathrm{K})$ *du complexe* K. *En particulier, on a une suite exacte naturelle*

$$\mathrm{Hom}(\mathrm{G}, \mathrm{G}(1)) \to \mathrm{T}_s\mathrm{Higgs}(c) \to \mathrm{Ext}^1(\mathrm{G}, \mathrm{G}) \to \mathrm{Ext}^1(\mathrm{G}, \mathrm{G}(1)) \to 0.$$

Ce lemme est une conséquence de la construction de l'espace de modules des paires de Higgs. La différentielle de Petri est bien entendu la flèche composée

$$\begin{array}{ccc} \mathrm{T}_s\mathrm{Higgs}(c) & & \\ \downarrow & \searrow & \\ \mathrm{Ext}^1(\mathrm{G}, \mathrm{G}) & \to & \mathrm{L}(\mathrm{H}^0(\mathrm{G}), \mathrm{H}^1(\mathrm{G})) \end{array}$$

où la flèche horizontale est l'accouplement naturel. Il résulte du lemme ci-dessus que si $\mathrm{Ext}^1(\mathrm{G}, \mathrm{G}(1)) = 0$, la flèche verticale est surjective : c'est le cas en un point s de la strate de codimension 1, et dans ce cas la flèche horizontale est un isomorphisme. Par suite, le diviseur Σ est génériquement lisse, et irréductible donc intègre. Ceci achève la démonstration. □

3.6. Fin du calcul, quand c est de dimension $d \geq 1$.

Sous les hypothèses du théorème, l'espace de modules n'est pas réduit à un point, donc $< c^*, c > \leq 0$. Soit P le polynôme de Hilbert de c, défini par

$$\mathrm{P}(m) = < c, (1-h)^{-m} > .$$

Puisque la dimension de c est ≥ 1, ce polynôme est de degré 1 ou 2, et prend des valeurs < 0. C'est évident si $\dim c = 1$, et résulte par exemple de la formule donnant $< c^*, c >$ (*cf.* §3.1) quand $\dim c = 2$. On peut donc quitte à remplacer c par $c(m)$ pour un entier m convenable, supposer que $\mu > -3$, $\mathrm{P}(0) > 0$, $\mathrm{P}(-1) \geq 0$ et $\mathrm{P}(-2) < 0$. On désigne alors par U l'ouvert de $\mathrm{M}_{\mathbf{P}_2}(c)$ correspondant aux points représentant des faisceaux polystables F tels que $\mathrm{H}^1(\mathrm{F}) = \mathrm{H}^1(\mathrm{F}(-1)) = \mathrm{H}^0(\mathrm{F}(-2)) = 0$. Soit K_i un espace vectoriel de dimension $|\mathrm{P}(-i)|$. On désigne par Q^* le fibré universel de rang 2 sur $\mathbf{P}_2$, noyau du morphisme d'évaluation $\mathrm{H}^0(\mathscr{O}_{\mathbf{P}_2}(1)) \otimes \mathscr{O}_{\mathbf{P}_2} \to \mathscr{O}_{\mathbf{P}_2}(1)$. La suite spectrale de Beilinson montre que si un faisceau polystable F représente un point de U, le faisceau F est le conoyau d'un morphisme génériquement injectif

$$\mathscr{V} = \mathrm{K}_2 \otimes \wedge^2\mathrm{Q}^* \oplus \mathrm{K}_1 \otimes \mathrm{Q}^* \to \mathscr{W} = \mathrm{K}_0 \otimes \mathscr{O}_{\mathbf{P}_2}$$

Réciproquement, considérons l'ouvert $\mathcal{M} \subset \text{Hom}(\mathcal{V}, \mathcal{W})$ des morphismes $s : \mathcal{V} \to \mathcal{W}$ injectifs en dehors d'un nombre fini de points quand $d = 2$, et génériquement injectifs quand $d = 1$. Le conoyau $\mathcal{F}_s$ d'un tel morphisme est alors pur de dimension d d'après le critère de Serre. Il est facile de voir, en utilisant le fait que le fibré $\underline{\text{Hom}}(\mathcal{V}, \mathcal{W})$ est engendré par ses sections, que le complémentaire de $\mathcal{M}$ est de codimension ≥ 2 dans l'espace vectoriel de tous les morphismes.

Considérons d'autre part l'ouvert $\mathcal{M}^{ss}$ de $\mathcal{M}$ des morphismes s dont le conoyau $\mathcal{F}_s = \text{coker}\ s$ est semi-stable et tels que le gradué de Jordan-Hölder appartienne à l'ouvert U. On obtient une famille universelle paramétrée par $\mathcal{M}^{ss}$ et par suite un morphisme

$$\pi : \mathcal{M}^{ss} \to \mathrm{U}$$

qui associe à s le point défini par le faisceau $\mathcal{F}_s$. Sur $\mathcal{M}^{ss}$, le groupe $\mathrm{G} = \text{Aut}(\mathcal{V}) \times \text{Aut}(\mathcal{W})$ opère par conjugaison ; la famille universelle $\mathcal{F}$ est aussi munie d'une action de G, et le morphisme π est équivariant pour cette action. Cette action sur $\mathcal{M}$ se factorise en fait au groupe $\overline{\mathrm{G}} = \mathrm{G}/\mathbf{C}^*$ quotient de G par son centre.

Lemme 3.14. — (i) *Le morphisme π fait de l'ouvert* U *un bon quotient de $\mathcal{M}^{ss}$ par l'action de* G.

(ii) *Sous l'une des hypothèse suivantes*

— $d = 1$, *et* $c_1 \geq 3$,

— $d = 2$ *et* $\delta(c) < 0$,

le complémentaire de $\mathcal{M}^{ss}$ dans $\mathcal{M}$ est de codimension ≥ 2.

Ce lemme généralise un énoncé analogue obtenu dans [16] quand c est de dimension $d = 2$. L'assertion (i) repose sur les propriétés de transitivité des bons quotients, en se ramenant au schéma de Hilbert. L'assertion (ii) repose sur le fait que l'on peut stratifier le complémentaire de l'ouvert des points semi-stables en utilisant les filtrations de Harder-Narasimhan ; on peut calculer la codimension des strates de Harder-Narasimhan dans une famille complète de faisceaux purs. C'est le cas pour la famille $\mathcal{F}$ paramétrée par $\mathcal{M}$. Pour obtenir l'énoncé, on considère l'ouvert $\mathcal{M}^s \subset \mathcal{M}^{ss}$ des points s tels que $\mathcal{F}_s$ soit stable ; sous l'une des hypothèses (ii) le complémentaire de cet ouvert est aussi de codimension 2 dans l'ouvert de $\mathcal{M}$ des points semi-stables.

Lemme 3.15. — *Considérons la famille universelle $\mathcal{F}$ de faisceaux de classe c paramétrée par $\mathcal{M}$.*

(i) *Le morphisme canonique*

$$\lambda_{\mathcal{F}} : \mathrm{K}(\mathbf{P}_2) \to \mathrm{Pic}^{\mathrm{G}}(\mathcal{M})$$

est surjectif; c'est un isomorphisme si $\mathrm{P}(-1) > 0$.

(ii) *Cet morphisme induit un morphisme surjectif*

$$c^{\perp} \to \mathrm{Pic}^{\overline{\mathrm{G}}}(\mathcal{M}).$$

Cet épimorphisme est un isomorphisme si $\mathrm{P}(-1) > 0$.

L'assertion (i) résulte du fait que $\mathcal{O}^*(\mathcal{M}) = \mathbb{C}^*$, et que $\mathrm{Pic}(\mathcal{M}) = 0$. Ainsi, le groupe $\mathrm{Pic}^{\mathrm{G}}(\mathcal{M})$ s'identifie au groupe Char(G). Ce groupe est $\mathbb{Z}^2$ ou $\mathbb{Z}^3$ suivant que $\mathrm{P}(-1) = 0$ ou $\mathrm{P}(-1) \neq 0$. Il suffit de constater que l'image de la base de $\mathrm{K}(\mathbf{P}_2)$ donnée par les classes e_{-i} des fibrés inversibles $\mathcal{O}(-i)$ pour $i = 0, 1, 2$ par $\lambda_{\mathcal{F}}$ fournit un système de générateurs de Char(G), ce qui est évident. Pour l'assertion (ii), on observe que l'on a un diagramme commutatif

$$\begin{array}{ccc} \mathrm{K}(\mathbf{P}_2) & \to & \mathrm{Char(G)} \\ & {\scriptstyle (1)}\searrow & \downarrow {\scriptstyle (2)} \\ & & \mathbb{Z} \end{array}$$

dans lequel la flèche (1) est donnée par $u \to < c, u >$ et la flèche verticale est induite par l'inclusion du centre $\mathbb{C}^* \hookrightarrow \mathrm{G}$. Ceci se vérifie immédiatement sur les éléments de la base e_{-i}. Le noyau de (1) est par définition $c^{\perp}$ et le noyau de (2) est le groupe $\mathrm{Char}(\overline{\mathrm{G}})$. D'où l'énoncé.

Fin de la preuve du théorème 3.10

Considérons le diagramme

$$\begin{array}{ccc} \mathcal{M}^{ss} & \hookrightarrow & \mathcal{M} \\ {\scriptstyle \pi}\downarrow & & \\ \mathrm{U} & \hookrightarrow & \mathrm{M} \end{array}$$

De la propriété universelle de $\lambda_M(u)$, il résulte que ce diagramme induit un diagramme commutatif

$$\begin{array}{ccccc} c^\perp & & \overset{\lambda_{\mathscr{F}}}{\longrightarrow} & & \mathrm{Pic}^{\overline{G}}(\mathscr{M}) \\ \lambda_M \downarrow & & & & \downarrow (3) \\ \mathrm{Pic}(M) & \overset{\rho}{\longrightarrow} & \mathrm{Pic}(U) & \overset{(4)}{\longrightarrow} & \mathrm{Pic}^{\overline{G}}(\mathscr{M}^{ss}) \end{array}$$

Surjectivité

On sait d'après le lemme précédent que le morphisme $\lambda_{\mathscr{F}}$ est surjectif, et le morphisme (3) est évidemment surjectif. De plus, le fait que U est un bon quotient de $\mathscr{M}^{ss}$ implique que la flèche (4) est injective. Il en résulte que (4) est un isomorphisme.

— Si $P(-1) \neq 0$, il résulte de Brill-Noether que le complémentaire de U est de codimension ≥ 2, donc, M étant normale, la flèche ρ est injective. C'est donc un isomorphisme, et par suite λ_M est surjective.

— Si $P(-1) = 0$, alors $\ker \lambda_{\mathscr{F}} = (e_{-1})$ et le noyau de ρ est justement $\lambda_M(e_{-1})$ d'après le théorème 3.12. On a alors un diagramme commutatif de suites exactes

$$\begin{array}{ccccccccc} 0 \rightarrow & (e_{-1}) & \rightarrow & c^\perp & \rightarrow & c^\perp/(e_{-1}) & \rightarrow 0 \\ & (5) \downarrow & & \lambda_M \downarrow & & \downarrow (6) & \\ 0 \rightarrow & \lambda_M(e_{-1}) & \rightarrow & \mathrm{Pic}(M) & \rightarrow & \mathrm{Pic}(U) & \rightarrow 0 \end{array}$$

dans lequel les deux flèches verticales extrêmes (5) et (6) sont surjectives, donc aussi la flèche λ_M.

Injectivité

Ici on doit se placer sous l'une des hypothèses

— $d = 1$ et $c_1 \geq 3$

— $d = 2$ et $\delta(c) < 0$.

Sous l'une de ces hypothèses, le lemme 3.14 a montré que $\mathscr{M}^{ss}$ est un ouvert de $\mathscr{M}$ dont le complémentaire est de codimension ≥ 2 : dans le diagramme ci-dessus la flèche (3) est alors un isomorphisme. Ceci entraîne l'injectivité de λ_M si $P(-1) \neq 0$. Si $P(-1) = 0$, la flèche (6) est un isomorphisme et on est ramené à vérifier que le fibré $\lambda_M(e_{-1})$ n'est pas trivial. Ce fibré possède une section canonique : il s'agit donc de vérifier qu'il existe un faisceau semi-stable F de classe c tel que $H^1(F(-1)) \neq 0$. On renvoie à [16] ou [17] pour le cas où $d = 2$. Dans le cas $d = 1$ c'est évident :

il suffit de prendre sur une courbe lisse C de degré c_1 un point L du diviseur thêta, et de considérer le faisceau $F = L(1)$.

Les cas particuliers

Sous l'une des hypothèses

— $d = 1$ et $c_1 \leq 2$,

— $d = 2$ et $\delta(c) = 0$

on a défini une classe exceptionnelle a associée à c : on a alors $< a, c >= 0$. Si E_a est un fibré exceptionnel de classe a, pour tout famille plate de faisceaux semi-stables $\mathscr{G}$ de classe c paramétrée par une variété algébrique S les images directes $R^i pr_{1*}(\mathscr{G} \otimes E_a)$ sont en fait toutes nulles, et par suite $\lambda_{\mathscr{G}}(a) = 0$. Par suite, a appartient au noyau de λ_M. Ainsi, λ_M induit un épimorphisme

$$c^{\perp}/(a) \to \mathrm{Pic}(M)$$

Le premier membre est un groupe cyclique infini ; puisque M est une variété projective, c'est un isomorphisme.

Factorialité locale

Il suffit de remplacer dans la démonstration de la surjectivité ci-dessus M par l'ouvert des points lisses M_{reg} de M, l'ouvert U par l'ouvert U_{reg} des points lisses de U et $\mathscr{M}^{ss}$ par l'image réciproque $\mathscr{M}^{ss}_{reg}$ de U_{reg}. Le morphisme $\mathscr{M}^{ss}_{reg} \to U_{reg}$ est encore un bon quotient. L'argument utilisé ci-dessus montre que le morphisme $c^{\perp} \to \mathrm{Pic}(M_{reg})$ est surjectif, et par suite le morphisme de restriction

$$\mathrm{Pic}(M) \to \mathrm{Pic}(M_{reg})$$

est surjectif.

4. Systèmes cohérents

On introduit dans cette section l'espace de modules $\mathrm{Syst}_X(c)$ des systèmes cohérents α−semi-stables de classe c, et on définit pour cet espace un opérateur λ_S analogue à l'homomorphisme λ_M de la section 2.

4.1. Motivation

Soit n un entier ≥ 2. Soit M_n l'espace de modules des faisceaux semi-stables de rang 2 et de classes de Chern $c_1 = 0, c_2 = n$ sur le plan projectif $\mathbf{P}_2$. Comme on l'a vu, c'est une variété projective irréductible de dimension $4n - 3$. D'après le théorème de Grauert-Mülich, la restriction d'un tel faisceau F à une droite générique ℓ est le fibré trivial de rang 2. Une droite de saut pour un tel faisceau est une droite ℓ telle que la restriction $F|_\ell$ n'est pas triviale : les droites de saut de F sont donc celles qui satisfont à la condition cohomologique $h^1(F(-1)|_\ell) \neq 0$. Si on considère la variété d'incidence D des couples (droites, points), munie des projections canoniques

$$\begin{array}{ccc} D & \xrightarrow{pr_2} & \mathbf{P}_2 \\ {\scriptstyle pr_1}\downarrow & & \\ \mathbf{P}_2^* & & \end{array}$$

l'ensemble des droites de saut est le support du faisceau cohérent sur le plan projectif dual $\Theta = R^1 pr_{1*}(pr_2^*(F(-1))$. Pour $p = -1$ et -2 les nombres de Hodge $h^q(F(p))$ sont nuls si $q \neq 1$ et on a $h^1(F(-1)) = h^1(F(-2)) = n$; la résolution standard $0 \to \mathcal{O}(-1,-1) \to \mathcal{O} \to \mathcal{O}_D \to 0$ de D dans $\mathbf{P}_2^* \times \mathbf{P}_2$, montre alors que ce faisceau Θ est le conoyau du morphisme canonique sur $\mathbf{P}_2^*$

$$H^1(F(-2)) \otimes \mathcal{O}_{\mathbf{P}_2^*}(-1) \xrightarrow{\varphi} H^1(F(-1)) \otimes \mathcal{O}_{\mathbf{P}_2^*}.$$

Ainsi, Θ est un faisceau pur de dimension 1 dont le support schématique est une courbe dont l'équation est le déterminant du morphisme φ. Cette courbe γ_F de degré n est appelée courbe des droites de saut de F. L'application qui associe à la classe de F la courbe γ_F définit un morphisme $F \mapsto \gamma_F$:

$$M_n \to |\mathcal{O}_{\mathbf{P}_2^*}(n)|$$

On sait que le morphisme γ est génériquement fini sur son image, et que cette image contient des courbes lisses [1] . En fait, sur l'ouvert des classes de faisceaux localement libres, ce morphisme est à fibres finies [13] . Pour $n = 2$, c'est un isomorphisme; pour $n = 3$, il est surjectif et de degré 3. Pour $n \geq 4$, la dimension du système linéaire $|\mathcal{O}_{\mathbf{P}_2^*}(n)|$ devient supérieure à $4n - 3$.

Problème

(i) *Le morphisme* γ *est-il de degré* 1 *sur son image pour* $n \geq 4$?

(ii) *Quel est le degré de l'image de* γ?

Il est facile de voir que $\gamma^*(\mathcal{O}(1)) = \mathcal{D}$, où $\mathcal{D} = \lambda_{\mathrm{M}}(-h + h^2)$ est le fibré déterminant de Donaldson, associé à la classe de $-\mathcal{O}_{\ell}(-1) = -h + h^2$ dans $\mathrm{K}(\mathbf{P}_2)$. Il en résulte si la réponse à la question (i) est affirmative, le degré de l'image est donné par l'intégrale de Donaldson

$$q_{4n-3}(\mathbf{P}_2) = \int_{\mathrm{M}_n} (c_1(\mathcal{D}))^{4n-3}.$$

Nous n'avons pas de réponse générale à ces questions. Pour $n = 4$, l'image de M_4 est une hypersurface $\mathcal{L} \subset \mathbf{P}_{14}$ appelée hypersurface des quartiques de Lüroth : nous verrons dans la section 5 que dans l'ouvert des quartiques lisses, cette hypersurface correspond aux quartiques qui sont circonscrites à un pentagone, ce qui est la définition historique des quartiques de Lüroth.

Théorème 4.1. —

(i) *Le morphisme* $\gamma = \mathrm{M}_4 \to \mathrm{Im}\gamma$ *est de degré* 1.

(ii) *L'hypersurface des quartiques de Lüroth est de degré* 54.

L'assertion (ii) a aussi été obtenue par une voie différente des notres par A. Tyurin et A. Tikhomirov [26] . Récemment, un calcul plus direct de l'invariant de Donaldson $q_{13}(\mathbf{P}_2)$ a été mené à son terme par Wei-Ping Li et Zhenbo Qin [18] . Nous nous limiterons ici à des indications sur la démonstration de l'assertion (i) ; cette assertion nous semble la plus délicate et nécessite une description précise de l'espace de modules M_4. On peut construire [12] une transformation birationnelle $\mathrm{M}_4 - \to \mathbf{S}$ sur la variété $\mathbf{S}$ des systèmes linéaires de degré 5 et de dimension projective 1 sur les coniques lisses du plan projectif : en effet si F est un fibré stable générique de rang 2, de classes de Chern $(0,4)$, on a $h^0(\mathrm{F}(1)) = 2$; le déterminant du morphisme d'évaluation

$$ev : \mathrm{H}^0(\mathrm{F}(1)) \otimes \mathcal{O}_{\mathbf{P}_2} \to \mathrm{F}(1)$$

est non nul et définit une conique lisse C. Le conoyau $\mathrm{G} = \mathrm{coker}\, ev$ est alors un faisceau inversible de degré -1 sur C muni d'un sous-espace vectoriel $\Gamma \subset \mathrm{Ext}^1(\mathrm{G}, \mathcal{O}_{\mathbf{P}_2}) = \mathrm{H}^0(\underline{\mathrm{Ext}}^1(\mathrm{G}, \mathcal{O}_{\mathbf{P}_2}))$ de dimension 2. Le faisceau $\mathrm{G}^* = \mathrm{Ext}^1(\mathrm{G}, \mathcal{O}_{\mathbf{P}_2})$ est un fibré inversible de degré 5 sur C. Réciproquement, la donnée de la paire (Γ, G^*) permet de reconstruire F. Malheureusement, l'étude de la situation générique n'est pas suffisante pour obtenir le résultat attendu sur le morphisme γ.

Problème

Compactifier la variété des systèmes linéaires **S** *ci-dessus et étendre la correspondance birationnelle décrite ci-dessus.*

C'est l'étude de cet exemple qui nous a incité en 1990 à introduire la notion de système cohérent semi-stable dans un contexte très général ; il se trouve que cette notion est utile dans d'autres situations. Notamment, elle a déjà été utilisée sur les courbes sous le nom de paires semi-stables par d'autres auteurs : S. Bradlow, A. Bertram, O. Garcia-Prada, M. Thaddeus (*cf.* par exemple [24]), N. Raghavendra et P. A. Vishwinath [22], et sous le nom de paires de Brill-Noether par A. King et P. Newstead [11]. Toutefois, même sur les courbes, on peut donner une notion de semi-stabilité différente de celle qu'ont introduite ces auteurs, et qui s'avère elle aussi utile ; comme illustration, nous donnerons par exemple une variante du calcul de Drézet et Narasimhan du groupe de Picard des espaces de modules de fibrés vectoriels semi-stables de déterminant fixé sur les courbes.

4.2. L'espace de modules des systèmes cohérents

Soit X une variété algébrique projective et lisse de dimension n.

Définition 4.2. — *Un système cohérent* (Γ, F) *de dimension* d *sur* X *est la donnée d'une paire formée d'un faisceau algébrique cohérent* F *de dimension* d *sur* X, *et d'un sous-espace vectoriel de l'espace vectoriel* $\mathrm{H}^0(\mathrm{F})$ *des sections de* F.

Un morphisme $f : (\Gamma, \mathrm{F}) \to (\Gamma', \mathrm{F}')$ *est un morphisme de faisceaux cohérents* $\mathrm{F} \to \mathrm{F}'$ *tel que* $f(\Gamma) \subset \Gamma'$.

Un système cohérent (Γ, F) *est appelé structure de niveau si* F *est de dimension* n *et si* $\dim \Gamma = \mathrm{rg}(\mathrm{F})$.

On obtient ainsi une catégorie additive.

Définition 4.3. — *Un morphisme de systèmes cohérents* $f : (\Gamma, \mathrm{F}) \to (\Gamma', \mathrm{F}')$ *est appelé strict si* $f(\Gamma) = \Gamma' \cap \mathrm{H}^0(\mathrm{Im} f)$.

La notion d'image, de noyau et de conoyau d'un morphisme strict de systèmes cohérents est définie sans ambiguité.

On se fixe un polynôme α strictement positif à coefficients rationnels. Etant donné un système cohérent (Γ, F) on désigne par r la multiplicité de F; on appelle polynôme de Hilbert réduit de (Γ, F) le polynôme

$$p_{(\Gamma,\mathrm{F})} = \frac{\dim \Gamma}{r}\alpha + p_{\mathrm{F}}.$$

Définition 4.4. — *On dit qu'un système cohérent* (Γ, F) *de dimension* d *est* α*-semi-stable (resp.* $\alpha-$*stable) si*

(i) *le faisceau* F *est pur de dimension* d;

(ii) *pour tout sous-faisceau cohérent* $\mathrm{F}' \subset \mathrm{F}$, *de multiplicité* $0 < r' < r$ *on a, en posant* $\Gamma' = \Gamma \cap \mathrm{H}^0(\mathrm{F}')$

$$p_{(\Gamma',\mathrm{F}')} \leq p_{(\Gamma,\mathrm{F})} \quad (\text{resp. } <).$$

Dans la catégorie des systèmes cohérents $\alpha-$semi-stables de même polynôme de Hilbert réduit fixé, les morphismes sont stricts, et cette catégorie est une catégorie abélienne. En particulier, les notions de filtrations de Jordan-Hölder, de gradué de Jordan-Hölder, et de S$-$équivalence ont encore un sens. Un système cohérent est dit polystable s'il est somme directe de système cohérents stables de même polynôme de Hilbert réduit.

Exemples.

(i) Le système cohérent $(0, \mathrm{F})$ est $\alpha-$semi-stable si et seulement si le faisceau F est semi-stable.

(ii) Supposons que α soit un polynôme de degré $\geq d$. Un système cohérent de dimension d est semi-stable si et seulement si F est pur de dimension d et si pour tout sous-faisceau cohérent $\mathrm{F}' \subset \mathrm{F}$ de multiplicité $0 < r' < r$ on a

$$\frac{\dim \Gamma'}{r'} \leq \frac{\dim \Gamma}{r} \text{ et en cas d'égalité } p_{\mathrm{F}'} \leq p_{\mathrm{F}}.$$

C'est la définition que nous avions adoptée dans [14] , article dans lequel nous avions choisi pour polynôme α le polynôme de Hilbert p_{X} de $\mathcal{O}_{\mathrm{X}}$.

Lemme 4.5. — *Soit* (Γ, F) *un système cohérent* p_X*-semi-stable de dimension* d*. Alors si* $\Gamma \neq 0$, *le conoyau du morphisme d'évaluation*

$$ev : \Gamma \otimes \mathcal{O}_X \to F$$

est de dimension $< d$.

Cet énoncé résulte immédiatement de la définition, en l'appliquant au faisceau cohérent F' image du morphisme d'évaluation. Il en résulte que si (Γ, F) est un système cohérent p_X-semi-stable de dimension n, et si $\Gamma \neq 0$, le morphisme d'évaluation est génériquement surjectif. En particulier l'existence d'un tel système cohérent semi-stable impose l'inégalité

$$\dim \Gamma \geq \mathrm{rg}(F).$$

(iii) Soit (Γ, F) un système cohérent de dimension n ; on suppose que $\dim \Gamma = 1$ et $\mathrm{rg}(F) = 2$. Un tel système cohérent est α-semi-stable si pour tout sous-faisceau cohérent $F' \subset F$ de rang 1 on a

$$p_{F'} \leq p_F + \frac{\alpha}{2\deg(X)}$$

si $\Gamma \cap H^0(F') = \{0\}$, et

$$p_{F'} \leq p_F - \frac{\alpha}{2\deg(X)}.$$

si $\Gamma \subset H^0(F')$. Si on applique cette définition au sous-faisceau cohérent engendré par le morphisme d'évaluation, on voit qu'un tel système cohérent semi-stable ne peut exister que si

$$0 < \alpha \leq 2\deg(X)(p_F - p_X).$$

On a bien entendu un énoncé analogue pour la stabilité, en remplaçant les inégalités ci-dessus par des inégalités strictes.

Lemme 4.6. — *Soit* (Γ, F) *un système cohérent de dimension* n *; on suppose* $\dim \Gamma = 1$ *et* $\mathrm{rg}(F) = 2$*. Si* α *est suffisamment petit, les assertions suivantes sont équivalentes :*

(1) *le faisceau cohérent* F *est stable*

(2) *le système cohérent* (Γ, F) *est* α*-stable.*

Pour les grandes valeurs de α, on a aussi l'énoncé suivant :

Lemme 4.7. — *On se place sous les mêmes hypothèses, et on prend* $\alpha = 2\deg(\mathrm{X})(p_{\mathrm{F}} - p_{\mathrm{X}})$. *Alors le système cohérent* (Γ, F) *est* α*−semi-stable si et seulement si le conoyau* G *du morphisme d'évaluation*

$$ev : \Gamma \otimes \mathcal{O}_{\mathrm{X}} \to \mathrm{F}$$

est un faisceau cohérent de rang 1 sans torsion.

Lorsque X est une courbe, la notion de α−semi-stabilité coïncide avec celle des paires semi-stables introduites par des méthodes différentielles par S. Bradlow, A. Bertram, O. Garcia-Prada,... et utilisées dans un cadre algébrique par N. Raghavendra et P.A. Vishwanath [22] pour désingulariser certaines sous-variétés de l'espace de modules des fibrés semi-stables et par M. Thaddeus pour le calcul de la formule de Verlinde (*cf.* par exemple [24]).

Familles de systèmes cohérents

Soit S une variété algébrique. On considère les projections canoniques

$$\begin{array}{ccc} \mathrm{S} \times \mathrm{X} & \xrightarrow{pr_2} & \mathrm{X} \\ {\scriptstyle pr_1}\downarrow & & \\ \mathrm{S} & & \end{array}$$

On désigne par ω_{X} le faisceau canonique sur X, et par $\omega_{\mathrm{S}\times\mathrm{X}/\mathrm{S}} = pr_2^*(\omega_{\mathrm{X}})$ le faisceau canonique relatif pour la projection pr_1. Si $\mathcal{F}$ un faisceau algébrique cohérent S-plat sur S × X l'image directe $pr_{1*}(\mathcal{F})$ se comporte mal par changement de base, sauf si les images directes supérieures sont nulles. Ceci introduit quelques difficultés dans la définition des familles de systèmes cohérents. Pour cette raison, nous assimilerons un système cohérent sur X à la donnée d'une paire (Γ, F) formée d'un faisceau algébrique cohérent F sur X et d'un espace vectoriel quotient de l'espace vectoriel $\mathrm{Ext}^n(\mathrm{F}, \omega_{\mathrm{X}})$. En vertu du théorème de dualité de Serre-Grothendieck, les espaces vectoriels $\mathrm{Ext}^n(\mathrm{F}, \omega_{\mathrm{X}})$ et $\mathrm{H}^0(\mathrm{F})$ sont duaux, et cela revient donc au même. Ceci conduit à la définition suivante :

Définition 4.8. — *Soit* $c \in \mathrm{K}_{num}(\mathrm{X})$ *de dimension* d. *On appelle famille de systèmes cohérents sur* X*, de classe* c*, paramétrée par la variété* *algébrique* S *la donnée d'une paire* $(\mathscr{F}, \mathscr{V})$, *où* $\mathscr{F}$ *est une famille* S$-$*plate de faisceaux algébriques cohérents sur* X *de classe* c, *et* $\mathscr{V}$ *un faisceau localement libre sur* S *quotient du faisceau cohérent* $\underline{\mathrm{Ext}}^n_{pr_1}(\mathscr{F}, \omega_{\mathrm{S}\times\mathrm{X}/\mathrm{S}})$.

En associant à S l'ensemble des classes d'isomorphisme de familles de systèmes cohérents $\alpha-$semi-stables de classe c sur X on définit un foncteur contravariant $\mathrm{S} \mapsto \underline{\mathrm{Syst}}_{\mathrm{X},\alpha}(c)(\mathrm{S})$: ceci résulte du bon comportement du foncteur $\underline{\mathrm{Ext}}^n_{pr_1}(\mathrm{F}, \omega_{\mathrm{S}\times\mathrm{X}/\mathrm{S}})$ par changement de base. Examinons par exemple la fibre au-dessus d'un point fermé $s \in \mathrm{S}$. Notons $\mathscr{F}_s$ le faisceau induit sur la fibre au-dessus de s.

Lemme 4.9. — *Soit* $\mathscr{F}$ *un faisceau algébrique cohérent* S-*plat sur* S$\times$X. *Alors pour tout point fermé* $s \in \mathrm{S}$, *on a un isomorphisme canonique*

$$\underline{\mathrm{Ext}}^n_{pr_1}(\mathscr{F}, \omega_{\mathrm{S}\times\mathrm{X}/\mathrm{S}})_s \simeq \mathrm{Ext}^n(\mathscr{F}_s, \omega_{\mathrm{X}})$$

Ainsi, la donnée d'une famille de systèmes cohérents $(\mathscr{F}, \mathscr{V})$ paramétrée par S fournit en chaque point $s \in \mathrm{S}$ un espace vectoriel quotient $\mathscr{V}_s$ de $\mathrm{Ext}^n(\mathscr{F}_s, \omega_{\mathrm{X}})$ et donc un sous-espace vectoriel $\mathscr{V}_s^*$ de $\mathrm{H}^0(\mathscr{F}_s)$; on obtient alors un système cohérent $(\mathscr{V}_s^*, \mathscr{F}_s)$ de dimension d.

Théorème 4.10. — *Soit* P *un polynôme. La famille des systèmes cohérents* $\alpha-$*semi-stables* (Γ, F) *tels que* $\mathrm{P_F} = \mathrm{P}$ *est limitée.*

Théorème 4.11. — *Soit* $c \in \mathrm{K}_{num}(\mathrm{X})$ *une classe effective de dimension* d. *Il existe pour le foncteur* $\underline{\mathrm{Syst}}_{\mathrm{X},\alpha}(c)$ *un espace de modules grossier,* $\mathrm{Syst}_{\mathrm{X},\alpha}(c)$. *C'est une variété projective dont les points fermés correspondent aux classes de* S$-$*équivalence de systèmes cohérents semi-stables de classe* c.

Bien entendu, on a une décomposition en réunion disjointe d'ouverts

$$\mathrm{Syst}_{\mathrm{X},\alpha}(c) = \coprod_k \mathrm{Syst}_{\mathrm{X},\alpha}(c, k)$$

où $\mathrm{Syst}_{\mathrm{X},\alpha}(c, k)$ désigne la composante des systèmes cohérents (Γ, F) de classe c tels que $\dim \Gamma = k$. Dans le cas $\alpha = p_{\mathrm{X}}$, on notera plus simplement $\mathrm{Syst}_{\mathrm{X}}(c)$ l'espace de modules ci-dessus.

La démonstration de ces deux énoncés est donnée dans le cas où $\alpha = p_X$ (ou ce qui revient au même $\deg \alpha \geq d$) dans [14]. Dans le cas général énoncé ci-dessus, on a seulement besoin de quelques variantes de cette démonstration. Soit P le polynôme de Hilbert, défini $P(m) =< c, (1-h)^{-m} >$. On considère comme au §2.2 pour un entier m suffisamment grand, un espace vectoriel H de dimension $P(m)$, et le schéma de Hilbert-Grothendieck $R = \mathbf{Hilb}(B, c)$ des faisceaux cohérents de classe c quotients du fibré vectoriel $B = H \otimes \mathcal{O}_X(-m)$. Ce schéma paramètre un faisceau quotient universel $\mathcal{F}$, et on peut introduire la grassmannienne relative de Grothendieck

$$\mathfrak{G} = \mathrm{Grass}(\underline{\mathrm{Ext}}^n_{pr_1}(\mathcal{F}, \omega_{R\times X})) \to R$$

dont les points correspondent aux paires (s, V), où $s \in R$ et où V est un espace vectoriel quotient de $\mathrm{Ext}^n(\mathcal{F}_s, \omega_X)$. Sur $\mathfrak{G}$ le groupe GL(H) opère de manière naturelle. Soit $\mathfrak{G}^{ss}$ l'ouvert des points (s,V) satisfaisant aux conditions suivantes

— le morphisme canonique $H \to H^0(\mathcal{F}_s(m))$ est un isomorphisme;

— le système cohérent $(V^*, \mathcal{F}_s)$ est semi-stable.

Alors $\mathfrak{G}^{ss}$ est un ouvert GL(H)−invariant qui possède alors un bon quotient : c'est la variété de modules attendue. Soit A_m l'espace vectoriel des sections $H^0(\mathcal{O}_X(m))$. Comme nous l'avions pressenti dans l'introduction de [14] , cet énoncé résulte de l'identification de cet ouvert $\mathfrak{G}^{ss}$ avec un ouvert de la variété produit

$$\mathfrak{P} = \mathbf{Hilb}_X(c) \times \mathrm{Grass}(H^* \otimes A_m),$$

d'un choix convenable d'une polarisation sur ce produit, et du calcul des points semi-stables relatifs à cette polarisation pour l'action GL(H) de $\mathfrak{P}$. Le schéma de Hilbert $\mathbf{Hilb}(B, c)$ étant polarisé par le plongement de Grothendieck du schéma de Hilbert $\mathbf{Hilb}(B, c)$ dans la grassmannienne $\mathrm{Grass}(H \otimes A_{\ell-m}, P(\ell))$ des espaces vectoriels quotients de dimension $P(\ell)$, pour $\ell > m$ suffisamment grand, la polarisation qui convient ici sur la composante définie par $\dim \Gamma = k$ est un multiple entier > 0 de l'élément de $\mathrm{Pic}(\mathfrak{P}) \otimes \mathbf{Q}$ défini par

$$= \mathcal{O}(\dim H + \alpha(m)k, \alpha(m)P(\ell)).$$

Nous ne donnerons pas ici la démonstration détaillée, qui s'appuie comme dans [14] sur un critère de semi-stabilité des systèmes cohérents comparable au critère 4.13 de [14] .

Etude infinitésimale et extensions

Considérons deux systèmes cohérents (Γ', F') et (Γ, F) sur X. Pour toute résolution injective $\epsilon : F \to R^{\cdot}$ on considère le complexe $H^0(R^{\cdot})/\Gamma$ défini par $0 \to H^0(R^0)/\Gamma \to H^0(R^1) \to H^0(R^2) \to \ldots$. On a alors un morphisme naturel de complexes d'espaces vectoriels

$$\varphi : \mathrm{Hom}_{\mathcal{O}_X}(F', R^{\cdot}) \to L(\Gamma', H^0(R^{\cdot})/\Gamma)$$

où $L(\Gamma', H^0(R^{\cdot})/\Gamma)$ est le complexe des applications $\mathbb{C}$−linéaires ; en considérant le mapping cône de φ on obtient une suite exacte de complexes

$$0 \to SL(\Gamma', H^0(R^{\cdot})/\Gamma) \to M^{\cdot}(\varphi) \to \mathrm{Hom}_{\mathcal{O}_X}(F', R^{\cdot}) \to 0$$

où la lettre S indique ici la suspension, c'est-à-dire le complexe décalé d'un degré vers la droite. On pose

$$\mathrm{Ext}^q((\Gamma', F'), (\Gamma, F)) = H^q(M^{\cdot}(\varphi))$$

Visiblement cet espace vectoriel ne dépend pas du choix de la résolution $R^{\cdot}$, et on a une suite exacte longue

$$\begin{aligned} 0 \to \mathrm{Hom}((\Gamma', F'), (\Gamma, F)) \to \mathrm{Hom}(F', F) \to L(\Gamma', H^0(F)/\Gamma) \to \\ \mathrm{Ext}^1((\Gamma', F'), (\Gamma, F)) \to \mathrm{Ext}^1(F', F) \to L(\Gamma', H^1(F)) \to \ldots \end{aligned}$$

La proposition suivante étend aux variétés de dimension quelconque un énoncé de Thaddeus [24] :

Proposition 4.12. — *Soit* (Γ, F) *un système cohérent* α*−stable de classe* c*, et* p *le point qu'il définit dans l'espace de modules* $\mathrm{Syst}_{X,\alpha}(c)$.

(i) *Si* $\mathrm{Ext}^2((\Gamma, F), (\Gamma, F)) = 0$*, l'espace de modules* $\mathrm{Syst}_{X,\alpha}(c)$ *est lisse au voisinage du point* p*. Cette condition est satisfaite en particulier si* $\mathrm{Ext}^2(F, F) = 0$ *et si la flèche naturelle* $\mathrm{Ext}^1(F, F) \to L(\Gamma, H^1(F))$ *est surjective.*

(ii) *L'espace tangent de Zariski en* p *est isomorphe à* $\mathrm{Ext}^1((\Gamma, F), (\Gamma, F))$.

Ces espaces vectoriels interviennent aussi pour classer les extensions strictes. On a par exemple l'énoncé suivant, qui précise le lemme 4.21 de [14] :

Proposition 4.13. — *L'ensemble des classes d'isomorphisme d'extensions strictes*

$$0 \to (\Gamma', \mathrm{F}') \to (\Gamma, \mathrm{F}) \to (\Gamma'', \mathrm{F}'') \to 0$$

est isomorphe à l'espace vectoriel $\mathrm{Ext}^1((\Gamma'', \mathrm{F}''), (\Gamma', \mathrm{F}'))$ *et cet isomorphisme est compatible avec l'action naturelle du groupe* $\mathrm{Aut}(\Gamma', \mathrm{F}') \times \mathrm{Aut}(\Gamma'', \mathrm{F}'')$.

4.3. La notion duale

Définition 4.14. — *Un cosystème cohérent de codimension d sur* X *est la donnée d'une paire* (Γ, F) *où* F *est un faisceau algébrique cohérent de dimension d, et* $\Gamma \subset \mathrm{Ext}^{n-d}(\mathrm{F}, \mathcal{O}_\mathrm{X})$ *un sous-espace vectoriel.*

Etant donné un tel cosystème cohérent, on introduit encore la notion de polynôme de Hilbert réduit pour un cosystème cohérent (Γ, F) en posant

$$p_{(\Gamma,\mathrm{F})} = \frac{\dim \Gamma}{r} p_\mathrm{X} - p_\mathrm{F}.$$

où r est la multiplicité de F.

Définition 4.15. — *Un cosystème cohérent de dimension d est dit semi-stable si*

— *le faisceau* F *est pur de dimension d*

— *pour tout faisceau quotient* $\mathrm{F} \to \mathrm{F}'$ *pur de dimension d on a , en posant* $\Gamma' = \Gamma \cap \mathrm{Ext}^1(\mathrm{F}', \mathcal{O}_\mathrm{X})$

$$p_{(\Gamma',\mathrm{F}')} \leq p_{(\Gamma,\mathrm{F})}$$

On pourrait naturellement songer à remplacer p_X par un polynôme positif quelconque α dans la définition ci-dessus ; cependant la notion que nous venons de décrire sera suffisante pour les applications que nous avons en vue. Ici encore la catégorie des cosystèmes cohérents semi-stables de polynôme de Hilbert réduit fixé est abélienne, et on peut définir la notion de filtration de Jordan-Hölder, et de S−équivalence.

Familles de cosystèmes cohérents

Soit S une variété algébrique. On considère à nouveau le diagramme

$$\begin{array}{ccc} S \times X & \xrightarrow{pr_2} & X \\ {\scriptstyle pr_1}\downarrow & & \\ S & & \end{array}$$

Définition 4.16. — *Soit* $c \in K_{num}(X)$ *une classe effective de dimension* d. *Une famille (plate) de cosystèmes cohérents de classe* c *paramétrée par* S *est la donnée d'une famille plate* $\mathscr{F}$ *de faisceaux algébriques cohérents de classe* c*, paramétrée par* S *et d'un fibré vectoriel quotient* $R^d pr_{1*}(\mathscr{F} \otimes \omega_{S\times X/S}) \to \mathscr{V}$.

Au-dessus de chaque point $s \in S$, une telle famille définit par dualité un sous-espace $\mathscr{V}_s^* \subset \mathrm{Ext}^{n-d}(\mathscr{F}_s, \mathscr{O}_X)$ c'est-à-dire un cosystème cohérent. Si on associe à la variété S l'ensemble des classes d'isomorphisme de cosystèmes cohérents semi-stables de classe c, on obtient encore un foncteur contravariant noté $\underline{\mathrm{Cosyst}}_X(c)$.

Théorème 4.17. — *Soit* P *un polynôme. La famille des cosystèmes cohérents* (Γ, F) *tels que* $P_F = P$ *est limitée.*

Théorème 4.18. — *Soit* $c \in K_{num}(X)$ *une classe effective de dimension* d. *On se place sous l'une des hypothèse suivantes :*

— $d = 0, n-1$ *ou* n

— $d = 1$ *et* $\mathscr{O}_X(1)$ *et* ω_X *sont numériquement liés.*

Alors il existe pour le foncteur $\underline{\mathrm{Cosyst}}_X(c)$ *un espace de modules grossier, noté* $\mathrm{Cosyst}_X(c)$, *dont les points sont les classes de* S*−équivalence de cosystèmes cohérents semi-stables de classe* c.

Naturellement, on a encore une décompostion en réunion disjointe d'ouverts

$$\mathrm{Cosyst}_X(c) = \coprod_k \mathrm{Cosyst}_X(c, k)$$

où $\mathrm{Cosyst}_X(c, k)$ désigne l'ouvert correspondant aux cosystèmes cohérents (Γ, F) tels $\dim \Gamma = k$.

Ces deux énoncés sont démontrés dans [14] . Le théorème 4.18 devrait être vrai pour tout d ; mais faute de motivation nous ne l'avons vérifié que sous les hypothèses énoncées, en nous ramenant, au moins dans les cas les plus importants, au cas des systèmes cohérents.

Relations avec les systèmes cohérents

(i) Pour $d = 0$, ou $d = 1$, si (Γ, F) est un cosystème cohérent semi-stable, le faisceau F est de Cohen-Macaulay. La classe dans $\mathrm{K}_{num}(\mathrm{X})$

$$c(\underline{\mathrm{Ext}}^{n-d}(\mathrm{F}, \mathcal{O}_\mathrm{X})) = (-1)^{n-d} c^*$$

est alors déterminée par la classe c de F, et

$$\mathrm{Ext}^{n-d}(\mathrm{F}, \mathcal{O}_\mathrm{X}) \simeq \mathrm{H}^0(\underline{\mathrm{Ext}}^{n-d}(\mathrm{F}, \mathcal{O}_\mathrm{X})).$$

Ainsi, un tel cosystème cohérent de classe c définit un système cohérent de classe $(-1)^{n-d}c^*$, et réciproquement ; si $d = 0$ ou si $d = 1$ et $\mathcal{O}_\mathrm{X}(1)$ et ω_X numériquement liés, on constate sans difficultés que les notions de semi-stabilité coïncident (en prenant $\alpha = p_\mathrm{X}$). Ceci passe aux familles. On obtient ainsi

Théorème 4.19. — *Soit $c \in \mathrm{K}_{num}(\mathrm{X})$, satisfaisant à l'une des conditions suivantes :*

— *c est de dimension $d = 0$;*

— *c est de dimension $d = 1$ et $\mathcal{O}_\mathrm{X}(1)$ et ω_X sont numériquement liés.*

Alors on a un isomorphisme canonique

$$\mathrm{Cosyst}_\mathrm{X}(c) \simeq \mathrm{Syst}_\mathrm{X}((-1)^{n-d}c^*)$$

Cet énoncé ne s'étend pas au cas $d > 1$, car F n'est pas obligatoirement de Cohen-Macaulay, et la classe $\underline{\mathrm{Ext}}^i(\mathrm{F}, \mathcal{O}_\mathrm{X})$ peut être non nulle pour $i \neq\neq n - d$.

(ii) On considère une structure de niveau (Γ, F) semi-stable relativement à $\alpha = p_\mathrm{X}$. Ainsi, F est un faisceau sans torsion, dont on désigne par r le rang : on a donc $\dim \Gamma = r$; on suppose que la classe c de F est $\neq r$. Considérons le morphisme d'évaluation $ev : \Gamma \otimes \mathcal{O}_\mathrm{X} \to \mathrm{F}$ alors :

— le morphisme ev est génériquement injectif ;

— soit Θ le conoyau de ev. Alors si Θ n'est pas nul (ce qui signifie que $c \neq r$) c'est un faisceau pur de dimension $n - 1$, de classe $c - r$ dans

$K_{num}(X)$. De plus, la propriété de semi-stabilité implique $\mathrm{Hom}(F, \mathcal{O}_X) = 0$, et par application du foncteur $\mathrm{Hom}(-, \mathcal{O}_X)$ à la suite exacte

$$0 \to \Gamma \otimes \mathcal{O}_X \to F \to \Theta \to 0$$

on obtient une application linéaire injective : $\Gamma^* \to \mathrm{Ext}^1(\Theta, \mathcal{O}_X)$. Ainsi, l'image V de Γ^* définit un cosystème cohérent de dimension $n-1$, dont on vérifie sans difficulté qu'il est semi-stable.

Réciproquement, la donnée d'un tel cosystème cohérent semi-stable (V, Θ), de classe $u \in K_{num}(X)$ définit un élément canonique $\omega \in \mathrm{Ext}^1(\Theta, V^* \otimes \mathcal{O}_X)$ ce qui fournit une extension

$$0 \to V^* \otimes \mathcal{O}_X \to F \to \Theta \to 0.$$

L'image de V^* dans $H^0(F)$ définit un système cohérent (Γ, F) de classe $c \neq r$. On vérifie encore que le système cohérent (Γ, F) est encore p_X-semi-stable. La correspondance obtenue s'étend aux familles, et fournit en fait sur les espaces de modules un isomorphisme

Théorème 4.20. — *Soit $c \in K_{num}(X)$ de dimension n et rang r. La correspondance décrite ci-dessus induit un isomorphisme*

$$\mathrm{Syst}_X(c, r) \simeq \mathrm{Cosyst}_X(c - r, r).$$

En conjuguant les deux énoncés ci-dessus, on obtient dans le cas des courbes et des surfaces :

Corollaire 4.21. — *Soit X une courbe ; si $c \in K_{num}(X)$ de dimension 1 et de rang r on a un isomorphisme canonique*

$$\mathrm{Syst}_X(c, r) \simeq \mathrm{Syst}_X(c - r, r)$$

Corollaire 4.22. — *Soit X une surface, telle que $\mathcal{O}_X(1)$ et ω_X soient numériquement liés. Soit $c \in K_{num}(X)$ de dimension 2 et de rang r. Alors*

$$\mathrm{Syst}_X(c, r) \simeq \mathrm{Syst}_X(r - c^*, r)$$

4.4. Construction de fibrés inversibles sur $\mathrm{Syst}_{X,\alpha}(c)$

On se fixe un polynôme α positif à coefficients rationnels. On se propose d'étendre la construction du §2.3 aux espaces de modules de systèmes cohérents α–semi-stables. On pose $\tilde{K}(X) = K(X) \oplus \mathbb{Z}$. Ce groupe sera muni de la forme quadratique $\tilde{q}$ définie par $\tilde{q}(u,m) = \chi(u^2) - m^2$. Pour cette forme quadratique, l'orthogonal de (u,m) ne dépend que de la classe de u dans $K_{num}(X)$ et de m.

Soit S une variété algébrique, munie d'une action d'un groupe algébrique G. Considérons une famille G–équivariante $(\mathscr{F}, \mathscr{V})$ de systèmes cohérents de classe $c \in K_{num}(X)$, telle que $\mathrm{rg}(\mathscr{V}) = k$: ceci signifie que la famille de faisceaux $\mathscr{F}$ est munie d'une action de G au-dessus de l'action donnée sur S, ainsi que le faisceau localement libre $\mathscr{V}$ sur S et que le morphisme canonique

$$\underline{\mathrm{Ext}}^n_{pr_1}(\mathscr{F}, \omega_{S\times X/S}) \to \mathscr{V}$$

est équivariant. On définit alors un homomorphisme de groupes abéliens

$$\lambda_{\mathscr{F},\mathscr{V}} : \tilde{K}(X) \to \mathrm{Pic}^G(S)$$

en posant $\lambda_{\mathscr{F},\mathscr{V}}(u,m) = \lambda_{\mathscr{F}}(u) \otimes \det \mathscr{V}^{\otimes m}$ pour $(u,m) \in \tilde{K}(X)$. Ce morphisme possède des propriétés fonctorielles analogues à celles de $\lambda_{\mathscr{F}}$:

Lemme 4.23. —

(i) *La formation du faisceau inversible $\lambda_{\mathscr{F},\mathscr{V}}$ est compatible aux changements de base G-équivariants.*

(ii) *Soit A un G–faisceau inversible sur S. Soit $(\mathscr{F}, \mathscr{V})$ une famille G–équivariante de systèmes cohérents de classe $c \in K_{num}(X)$, telle que $\mathrm{rg}(\mathscr{V}) = k$. Alors dans $\mathrm{Pic}^G(S)$ on a*

$$\lambda_{\mathscr{F}\otimes pr_1^*(A),\mathscr{V}\otimes A^*}(u,m) = \lambda_{\mathscr{F},\mathscr{V}}(u,m) \otimes A^{\otimes <c,u> - km}$$

(iii) *Etant donnée une suite exacte stricte de familles G–équivariantes de systèmes cohérents paramétrées par S*

$$0 \to (\mathscr{F}', \mathscr{V}') \to (\mathscr{F}, \mathscr{V}) \to (\mathscr{F}'', \mathscr{V}'') \to 0$$

on a $\lambda_{\mathscr{F},\mathscr{V}}(u,m) = \lambda_{\mathscr{F}',\mathscr{V}'}(u,m) \otimes \lambda_{\mathscr{F}'',\mathscr{V}''}(u,m)$ dans $\mathrm{Pic}^G(X)$.

On peut maintenant appliquer cette construction à la composante $\mathfrak{G}^{ss}(k)$ de l'ouvert $\mathfrak{G}^{ss}$ (*cf.* §4.2) des paires (s, V) telles que $\dim \mathrm{V} = k$. Ce schéma est muni d'une famille universelle de systèmes cohérents α−semi-stables $(\mathscr{F}, \mathscr{V})$, équivariante pour l'action de $\mathrm{G} = \mathrm{GL}(\mathrm{H})$, ce qui fournit un homomorphisme

$$\lambda_{\mathscr{F},\mathscr{V}} : \tilde{\mathrm{K}}(\mathrm{X}) \to \mathrm{Pic}^{\mathrm{G}}(\mathfrak{G}^{ss}(k)).$$

On désigne par $\tilde{\mathbf{H}}$ le sous-espace vectoriel de $\tilde{\mathrm{K}}(\mathrm{X}) \otimes \mathbf{Q}$ engendré les éléments

$$\tilde{e}_i = ((1-h)^{-i}, -\alpha(i))$$

et par $(c,k)^\perp$ l'orthogonal de (c,k) pour la forme quadratique $\tilde{q}$. Par application du lemme de descente de Kempf, Drézet et Narasimhan, on obtient comme au §2.3 :

Théorème 4.24. — *Soit $c \in \mathrm{K}_{num}(\mathrm{X})$ et k un entier ≥ 0.*

(i) *L'homomorphisme $\lambda_{\mathscr{F},\mathscr{V}} : \tilde{\mathbf{H}}^{\perp\perp} \cap (c,k)^\perp \to \mathrm{Pic}^{\mathrm{G}}(\mathfrak{G}^{ss}(k))$ se factorise de manière unique suivant le diagramme*

$$\begin{array}{ccc} \tilde{\mathbf{H}}^{\perp\perp} \cap (c,k)^\perp & \longrightarrow & \mathrm{Pic}^{\mathrm{G}}(\mathfrak{G}^{ss}(k)) \\ & {}_{\lambda_{\mathrm{S}}}\searrow & \uparrow \\ & & \mathrm{Pic}(\mathrm{Syst}_{\mathrm{X},\alpha}(c,k)) \end{array}$$

(ii) *Soit $(u,m) \in \tilde{\mathbf{H}}^{\perp\perp} \cap (c,k)^\perp$; le fibré inversible ainsi défini $\lambda_{\mathrm{S}}(u,m) \in \mathrm{Pic}(\mathrm{Syst}_{\mathrm{X},\alpha}(c,k))$ est caractérisé par la propriété universelle suivante : pour toute famille $(\mathscr{F}', \mathscr{V}')$ de systèmes cohérents semi-stables de classe c telle que $\mathrm{rg}\,\mathscr{V}' = k$, paramétrée par une variété algébrique S, on a dans $\mathrm{Pic}(\mathrm{S})$*

$$\lambda_{(\mathscr{F}',\mathscr{V}')}(u,m) = f^*_{(\mathscr{F}',\mathscr{V}')}(\lambda_{\mathrm{S}}(u,m)),$$

où $f_{(\mathscr{F}',\mathscr{V}')} : \mathrm{S} \to \mathrm{Syst}_{\mathrm{X},\alpha}(c,k))$ désigne le morphisme modulaire associé à cette famille.

Soient $c \in \mathrm{K}_{num}(\mathrm{X})$ une classe effective de dimension d, de multiplicité r, et k un entier ≥ 0. L'examen des systèmes cohérents α−polystables montre en fait que le fibré $\lambda_{\mathrm{S}}(u,m)$ a un sens dans $\mathrm{Pic}(\mathrm{Syst}_{\mathrm{X},\alpha}(c,k))$ dès que la propriété suivante est satisfaite : (u,m) est orthogonal à tous les couples $(c',k') \in \tilde{\mathrm{K}}(\mathrm{X})$ satisfaisant aux conditions suivantes :

(1) la classe c' est effective de dimension d, de multiplicité $r' = < c', h^d >$;

(2) l'élément de $\tilde{\mathrm{K}}(\mathrm{X}) \otimes \mathbf{Q}$ défini par $\frac{1}{r}(c,k) - \frac{1}{r'}(c',k')$ est orthogonal à $\tilde{\mathbf{H}}$.

Dans le cas où α est de degré $\geq d$, cette condition est équivalente à

$$\frac{c}{r} - \frac{c'}{r'} \in \mathbf{H}^{\perp} \quad \text{et} \quad \frac{k}{r} = \frac{k'}{r'}.$$

Ainsi, dans ce cas particulier $\lambda_{\mathrm{S}}(u,m)$ a un sens dès que $u \in \mathbf{H}^{\perp\perp}$ et $(u,m) \in (c,k)^{\perp}$. En particulier, si X est une courbe ou le plan projectif, le fibré inversible $\lambda_{\mathrm{S}}(u,m)$ a un sens dans $\mathrm{Pic}(\mathrm{Syst}_{\mathrm{X}}(c,k))$ dès que $(u,m) \in (c,k)^{\perp}$.

5. Exemples et applications

Nous nous limitons ici à deux applications de la théorie des systèmes cohérents. La première application est une variante du calcul de Drézet et Narasimhan du groupe de Picard des espaces de modules de fibrés vectoriels de déterminant fixé sur une courbe. La seconde permet de comprendre la structure de l'espace de modules des faisceaux semi-stables M_4 sur le plan projectif, introduit au §4.1, et d'aborder l'étude du morphisme γ qui associe à la classe d'un faisceau F la quartique des droites de saut de F. Cette notion de système cohérent est aussi utile sur des variétés de dimension supérieure; par exemple, sur l'espace projectif $\mathbf{P}_3$, elle permet d'étudier les composantes irréductibles de l'espace de modules $\mathrm{M}_{\mathbf{P}_3}(2;0,2,0)$ des classes de faisceaux semi-stables de rang 2 et classes de Chern $(0,2,0)$ (*cf.* [14]) .

5.1. Systèmes cohérents sur une courbe.

Soit X une courbe projective lisse et irréductible de genre $g \geq 2$. On se fixe une classe $c \in \mathrm{K}(\mathrm{X})$ de dimension 0, et on considère l'espace de modules $\mathrm{Syst}_{\mathrm{X}}(c,r)$ des systèmes cohérents (Γ, F) de classe c et tels que $\dim \Gamma = r$. Si $\bar{c}$ est l'image de c dans $\mathrm{K}_{num}(\mathrm{X}) = \mathbf{Z}^2$, l'espace de modules $\mathrm{Syst}_{\mathrm{X}}(c,r)$ est la fibre au-dessus du point $\mathrm{L} = \det c$ du morphisme déterminant $\mathrm{Syst}_{\mathrm{X}}(\bar{c},r) \to \mathrm{Pic}(\mathrm{X})$. Comme on l'a vu dans la section précédente, le morphisme

$$\lambda_{\mathrm{S}} : (c,r)^{\perp} \to \mathrm{Pic}(\mathrm{Syst}_{\mathrm{X}}(c,r))$$

a un sens ; si $\ell = \deg c$, l'orthogonal de (c,r) est donné par le groupe des classes $(u,m) \in \tilde{\mathrm{K}}(\mathrm{X})$ telles que

$$\mathrm{rg}(u)\ell - mr = 0$$

On posera $\delta = \mathrm{pgcd}(\ell, r)$.

Si $\mathrm{Z}^1(\mathrm{X}) \subset \mathrm{K}(\mathrm{X})$ désigne le groupe des classes 0−cycles de degré 0, cet homomorphisme s'annule sur $\mathrm{Z}^1(\mathrm{X}) \subset (c,r)^{\perp}$.

Théorème 5.1. — *Soit r un entier > 1. Soit $c \in \mathrm{K}(\mathrm{X})$ de dimension* 0 *de degré $\ell > 2rg$.*

(i) *L'espace de modules $\mathrm{Syst}_{\mathrm{X}}(c,r)$ est une variété non vide, irréductible et normale de dimension $r(\ell - r) + 1 - g$, lisse sur l'ouvert des points stables.*

(ii) *Le morphisme ci-dessus*

$$\lambda_{\mathrm{S}} : (c,r)^{\perp} \to \mathrm{Pic}(\mathrm{Syst}_{\mathrm{X}}(c,r))$$

est un épimorphisme de noyau $\mathrm{Z}^1(\mathrm{X})$.

(iii) *La variété $\mathrm{Syst}_{\mathrm{X}}(c,r)$ est localement factorielle.*

On pose $\mathrm{L} = \det c$, et on suppose désormais $r > 1$ et $\ell > 2rg$. Pour voir que cet espace de modules n'est pas vide, on considère les systèmes cohérents de la forme $(\Gamma, \mathcal{O}_{\mathrm{D}})$, où D est un diviseur obtenu comme schéma des zéros d'une section de L. Autrement dit, soit $\mathrm{P} = \mathbf{P}(\mathrm{H}^0(\mathrm{L}))$ le système linéaire des sections de L, et $\Sigma \subset \mathrm{P} \times \mathrm{X}$ le diviseur universel. On considère l'image directe $\mathscr{A} = pr_{1*}(\mathcal{O}_{\Sigma})$; c'est un faisceau localement libre de rang ℓ, qu'on identifie au faisceau des sections d'un fibré vectoriel A sur P; le schéma A est en fait muni d'une structure de P−algèbre. On considère la grassmannienne relative des sous-espaces de dimension r :

$$\mathrm{Grass}(r, \mathrm{A}) \to \mathrm{P}$$

Lemme 5.2. — *Dans $\mathrm{Grass}(r,\mathrm{A})$ le complémentaire de l'ouvert des systèmes cohérents stables est de codimension ≥ 2.*

Cet énoncé ne serait pas vrai pour $r = 1$.

Sur $\mathrm{Grass}(r,\mathrm{A})$, le P−schéma en groupes des éléments inversibles A^* de A opère de manière naturelle. Soit $\mathrm{Grass}^s(r,\mathrm{A})$ l'ouvert des systèmes cohérents stables. On a alors un morphisme A^*-équivariant

$$\pi : \mathrm{Grass}^s(r, \mathrm{A}) \to \mathrm{Syst}_{\mathrm{X}}(c, r)$$

dont l'image est un ouvert U non vide. L'image U est en fait partout dense, comme le prouve l'énoncé suivant :

Lemme 5.3. — *Dans* $\mathrm{Syst}_{\mathrm{X}}(c,r)$ *le complémentaire de l'ouvert* U *des classes de systèmes cohérents stables de la forme* $(\Gamma, \mathcal{O}_{\mathrm{D}})$, *avec* $\dim \Gamma = r$ *et* $c(\mathcal{O}_{\mathrm{D}}) = c$ *est de codimension* ≥ 2.

Ceci entraîne l'irréductibilité; la normalité résulte du fait que la composante $\mathfrak{G}^{ss}(r)$ de l'ouvert $\mathfrak{G}^{ss}$ utilisé dans la construction de $\mathrm{Syst}_{\mathrm{X}}(c,r)$ est une variété lisse. La lissité de l'ouvert des points stables résulte du fait que sur l'ouvert $\mathfrak{G}^s$ l'action du groupe $\mathrm{GL(H)}/\mathbb{C}^*$ est libre.

Pour terminer la démonstration il faut encore étendre à cette situation relative ce qu'on a énoncé au §3.3.

Le groupe $\mathrm{Pic}^{\mathrm{A}^*/\mathbb{C}^*}(\mathrm{Grass}(r, \mathrm{A}))$

Soit S une variété intègre, et $\pi : \mathrm{Y} \to \mathrm{S}$ une S−variété intègre. Soit $\mathrm{G} \to \mathrm{S}$ un schéma en groupes algébriques opérant sur Y, c'est-à-dire que l'on a un S−morphisme

$$\mathrm{G} \times_{\mathrm{S}} \mathrm{Y} \to \mathrm{Y}$$

induisant sur chaque fibre Y_s une action du groupe G_s. Dans cette situation relative, les notions de bon quotient, de G−fibrés vectoriels, de fibrés vectoriels G−invariants ont un sens.

Un morphisme croisé $\varphi : \mathrm{G} \times_{\mathrm{S}} \mathrm{Y} \to \mathbb{C}^*$ est un morphisme induisant au-dessus de chaque point de S un morphisme croisé. Un G−morphisme croisé principal est un morphisme croisé de la forme

$$\varphi(g,x) = \frac{\psi(gx)}{\psi(x)}$$

pour $(g,x) \in \mathrm{G} \times_{\mathrm{S}} \mathrm{Y}$, où $\psi : \mathrm{Y} \to \mathbb{C}^*$ est un morphisme. On pose

$$\mathrm{H}^1(\mathrm{G}, \pi_*(\mathcal{O}_{\mathrm{Y}}^*)) = \frac{\{\text{Morphismes croisés } \mathrm{G} \times_{\mathrm{S}} \mathrm{Y} \to \mathbb{C}^*\}}{\{\text{Morphismes croisés principaux}\}}$$

Dans cette situation, on a encore une suite exacte

$$0 \to \mathrm{H}^1(\mathrm{G}, \pi_*(\mathcal{O}_{\mathrm{Y}}^*)) \to \mathrm{Pic}^{\mathrm{G}}(\mathrm{Y}) \to \mathrm{Pic}(\mathrm{Y})^{\mathrm{G}} \qquad (*)$$

Définition 5.4. — *Un caractère de* G *est un morphisme* $\mathrm{G} \to \mathbf{C}^*$ *induisant sur chaque fibre* G_s *un caractère.*

Les caractères de G constituent un sous-groupe Char(G) du groupe des morphismes croisés $\mathrm{G} \times_{\mathrm{S}} \mathrm{Y} \to \mathbf{C}^*$.

On applique ceci avec $\mathrm{G} = \mathrm{A}^*$.

Les éléments non inversibles de A constituent une hypersurface irréductible de A, plate au-dessus de P : en effet le noyau du morphisme d'évaluation $\mathrm{A} \otimes \mathcal{O}_\Sigma \to \mathcal{O}_\Sigma$ définit un sous-fibré vectoriel K de rang $\ell - 1$ de $\mathrm{A} \times_{\mathrm{P}} \Sigma$ et l'image de K par le morphisme fini $\mathrm{A} \times_{\mathrm{P}} \Sigma \to \mathrm{A}$ est exactement le fermé des éléments non inversibles. Comme le fibré L est engendré par ses sections, Σ est irréductible (et lisse), et donc les éléments non inversibles constituent une hypersurface Ξ irréductible de A. Le fibré inversible $\mathcal{L}$ sur A associé à cette hypersurface provient d'un fibré inversible sur P et est muni d'une section f qui ne s'annule pas sur A^*, bien définie à une constante multiplicative près. Si $s \mapsto 1_s$ désigne la section unité de A sur P, la section $s \to f(1_s)$ définit une section partout non nulle de $\mathcal{L}$, qui est donc un fibré trivial. Ainsi, Ξ est défini par une équation scalaire $f = 0$, et si on impose $f(1_s) = 1$, alors f est un caractère de A^*.

Lemme 5.5. — *Soit* $\mathrm{Y} \to \mathrm{P}$ *une* P*−variété munie d'une action de* A^* *; on suppose que par tout point de* Y *passe une section locale. Le groupe des morphismes croisés* $\varphi : \mathrm{A}^* \times_{\mathrm{P}} \mathrm{Y} \to \mathbf{C}^*$ *est engendré par le caractère* f.

En effet, soit $\varphi : \mathrm{A}^* \times_{\mathrm{P}} \mathrm{Y} \to \mathbf{C}^*$ un morphisme croisé. Soit $p : \mathrm{A}^* \to \mathrm{P}$ la projection canonique ; si σ est une section locale de Y au-dessus d'un ouvert V de P, la fonction $g \to \varphi(g, \sigma p(g))$ définit une fonction régulière inversible sur $\mathrm{W} = p^{-1}(\mathrm{V})$. Il existe donc un entier $k \in \mathbf{Z}$ et une fonction u inversible sur V tels que

$$\varphi(g, \sigma p(g)) = u(p(g)) f^k(g)$$

pour tout $g \in \mathrm{W}$. De la définition des morphismes croisés, il résulte que la fonction u doit être constante et égale à 1. Par suite, $\varphi(g, y) = f^k(g)$ pour tout $(g, y) \in \mathrm{A}^* \times_{\mathrm{P}} \mathrm{Y}$. Ceci entraîne le résultat. □

En particulier, le groupe des caractères Char(A^*) est un groupe cyclique, engendré par f. Sur un ouvert de P au-dessus duquel la fibre Σ_s est lisse, la fibre A^*_s est isomorphe à $(\mathbf{C}^*)^\ell$, et le caractère f est de la forme

$$(\xi_1, \ldots, \xi_\ell) \mapsto \xi_1 \ldots \xi_\ell.$$

Il en résulte l'énoncé suivant :

Lemme 5.6. — *L'inclusion* $P \times \mathbf{C}^* \hookrightarrow A^*$ *induit un homomorphisme injectif*

$$\mathrm{Char}(A^*) \hookrightarrow \mathrm{Char}(P \times \mathbf{C}^*) = \mathbf{Z}$$

On applique ce qui précède à $Y = \mathrm{Grass}(r, A)$, sur lequel le groupe relatif A^* opère. L'action de A^* sur Grass(r, A) se factorise à travers $\overline{G} = A^*/\mathbf{C}^*$. Sur Grass($r$, A) $\times$ X on dispose d'un système cohérent universel $(\mathcal{O}_\Sigma, \mathcal{V})$, équivariant pour l'action de A^*. Le caractère induit sur $\mathbf{C}^*$ par le fibré $\lambda_{\mathcal{O}_\Sigma, \mathcal{V}}(u, m)$ est donné par $v \mapsto v^{<c,u>-rm}$. D'après la suite exacte (*) et le lemme 5.6 le groupe $\mathrm{Pic}^{A^*/\mathbf{C}^*}(\mathrm{Grass}(r, A))$ est alors un groupe abélien libre de rang 2, d'indice ℓ/δ dans le groupe de Picard de Grass(r, A), et le morphisme canonique $\lambda_{\mathcal{O}_\Sigma, \mathcal{V}}$ induit un morphisme

$$\lambda_{\mathcal{O}_\Sigma, \mathcal{V}} : (c, r)^\perp \to \mathrm{Pic}^{A^*/\mathbf{C}^*}(\mathrm{Grass}(r, A))$$

qui est surjectif, et de noyau $Z^1(X)$; ceci résulte en effet de la formule $\lambda_{\mathcal{F}}(u) = \mathcal{O}_P(< L^*, u >)$, qu'on obtient en écrivant sur $P \times X$ la suite exacte

$$0 \to \mathcal{O}(-1) \boxtimes L^* \to \mathcal{O}_{P \times X} \to \mathcal{O}_\Sigma \to 0.$$

Fin de la démonstration du théorème 5.1

Démontrons l'assertion (ii) : elle est semblable à celle du §3.6.

L'ouvert U est un quotient géométrique de Y par $\overline{G}$. Compte-tenu de la propriété universelle de λ_S, le diagramme

$$\begin{array}{ccc} \mathrm{Grass}^s(r, A) & \hookrightarrow & \mathrm{Grass}(r, A) \\ \downarrow & & \\ U & \hookrightarrow & \mathrm{Syst}_X(c, r) \end{array}$$

induit un diagramme commutatif

$$\begin{array}{ccc} (c,r)^{\perp} & \xrightarrow{\lambda_{\mathcal{O}_\Sigma,\mathcal{V}}} & \mathrm{Pic}^{\overline{\mathrm{G}}}(\mathrm{Grass}(r,\mathrm{A})) \\ \lambda_{\mathrm{S}}\downarrow & & \downarrow (1) \\ \mathrm{Pic}(\mathrm{Syst}(c,r)) & \overset{(3)}{\longrightarrow} & \mathrm{Pic}(\mathrm{U}) \overset{(2)}{\longrightarrow} \mathrm{Pic}^{\overline{\mathrm{G}}}(\mathrm{Grass}^{s}(r,\mathrm{A})) \end{array}$$

Il résulte des lemmes 5.5 et 5.6, et de la suite exacte (*) que le groupe $\mathrm{Pic}^{\overline{\mathrm{G}}}(\mathrm{Grass}^s(r,\mathrm{A}))$ s'identifie à un sous-groupe de $\mathrm{Pic}(\mathrm{Grass}^s(r,\mathrm{A}))$; il résulte alors du lemme 5.2 que la flèche de restriction (1) est un isomorphisme. L'homomorphisme (2) est injectif parce que U est un bon quotient de $\mathrm{Grass}^s(r,\mathrm{A})$ et la restriction (3) est injective en raison du lemme 5.3, compte-tenu du fait que $\mathrm{Syst}(c,r)$ est une variété normale. Donc λ_{S} est un morphisme surjectif de noyau $\mathrm{Z}^1(\mathrm{X})$.

La factorialité locale se démontre par des arguments voisins, en se plaçant au-dessus d'un ouvert de $\mathrm{Syst}_{\mathrm{X}}(c,r)$. □

Structures de niveau sur une courbe.

Du théorème ci-dessus on déduit l'énoncé suivant :

Théorème 5.7. — *Soit* $c \in \mathrm{K}(\mathrm{X})$ *de dimension 1, de rang* $r > 1$ *et de degré* $\ell > 2rg$.

(i) *L'espace de modules* $\mathrm{Syst}_{\mathrm{X}}(c,r)$ *des structures de niveau de classe* c *est non vide, irréductible et normale de dimension* $r(\ell - r) + 1 - g$ *; elle est lisse en tout point stable.*

(ii) *Le morphisme* $\lambda_{\mathrm{S}} : (c,r)^{\perp} \to \mathrm{Pic}(\mathrm{Syst}_{\mathrm{X}}(c,r))$ *est surjectif, et de noyau* $\mathrm{Z}^1(\mathrm{X})$. *En particulier, le groupe de Picard de* $\mathrm{Syst}(c,r)$ *est un groupe abélien libre de rang 2.*

(iii) *C'est une variété localement factorielle.*

Le corollaire 4.21 fournit en effet un isomorphisme $\mathrm{Syst}_{\mathrm{X}}(c,r) \simeq \mathrm{Syst}_{\mathrm{X}}(c-r,r)$ ce qui entraîne évidemment les assertions (i) et (iii). Une structure de niveau $(\mathcal{F},\mathcal{V})$ de classe c définit une suite exacte sur $\mathrm{S}\times\mathrm{X}$

$$0 \to \mathcal{V}^* \to \mathcal{F} \to \Theta \to 0$$

Alors le système cohérent associé est $(\Theta^*, \mathcal{V}^*)$ où Θ^* est défini par $\Theta^* = \underline{\mathrm{Ext}}^1(\Theta, \mathcal{O}_{\mathrm{S}\times\mathrm{X}})$. Pour $u \in \mathrm{K}(\mathrm{X})$, posons $u^{\vee} = u^* \otimes \omega_{\mathrm{C}}$.

Lemme 5.8. — *Pour* $(u,m) \in \tilde{\mathrm{K}}(\mathrm{X})$, *on a dans* Pic(S)

$$\lambda_{(\Theta^*,\mathcal{V}^*)}(u,m) = \lambda_{\mathcal{F},\mathcal{V}}(-u\check{}, \chi(u) - m).$$

En effet, d'après le lemme 2.1, et la suite exacte ci-dessus, on a

$$\lambda_\Theta(u) = \lambda_{\mathcal{F}}(u) \otimes \det(\mathcal{V})^{\otimes \chi(u)}.$$

D'autre part, par dualité de Serre, $pr_{1!}(\Theta^*(u)) = (pr_{1!}(\Theta(u\check{})))^*$ et par conséquent $\lambda_{\Theta^*}(u) = \lambda_\Theta(-u\check{})$. Compte-tenu du fait que $\chi(u\check{}) = -\chi(u)$ ceci conduit à la formule attendue. □

Dès lors, l'involution de $\tilde{\mathrm{K}}(\mathrm{X})$ définie par $j : (u,m) \mapsto (-u\check{}, \chi(u) - m)$ induit un isomorphisme $(c-r,r)^\perp \to (c,r)^\perp$; la propriété universelle de λ_S montre que le diagramme

$$\begin{array}{ccc} (c-r,r)^\perp & \xrightarrow{\lambda_\mathrm{S}} & \mathrm{Pic}(\mathrm{Syst}_\mathrm{X}(c-r,r)) \\ j\downarrow & & \downarrow\wr \\ (c,r)^\perp & \xrightarrow{\lambda_\mathrm{S}} & \mathrm{Pic}(\mathrm{Syst}_\mathrm{X}(c,r)) \end{array}$$

dans lequel la flèche verticale de droite est induite par l'identification ci-dessus, est commutatif. La première flèche horizontale est surjective de noyau $\mathrm{Z}^1(\mathrm{X})$ d'après le théorème 5.1. Il en est de même de la seconde, ce qui achève la démonstration. □

Le théorème de Drézet et Narasimhan

On sait que l'espace de modules des fibrés semi-stables sur X de rang r et de classe c dans K(X) est une variété irréductible, normale, de dimension $(r^2-1)(g-1)$. En un point représentant un fibré stable F, la variété est lisse, et l'espace tangent de Zariski est isomorphe à $\mathrm{Ext}^1_0(\mathrm{F},\mathrm{F})$, où $\mathrm{Ext}^1_0(\mathrm{F},\mathrm{F}) = \mathrm{H}^1(\mathrm{X}, \mathrm{Hom}_0(\mathrm{F},\mathrm{F}))$ désigne l'espace vectoriel des classes de cohomologie à valeurs dans le fibré des endomorphismes de trace nulle. En ce qui concerne le groupe de Picard, le résultat ci-dessus permet d'obtenir une variante de la démonstration du théorème de Drézet et Narasimhan :

Corollaire 5.9. — (Drézet et Narasimhan [7]) *Soit* $c \in \mathrm{K}(\mathrm{X})$ *de dimension 1, de rang* $r > 1$. *Alors*

(i) *l'homomorphisme canonique*

$$\lambda_\mathrm{M} : c^\perp \to \mathrm{Pic}(\mathrm{M}_\mathrm{X}(c))$$

est surjectif, et a pour noyau $Z^1(X)$ *; en particulier, le groupe de Picard est un groupe cyclique.*

(ii) *la variété* $M_X(c)$ *est localement factorielle.*

En effet, on peut supposer que c est de degré $\ell > 2rg$, et considérer la variété $\mathrm{Syst}_X(c,r)$ des structures de niveau ci-dessus. La vérification du lemme ci-dessous est immédiate :

Lemme 5.10. — (i) *Si* (Γ, F) *est une structure de niveau semi-stable de classe* c*, on a* $H^1(F) = 0$.

(ii) *Dans* $\mathrm{Syst}_X(c,r)$*, le complémentaire de l'ouvert des points stables est de codimension* $\geq 4g-2$.

L'étude locale de $\mathrm{Syst}_X(c,r)$ (*cf.* proposition 4.12) montre que si $s \in \mathrm{Syst}_X(c,r)$ est un point stable représenté par un système cohérent (Γ, F), l'application linéaire canonique $T_s\mathrm{Syst}_X(c,r) \to \mathrm{Ext}^1_0(F,F)$ est surjective. L'étude de la stratification définie par la filtration de Harder-Narasimhan de F montre alors que le fermé des points représentant les structures de niveau (Γ, F) telles que F soit instable est de codimension ≥ 2. Soit V l'ouvert complémentaire. On a alors un morphisme

$$\pi : V \to M_X(c).$$

Soit $p \in M_X(c)$ un point représenté par un faisceau stable F et V_p la fibre de π au-dessus de p. Elle s'identifie à un ouvert de la grassmannienne $\mathrm{Grass}(r, H^0(F))$.

Lemme 5.11. — *Le complémentaire de l'ouvert* V_p *dans la grassmannienne* $\mathrm{Grass}(r, H^0(F))$ *des sous-espaces de dimension* r *est de codimension* ≥ 2.

En effet, pour tout diviseur effectif $D \subset X$ de degré 2, on a $H^1(F(-D)) = 0$ ce qui entraîne que le morphisme canonique $H^0(F) \otimes \mathcal{O}_D \to F|_D$ est surjectif. Si on prend D lisse, il en résulte que dans $\mathrm{Grass}(r, H^0(F))$ les points Γ tels que $\Gamma \otimes \mathcal{O}_D \to F|_D$ soit de rang $< r$ est l'intersection de deux hypersurfaces irréductibles distinctes, donc de codimension 2. Ce fermé contient le complémentaire de V_p, ce qui prouve le lemme. □

Le morphisme λ_S induit un diagramme commutatif

$$\begin{array}{ccccccc} 0 \to & c^{\perp} & \to & (c,r)^{\perp} & \to & \mathbf{Z} \\ & \lambda_M \downarrow & & \lambda_S \downarrow & & \downarrow (1) \\ 0 \to & \mathrm{Pic}(M_X(c)) & \xrightarrow{\pi^*} & \mathrm{Pic}(V) & \to & \mathrm{Pic}(V_p) \end{array}$$

dans lequel la première ligne est exacte, et la seconde une 0-suite. Il est facile de voir que la flèche π^* est injective en revenant aux schémas de Hilbert qui permettent la construction de ces espaces de modules. D'après le lemme 5.11 la flèche (1) est un isomorphisme; on sait que la flèche λ_S du milieu est surjective, de noyau $Z^1(X)$. Il en résulte que λ_M a même noyau, et qu'elle est aussi surjective. La propriété de factorialité locale de $M_X(c)$ résulte trivialement de celle de V. □

Soient χ la caractéristique d'Euler-Poincaré de c, et $\delta = \mathrm{pgcd}(r, \chi)$. Soit a un point de X. L'élément $u \in c^{\perp}$ défini par $\delta u = -r + \chi a$ fournit un générateur de $c^{\perp}$; le fibré inversible $\mathscr{D} = \lambda_M(u)$ associé est appelé classiquement fibré déterminant.

5.2. Systèmes cohérents sur $\mathbf{P}_2$

L'espace de modules $M_{\mathbf{P}_2}(2; 1, 2)$

Considérons l'espace de modules $M_{\mathbf{P}_2}(2; 1, 2)$ des faisceaux stables de rang 2, de classes de Chern $(1, 2)$, sur le plan projectif, *i.e.* de classe $c = 2 + h - h^2$ dans $K(\mathbf{P}_2)$. C'est une variété projective lisse de dimension 4. On sait que pour un tel faisceau stable $h^q(F) = 0$ pour $q = 1$ et 2 ; de la formule de Riemann-Roch on tire $h^0(F) = 2$.

Lemme 5.12. — *Soit* (Γ, F) *une structure de niveau de classe* $c = 2 + h - h^2$ *sur le plan projectif. Alors* (Γ, F) *est un système cohérent stable si et seulement si le faisceau* F *est stable.*

On a alors un isomorphisme

$$\mathrm{Syst}_{\mathbf{P}_2}(c, 2) \simeq M_{\mathbf{P}_2}(c)$$

en associant à la structure de niveau stable (Γ, F) le faisceau stable F sous-jacent. On a $2 - c^* = h + 2h^2$. On a d'autre part l'isomorphisme du corollaire

4.22

$$\mathrm{Syst}_{\mathbf{P}_2}(c,2) \simeq \mathrm{Syst}_{\mathbf{P}_2}(h+2h^2,2);$$

le membre de droite est la variété des systèmes linéaires de degré 2 et de dimension vectorielle 2 sur les droites du plan projectif : cette variété est évidemment isomorphe au schéma de Hilbert $\mathbf{Hilb}^2(\mathbf{P}_2)$ des sous-schémas finis de longueur 2 du plan projectif : il suffit en effet d'associer à un tel système linéaire $(\Gamma, \mathcal{O}_\ell(2))$ sur la droite ℓ les zéros des quadriques singulières de $\Gamma \subset \mathrm{H}^0(\mathcal{O}_\ell(2))$. Ainsi, on obtient l'isomorphisme bien connu :

Corollaire 5.13. — *Soit $c = 2 + h - h^2$. On a un isomorphisme canonique*

$$\mathrm{M}_{\mathbf{P}_2}(c) \simeq \mathbf{Hilb}^2(\mathbf{P}_2)$$

L'espace de modules $\mathrm{M}_{\mathbf{P}_2}(2;0,4)$

On considère maintenant l'espace de modules $\mathrm{M} = \mathrm{M}_{\mathbf{P}_2}(c)$ des classes de S−équivalence de faisceaux semi-stables de rang 2, de classes de Chern $(0,4)$, c'est-à-dire de classe $c = 2-4h^2$ dans $\mathrm{K}(\mathbf{P}_2)$. C'est une variété irréductible et normale de dimension 13 dont l'ensemble singulier est de dimension 8.

Si F est un faisceau semi-stable de classe c, on a $h^0(\mathrm{F}(1)) = 2$ ou 3 . Si $h^0(\mathrm{F}(1)) = 3$, on dit que F est *spécial*. Un tel faisceau est obligatoirement stable. Dans $\mathrm{M}_{\mathbf{P}_2}(c)$ les points correspondant aux faisceaux spéciaux forment une sous-variété lisse Σ de codimension 3 qui évite l'ensemble singulier.

Théorème 5.14. — [14] *On pose $c = 2-4h^2$.*

(i) *Soit* F *un faisceau cohérent de classe c, et $\Gamma \subset \mathrm{H}^0(\mathrm{F}(1))$ un sous-espace vectoriel de l'espace des sections. Alors la structure de niveau $(\Gamma, \mathrm{F}(1))$ est semi-stable si et seulement si* F *est semi-stable.*

(ii) *Le morphisme canonique*

$$\mathrm{Syst}_{\mathbf{P}_2}(c(1),2) \to \mathrm{M}_{\mathbf{P}_2}(c)$$

qui associe à la classe du système cohérent semi-stable $(\Gamma, \mathrm{F}(1))$ la classe du faisceau F *est l'éclatement $\mathrm{Bl}_\Sigma(\mathrm{M}_{\mathbf{P}_2}(c))$ de $\mathrm{M}_{\mathbf{P}_2}(c))$ le long de la sous-variété Σ.*

On a $2 - c(1)^* = 2h + 4h^2$, et les faisceaux cohérents de classes $2h + 4h^2$ sont les faisceaux de dimension 1, de multiplicité 2, de caractéristique d'Euler-Poincaré $\chi = 6$: ainsi le support schématique d'un tel faisceau est une conique. D'après le corollaire 4.22 on a un isomorphisme $\mathrm{Syst}_{\mathbf{P}_2}(c(1), 2) \simeq \mathrm{Syst}_{\mathbf{P}_2}(2h + 4h^2, 2)$. On obtient ainsi un diagramme de variétés projectives

$$\begin{array}{ccc} \mathrm{Syst}_{\mathbf{P}_2}(c(1),2) & \xrightarrow{\sigma} & \mathbf{P}_5 = |\mathcal{O}_{\mathbf{P}_2}(2)| \\ {\scriptstyle\pi}\downarrow & & \\ \mathrm{M}_{\mathbf{P}_2}(c) & & \end{array}$$

dans lequel le morphisme π est birationnel ; le morphisme σ est le morphisme qui associe à un point $\mathrm{Syst}_{\mathbf{P}_2}(2h + 4h^2, 2)$ représenté par un système cohérent (Γ, Θ) le support schématique de Θ. Au-dessus d'une conique lisse C, la fibre est isomorphe à la grassmanienne $\mathrm{Grass}(2, \mathrm{H}^0(\Theta))$, où Θ est l'unique faisceau semi-stable de rang 1 sur C de degré 5 sur C. On se propose dans ce qui suit de montrer comment cette description permet d'aborder la démonstration de l'assertion (i) du théorème 4.1.

Courbe des droites de saut

Le calcul de la quartique des droites de saut peut se faire en termes de systèmes cohérents : en effet, considérons le diagramme standard du §4.1 :

$$\begin{array}{ccc} \mathrm{D} & \xrightarrow{pr_2} & \mathbf{P}_2 \\ {\scriptstyle pr_1}\downarrow & & \\ \mathbf{P}_2^* & & \end{array}$$

Soit (Γ, Θ) est un système cohérent semi-stable de classe $2h + 4h^2$ et tel que $\dim \Gamma = 2$, et F le faisceau semi-stable de classe c associé dans la correspondance ci-dessus. Sur le plan projectif dual, le faisceau $pr_{1*}(pr_2^*(\Theta))$ est un faisceau cohérent sans torsion de rang 2, de classes de Chern $(4, 6)$; ce faisceau est non singulier en dehors des points correspondant aux droites contenues dans le support de Θ. On a alors un morphisme canonique sur le plan projectif dual

$$\Gamma \otimes \mathcal{O}_{\mathbf{P}_2^*} \to pr_{1*}(pr_2^*(\Theta))$$

dont il est facile de vérifier qu'il est génériquement injectif ; le conoyau est alors pur de dimension 1, et son support schématique est une quartique $\mathfrak{q}$

qui coïncide avec la quartique γ_F des droites de saut de F. En particulier, si ℓ est une droite non contenue dans le support de Θ, une telle droite définit un point de γ_F si l'application linéaire

$$\Gamma \to H^0(\Theta|_\ell)$$

n'est pas inversible.

Exemple

On voit en particulier que si le support de Θ est une conique lisse, et si Γ contient une section qui a 5 zéros distincts, la quartique γ_F est circonscrite au pentagone déterminé par ces 5 points. De plus, l'interprétation ci-dessus permet d'étudier les singularités de la courbe des droites de saut : ceci entraîne que si Γ ne contient pas de sections ayant 2 zéros doubles, la quartique γ_F est lisse (*cf.* Maruyama [19] ; Trautmann [25]). Réciproquement, étant donné un vrai pentagone dans le plan projectif dual, c'est-à-dire à 10 sommets, ce pentagone définit 5 points distincts sur une conique lisse C de $\mathbf{P}_2$; ces points sont les zéros d'une section s d'un fibré inversible Θ de degré 5 sur C. Les systèmes linéaires $\Gamma \subset H^0(\Theta)$ qui contiennent la section s constituent dans la grassmannienne $\mathrm{Grass}(2, H^0(\Theta))$ des sous-espaces de dimension 2 de $H^0(\Theta)$ un espace projectif de dimension 4, et le morphisme γ identifie cet espace projectif avec l'espace projectif des quartiques passant par les sommets du pentagone donné. Par suite, toute quartique γ de $\mathbf{P}_2^*$ passant par les 10 sommets d'un tel pentagone définit un point de l'image $\mathscr{L}$ de γ, autrement de l'hypersurface des quartiques de Lüroth.

Trois diviseurs dans $\mathrm{Syst}_{\mathbf{P}_2}(2h + 4h^2, 2)$

La description ci-dessus permet de mettre en évidence, outre le diviseur exceptionnel, trois diviseurs irréductibles $\partial S, D_1, D_2$ qui jouent un rôle important dans l'étude de la courbe des droites de saut :

1. Le diviseur ∂S des systèmes cohérents à point de base, *i.e.* tels que le morphisme d'évaluation $\Gamma \otimes \mathcal{O}_{\mathbf{P}_2} \to \Theta$ ne soit pas surjectif en au moins un point p. Alors la quartique associée $\mathfrak{q}$ est décomposée en une droite (correspondant à p) et une cubique. Pour un tel système cohérent, le faisceau semi-stable F associé est singulier en p.

2. Le diviseur D_1 est l'adhérence du sous-ensemble localement fermé des classes de systèmes cohérents stables (Γ, Θ) tels que le support C de Θ soit

lisse, et tels qu'il existe une section s de Γ ayant deux zéros doubles a et b. La quartique de Lüroth associée est singulière au point de $\mathbf{P}_2^*$ correspondant à la droite $\ell = ab$. Ces quartiques sont appelées quartiques de Lüroth singulières de type I.

3. Considérons dans $\mathrm{Syst}_{\mathbf{P}_2}(2h + 4h^2)$ l'image réciproque du diviseur Δ des coniques singulières par le morphisme σ. Ce diviseur est réduit et a deux composantes irréductibles dont l'une correspond via l'isomorphisme ci-dessus au diviseur exceptionnel. L'autre composante D_2 est l'adhérence de la variété des classes de structures de niveau (Γ, F) qui s'écrivent comme extension stricte

$$0 \to (0, \Theta') \to (\Gamma, \Theta) \to (\Gamma'', \Theta'') \to 0$$

où Θ' et Θ'' sont des faisceaux purs de classe $h + 2h^2$, donc de multiplicité 1 : ce sont des fibrés inversibles de degré 2 sur des droites ℓ' et ℓ'' respectivement. De telles suites exactes sont classées à isomorphisme près par un fibré en espaces projectif $\mathrm{P}(\mathscr{V})$ de rang 6 au-dessus de $\mathbf{P}_2^* \times \mathbf{Hilb}^2(\mathbf{P}_2)$ associé à un fibré vectoriel $\mathscr{V}$ de rang 7, (*cf.* proposition 4.13) et une telle extension est semi-stable, sauf peut-être si le système linéaire (Γ'', Θ'') a un point de base. En fait, cette extension définit un système cohérent semi-stable en dehors d'un fermé de dimension 5 de $\mathrm{P}(\mathscr{V})$; ce fermé ne rencontre pas la fibre générique de $\mathrm{P}(\mathscr{V})$. On désigne par $\mathrm{P}(\mathscr{V})^{ss}$ l'ouvert des points de $\mathrm{P}(\mathscr{V})$ qui fournissent un système cohérent semi-stable.

L'exemple traité ci-dessus montre que le système cohérent (Γ'', Θ'') définit un faisceau stable F'' de rang 2 et classes de Chern (1,2) ; le faisceau semi-stable F associé à (Γ, Θ) dans la correspondance ci-dessus s'insère dans une suite exacte

$$0 \to \mathrm{F}'' \to \mathrm{F}(1) \to \mathscr{O}_{\ell'}(-1) \to 0$$

ce qui implique que ℓ' est une droite de saut d'ordre 2 pour F, *i.e.* $h^1(\mathrm{F}(-1)|_{\ell'}) = 2$. Il en résulte que l'image par π du diviseur D_2 est le diviseur des classes de faisceaux semi-stables F qui ont au moins une droite de saut d'ordre ≥ 2. La quartique $\mathfrak{q} = \gamma_{\mathrm{F}}$ des droites de saut de F est encore singulière ; elle possède en outre des éléments géométriques intéressants que nous allons mettre en évidence :

Proposition 5.15. — *Soit* (Γ, Θ) *un point générique du diviseur* D_2, *et $\mathfrak{q}$ la quartique associée. On garde les notations ci-dessus et on désigne par O le point d'intersection des droites ℓ' et ℓ'', et par τ le deuxième zéro de la section $s \in \Gamma''$ qui s'annule en O. Alors*

(i) *la quartique $\mathfrak{q}$ a un et un seul point singulier, correspondant à ℓ';*

(ii) *la droite de $\mathbf{P}_2^*$ définie par τ est tangente en ℓ'' à $\mathfrak{q}$;*

(iii) *le cône tangent en ℓ' et la tangente τ se coupent sur $\mathfrak{q}$.*

On dispose en outre sur la droite ℓ'' d'une involution qui associe à un point générique $a \in \ell''$ le deuxième zéro $a^{\check{}}$ de l'unique forme quadratique de $\Gamma'' \subset \mathrm{H}^0(\Theta'')$ qui s'annule en a. Cette involution détermine évidemment Γ''. On peut en fait retrouver cette involution sur la géométrie de la quartique $\mathfrak{q}$ associée au point générique de $\mathrm{P}(\mathscr{V})$: en effet, dans ce cas le faisceau $\Theta|_{\ell'}$ est un fibré inversible de degré 3 sur ℓ', et Γ détermine aussi un système linéaire sans point de base sur ℓ'. L'unique section de Γ qui s'annule en a et $a^{\check{}}$ s'annule en trois points a_i sur ℓ'. Les droites $a_i a$ et $a_i a^{\check{}}$ définissent sur $\mathfrak{q}$ des points alignés. Par dualité, ceci se lit de la manière suivante sur la quartique : la droite a, qui passe par ℓ'', recoupe $\mathfrak{q}$ en trois autres points ; ceci détermine trois droites dans le pinceau des droites passant par ℓ', et ces droites recoupent la quartique en trois autres points appartenant à une même droite $a^{\check{}}$ passant par ℓ''.

Corollaire 5.16. — *Le morphisme $\gamma : \mathrm{P}(\mathscr{V})^{ss} \to \mathbf{P}_{14}$ est génériquement injectif.*

On vient de voir en effet en effet qu'en dehors d'un fermé de dimension 11, la donnée de la quartique $\mathfrak{q}$ détermine les systèmes cohérents Θ' et (Γ'', Θ''). Au-dessus d'un tel point, la fibre de $\mathrm{P}(\mathscr{V})^{ss} \to \mathbf{P}_2^* \times \mathbf{Hilb}^2(\mathbf{P}_2)$ est un espace projectif de dimension 6, et il suffit de constater que le morphisme $\gamma : \mathrm{P}(\mathscr{V})^{ss} \to \mathrm{P}_{14}$ est induit sur ces fibres par une application linéaire ; elle est donc injective.

On peut en fait prouver beaucoup plus :

Théorème 5.17. —

(i) *Le morphisme $\gamma : \mathrm{Syst}_{\mathbf{P}_2}(2h + 4h^2, 2) \to \mathbf{P}_{14}$ est génériquement injectif le long de D_2;*

(ii) *ce morphisme est non ramifié au point générique de D_2;*

(iii) *soit* $\mathscr{S} \subset \mathbf{P}_{14}$ *l'hypersurface des quartiques singulières, et* $\mathscr{L} = \operatorname{Im}\gamma$ *le diviseur des quartiques de Lüroth. Alors* $\mathscr{S} \cap \mathscr{L}$ *a deux composantes irréductibles*

$$\mathscr{S} \cap \mathscr{L} = \gamma(\mathrm{D}_1) \cup \gamma(\mathrm{D}_2).$$

Une quartique singulière de $\gamma(\mathrm{D}_2)$ est classiquement appelée quartique de Lüroth singulière de type II. Il résulte de cet énoncé que l'image réciproque d'un point représentant une quartique singulière de type II générique est schématiquement réduit à un point. Ceci conduit à l'assertion (i) du théorème 4.1.

L'existence des deux types de quartiques de Lüroth singulières semblait connue au début du siècle ([2] ,[21]). J'ignore si on peut caractériser par des propriétés géométriques analogues à celles de la proposition 5.15 le point générique de $\gamma(\mathrm{D}_1)$.

BIBLIOGRAPHIE

[1] W. BARTH, *Moduli of vector bundles on the projective plane,* Invent. Math. **42** (1977) 63–91.

[2] H. BATEMAN, *The quartic curve and its inscribed configurations,* Amer. J. Math. **36** (1914) 357–386.

[3] S. DONALDSON, *Polynomial invariants for smooth 4-manifolds,* Topology **29** (1990) 257–315.

[4] J.-M. DRÉZET, *Fibrés exceptionnels et suite spectrale de Beilinson généralisée,* Math. Annalen **275** (1986) 25–48.

[5] J.-M. DRÉZET, *Groupe de Picard des variétés de modules de faisceaux semi-stables sur* $\mathbf{P}_2(\mathbf{C})$, Ann. de l'Institut Fourier **38** (1988) 105–168.

[6] J.-M. DRÉZET ET J. LE POTIER, *Fibrés stables et fibrés exeptionnels sur le plan projectif,* Ann. scient. Ec. Norm. Sup. 4e **série, t.18** (1985) 193–244.

[7] J.-M. DRÉZET ET M. S. NARASIMHAN, *Groupe de Picard des variétés de modules de fibrés semi-stables sur les courbes algébriques,* Invent. Math. **97** (1989) 53–94.

[8] D. GIESEKER, *On the moduli of vector bundles on an algebraic surface,* Ann. Math. **106** (1977) 45–60.

[9] D. GIESEKER ET J. LI, *Moduli of vector bundles on a surface I,* Preprint (1993).

[10] L. GÖTTSCHE ET A. HIRSCHOWITZ, *Weak Brill-Noether for vector bundles on the projective plane,* Europroj 1993 (Proceedings of the Catania Conference), to appear.

[11] A.D. KING AND P.E. NEWSTEAD, *Moduli of Brill-Noether pairs on algebraic curves*, Preprint Liverpool (1994).

[12] J. LE POTIER, *Stabilité et amplitude sur* $\mathbf{P}_2$, in Vector Bundles and Differential Equations (Proceedings de la conférence de Nice), Progress in Maths Birkhaüser, 7 (1980) 145–182.

[13] J. LE POTIER, *Fibré déterminant et courbes de saut sur les surfaces algébriques*, in Complex projective Geometry (Proceedings de la Conférence de Bergen) London mathematical Society, Lecture Note Series 179 (1992) 213–240.

[14] J. LE POTIER, *Systèmes cohérents et structures de niveau,* Astérisque **214** (1993).

[15] J. LE POTIER, *Faisceaux semi-stables de dimension 1 sur le plan projectif,* Revue Roumaine de Mathématiques Pures et Appliquées, **38**, 7-8 (A la mémoire de C. Bănică) (1993) p. 635-678.

[16] J. LE POTIER, *Espace de modules de faisceaux semi-stables sur le plan projectif,* Ecole EUROPROJ et CIMPA «Vector bundles on Surfaces» Nice Sophia-Antipolis, juin 1993.

[17] J. LE POTIER, *Problème de Brill-Noether et groupe de Picard de l'espace de modules des faisceaux semi-stables sur le plan projectif,* Europroj 1993 (Proceedings of the Catania Conference), à paraître.

[18] W.-P. LI ET Z. QIN, *Lower-degree Donaldson polynomial invariants of rational surfaces,* J. Algebraic geometry **2** (1993) 413–442.

[19] M. MARUYAMA, *Moduli of stable sheaves, II,* J. Math. Kyoto University **18** (1978) 557–614.

[20] M. MARUYAMA, *Singularities of the curve of jumping lines of a vector bundle of rank 2 on* $\mathbf{P}_2$,Algebraic Geometry, Proc. of Japan-France Conf. (1982), Lectures Notes in Math., Springer **1016** (1983) 370–411.

[21] F. MORLEY, *On the Lüroth quartic curve,* Amer. J. Math. **36** (1918).

[22] N. RAGHAVENDRA AND P.A. VISHWANATH, *Moduli of pairs and generalized theta divisors,* Tohoku Math. J., to appear.

[23] C. T. SIMPSON, *Moduli of Representations of the Fundamental Group of a Smooth Variety,* Preprint, Princeton University (1990).

[24] M. THADDEUS, *Stable pairs, linear systems and the Verlinde formula,* Preprint (1992).

[25] G. TRAUTMANN, *Poncelet curves and associated theta characteristics,* Expositiones Mathematicae **6** (1988) 29–64.

[26] A. N. TYURIN, *The moduli spaces of vector bundles on threefolds, surfaces and curves,* Preprint Erlangen (1990).

THE COMBINATORICS OF THE VERLINDE FORMULAS

ANDRÁS SZENES

1. Introduction

In this short note we discuss the origin and properties of the Verlinde formulas and their connection with the intersection numbers of moduli spaces. Given a simple, simply connected Lie group G, the Verlinde formula is an expression $V_k^G(g)$ associated to this group depending on two integers k and g. For $G = \mathrm{SL}_2$ the formula is

$$V_k^{\mathrm{SL}_2}(g) = \sum_{j=1}^{k-1} \left(\frac{k}{2\sin^2 \frac{j\pi}{k}} \right)^{g-1}. \tag{1.1}$$

We describe V_k^G for general groups in §2. These formulas were first written down by E. Verlinde [23] in the context of conformal field theory. The interest towards them in algebraic geometry stems from the fact that they give the Hilbert function of moduli spaces of principal bundles over projective curves. More precisely, let C be a smooth projective curve of genus g, and let $\mathfrak{M}_C^G$ be the moduli space of principal G-bundles over C (cf. e.g. [16] and references therein). Then there is an ample line bundle $\mathcal{L}$ over $\mathfrak{M}_C^G$ such that

$$\dim \mathrm{H}^0(\mathfrak{M}_C^G, \mathcal{L}^k) = V_{k+h}^G(g), \tag{1.2}$$

where h is the dual Coxeter number of G. This statement requires some modifications for a general simple G, but it holds for SL_n ([3, 7, 6, 16]).

Proving (1.2) is important, but in this paper we will address a different question: what can be said about the moduli spaces knowing (1.2)? Accordingly, first we concentrate on understanding the formula.

Two rather trivial aspects of (1.2) are that

- $V_k^G(g)$ is integer valued,
- $V_k^G(g)$ is a polynomial in k.

Note that looking at the formula itself, none of this is obvious. Our goal is to explain these properties and connect them to the intersection theory of $\mathfrak{M}_C^G$.

The paper is structured as follows: in §2 we discuss some of the ideas of Topological Field Theory, which explain the structure of the formula for general G and show its integrality (cf. [13, 21, 8, 5]). In §3 we give our main result, a residue formula for V_k^G for $G = \mathrm{SL}_n$. Such a formula gives an explicit way of calculating the coefficients of V_k^G as a polynomial in k. Finally, in §4 we give an application of our formulas: a "one-line proof" of (1.2).

Research was partially supported by an NSF grant.

This paper is intended as an announcement and overview. As a result, few proofs will be given, and even most of those will be sketchy. A more complete treatment will appear separately.

Acknowledgements. I am grateful to Raoul Bott, my thesis advisor, for suggesting to me this circle of problems and helping me with advice and ideas along the way. I would like to thank Noam Elkies for useful discussions.

I am thankful to the organizers of the Durham Symposium on Vector bundles, in particular to Peter Newstead and Bill Oxbury for their help and for the opportunity to present my work.

2. Topological Field Theory and Fusion Algebras

This section is independent from the rest of the paper. It contains a quick and rather formal overview of the structure of Topological Field Theories [20, 1] and Verlinde's calculus [23].

Consider a finite dimensional vector space F (the space of fields) with a marked element $1 \in F$ (the vacuum). Assume that a number $F(g)_{v_1,v_2,\dots v_n}$ (correlation functions) is associated to every topological Riemann surface of genus g, with elements of the algebra $v_1, v_2, \dots v_n \in F$ inserted at n punctures, which satisfies the following axioms:

Normalization: $F(0)_{1,1,1} = 1$,
Invariance: $F(g)_{v_1,\dots} = F(g)_{1,v_1,\dots}$,
Linearity: $F(g)_{v_1,\dots}$ is linear in v_i.

Introduce the symmetric linear 3-form $\omega : F \otimes F \otimes F \to \mathbb{C}$ by $\omega(u,v,w) = F(0)_{u,v,w}$, the bilinear form $(u,v) = F(0)_{u,v}$ and the trace $\int u = F(0)_u$. Assume that $(,)$ is **non-degenerate**, and fix a pair of bases $\{u_i, u^i\}$ of F, dual with respect to this form, that is $(u_i, u^j) = \delta_{ij}$.

Verlinde's fusion rule: $F(g)_{v_1,\dots} = \sum_i F(g-1)_{u_i,u^i,v_1\dots}$.

One can extend F to disconnected surfaces by the axiom:

Multiplicativity: F is multiplicative under disjoint union.

Remark 2.1. These axioms serve as an algebraic model of certain relations among the Hilbert functions of various moduli spaces. The number $F(g)_{v_1,v_2,\dots v_n}$ represents the dimension of the space of sections of a certain line bundle over a moduli space of parabolic bundles with weights depending on the insertions $v_1, v_2, \dots v_n$. The fusion axiom describes how the space of sections of a line bundle decomposes over a family of curves degenerating to a nodal curve. (See [7, 22]; also the article by Ueno in the present volume.)

Lemma 2.1. *The axioms above define the structure of an associative and commutative algebra on F, by the formula $vw = \sum_i \omega(v,w,u^i)u_i$, compatible with $(,)$ and $\int$. Then if we denote the invariantly defined element $\sum_i u_i u^i \in F$ by α, we have*

$$F(g)_{v_1,v_2,\dots,v_n} = \int \alpha^g v_1 v_2 \dots v_n. \tag{2.1}$$

Now assume in addition that the algebra F is **semisimple**. Then it has the form $F \cong L^2(S,\mu)$, where $S = \operatorname{Spec} F$ is a finite set and the complex measure μ can be given via a function $\mu : S \to \mathbb{C}$.

The elements of F become functions on S and the trace $\int$ turns out to be the actual integral with respect to μ. Now take the following pair of dual bases: $\{\delta_s, \delta_s/\mu(s) |\ s \in S\}$, where $\delta_s(x) = \delta_{sx}$. We call this the spectral basis. Using this basis and (2.1), we obtain the following formula:

$$F(g) = \sum_{s\in S} \mu(s)^{1-g}. \tag{2.2}$$

This formula resembles (1.1), but what is the appropriate algebra?

2.1. Fusion algebras. Here we construct the fusion algebras for arbitrary simple, simply connected Lie groups. First we need to introduce some standard notation.

Notation. In this paragraph we will use the compact form of simple Lie groups, still denoting them by the same letter. Thus let G be a compact, simply connected, simple Lie group, $\mathfrak{g}$ its complexified Lie algebra, T a fixed maximal torus, and $\mathfrak{t}$ the complexified Lie algebra of T. Denote by Λ the unit lattice in $\mathfrak{t}$ and by $\mathcal{W} \subset \mathfrak{t}^*$ its dual over $\mathbb{Z}$, the weight lattice. Let $\Delta \subset \mathcal{W}$ be the set of roots and W the Weyl group of G. A fundamental domain for the natural action of the Weyl group on T is called an alcove; a fundamental domain for the associated action on $\mathfrak{t}^*$ is called a chamber. We will use the multiplicative notation for weights and roots, and think of them as characters of T. The element of $\mathfrak{t}^*$ corresponding to a weight λ under the exponential map will be denoted by L_λ.

Fix a dominant chamber $\mathfrak{C}$ in $\mathfrak{t}^*$ or a corresponding alcove $\mathfrak{a}$ in T. This choice induces a splitting of the roots into positive (Δ^+) and negative (Δ^-) ones. For a weight λ, denote its Weyl antisymmetrization by $\mathrm{A}\cdot\lambda = \sum_{w\in W} \sigma(w) w\cdot \lambda$, where $\sigma : W \to \pm 1$ is the standard character of W. According to the Weyl character formula, for a dominant weight λ, the character of the corresponding irreducible highest weight representation is $\chi_\lambda = \mathrm{A}\cdot\lambda\rho/\,\mathrm{A}\cdot\rho$, where ρ is the square root of the product of the positive roots.

The ring $R(G)$, the representation ring of G, can be identified with $R(T)^W$, the ring of Weyl invariant linear combinations of the weights. Denote by $d\mu_T$ the normalized Haar measure on T. If we endow T/W with the Weyl measure

$$d\mu_W = \mathrm{A}\cdot\rho\, \mathrm{A}\cdot\bar{\rho}\ d\mu_T,$$

then $R(G)$ becomes a pre-Hilbert space with orthonormal basis $\{\chi_\lambda\}$, i.e. one has
$\int_{T/W} \chi_\lambda \chi_\mu\, d\mu_W = \delta_{\lambda,\bar{\mu}}$. □

We need to introduce an integer parameter denoted by k called the *level*, which can be thought of as an element of $\mathrm{H}^3(G,\mathbb{Z}) \cong \mathrm{H}^4(BG,\mathbb{Z}) \cong \mathbb{Z}$, and in turn can be identified with a Weyl-invariant integral inner product on $\mathfrak{t}$.

The *basic* invariant inner product on $\mathfrak{t}$ corresponding to $k = 1$ is specified by the condition $(H_\theta, H_\theta) = 2$, where $H_\theta \in \mathfrak{t}$ is the coroot of the highest root L_θ. It has the following properties (see [13, §6],[19, Ch.4]):

- For the induced inner product on $\mathfrak{t}^*$, we have $(L_\theta, L_\theta) = 2$.
- For $\lambda \in \mathcal{W}$, the inner product (L_θ, L_λ) is an integer, and $(,)$ is the smallest inner product with this property.
- The Killing form is equal to $-2h(,)$, where $h = (L_\theta, L_\rho) + 1$ is the dual Coxeter number of G.

The basic inner product also gives an identification $\nu : \mathfrak{t}^* \to \mathfrak{t}$ between $\mathfrak{t}^*$ and $\mathfrak{t}$, by the formula $\beta(x) = (\nu(\beta), x)$.

2.2. The simply-laced subgroup. Let $\Delta_l \in \Delta$ be the set of long roots of G. Denote by $\mathcal{W}_r$ the lattice in $\mathfrak{t}^*$ generated by Δ, and by $\mathcal{W}_l$ the lattice generated by Δ_l. By definition Λ is the dual of $\mathcal{W}$ over $\mathbb{Z}$ with respect to the canonical pairing $\langle , \rangle$ between $\mathfrak{t}^*$ and $\mathfrak{t}$. The dual of $\mathcal{W}_r$ is the center lattice in $\mathfrak{t}$. Denote the dual of $\mathcal{W}_l$ by Λ_l.

The root system Δ_l corresponds to a subgroup G_l of G with maximal torus T and Weyl group $W_l \subset W$, which is generated by reflections corresponding to the elements of Δ_l. Denote the center of G_l by Z_l. Then Z_l can be described as the set of elements of T invariant under W_l, and we have $\exp^{-1} \Lambda_l = Z_l$. It is important to note that in view of the second property of $(,)$ above, $\nu : \mathcal{W}_l \to \Lambda$ is an isomorphism. Since $\mathcal{W}$ is paired to Λ and $\mathcal{W}_l$ is paired to Λ_l over $\mathbb{Z}$, it follows that $\nu : \mathcal{W} \to \Lambda_l$ is also an isomorphism. Then the map $\exp \cdot \nu : \{\alpha \in \mathfrak{C} |\, (L_\theta, \alpha) \leq 1\} \to \mathfrak{a}$ is a bijection.

Naturally, if G is simply laced, then $G_l = G$. For the non-simply laced groups one has the following subgroups:

- $\mathrm{Spin}_{2n} \subset \mathrm{Spin}_{2n+1}$
- $\mathrm{SU}_2^n \subset \mathrm{Sp}_n$
- $\mathrm{SU}_3 \subset \mathrm{G}_2$
- $\mathrm{Spin}_8 \subset \mathrm{F}_4$

2.3. The definition of the fusion algebra. We give a different definition from the standard one via co-invariants of infinite dimensional Lie algebras [22], but one which is very natural from the point of view of representation theory.

To motivate the construction, recall the procedure of holomorphic induction [4]: the flag variety $F = G/T$ has a complex structure and every character λ of T induces a holomorphic equivariant line bundle $\mathcal{L}_\lambda = G \times_\lambda \mathbb{C}$ over F. Then one can define the induction map $\mathcal{I} : R(T) \to R(G)$ as a homomorphism of additive groups by the formula $\lambda \mapsto \sum (-1)^i \mathrm{H}^i(F, \mathcal{L}_\lambda)$, where the cohomology groups in the latter expression are thought of as G-modules. The Borel-Weil-Bott theorem then says that

(2.3)

$$\text{for } \lambda \text{ dominant } \mathcal{I}(\lambda) = \chi_\lambda$$

and

(2.4)
$$\mathcal{I}(\lambda') = \sigma(w)\mathcal{I}(\lambda), \text{ whenever } \lambda'\rho = w(\lambda\rho) \text{ for some } w \in W.$$

Note that $\mathcal{I}$ is not expected to be a ring homomorphism.

This procedure applies to the loop group $\widehat{LG}$ as well ([19, 15, 17]). Once the action of the central elements is fixed as $c \mapsto c^k$, where $k \in \mathbb{N}$ is the level, again, we have a map $\bar{\mathcal{I}} : R(T) \to R_k(\widehat{LG})$. This last object $R_k(\widehat{LG})$ has only additive structure, since the tensor product of two level k representations has level $2k$. The role of the Weyl group is played by the affine Weyl group W_k obtained by adjoining to W the translation by $(k+h)L_\theta$. Again the Borel-Weil-Bott theorem applies, and (2.4), with W replaced with W_k, gives a description of the kernel of $\bar{\mathcal{I}}$.

Since $W \subset W_k$, the map $\bar{\mathcal{I}}$ factors through $\mathcal{I}$, and as a result we have a map $\mathcal{J} : R(G) \to R_k(\widehat{LG})$. It is easy to see from (2.4) that the set of characters: $\Xi_k = \{\chi_\lambda | (L_\theta, L_\lambda) \leq k\}$ forms a basis of $R_k(\widehat{LG})$ if we identify $\mathcal{J}(\chi_\lambda)$ with χ_λ.

Lemma 2.2. *The additive group $R_k(\widehat{LG})$ can be endowed with a ring structure F_k^G so that the map $\mathcal{J}$ becomes a homomorphism of rings.*

The algebra F_k^G is called the *fusion algebra* of G of level k. As noted above, we can consider Ξ_k to be a basis of F_k^G. Endow F_k^G with the trace function $\int$ by the formula $\int \chi_\lambda = 0$ except for the trivial character χ_1, which has trace equal to 1. Also note that since $\mathrm{Spec}(R(G)) = T/W$ and F_k^G is a quotient of $R(G)$, we expect $\mathrm{Spec}(F_k^G) \subset T/W$.

Lemma 2.3. *F_k^G can be identified with "L^2" of the finite normalized measure space $Z_k = \{t \in T | t^{k+h} \in Z_l, t \text{ is regular}\}/W$, with measure $d\mu_k$ given by the function*

(2.5)
$$\frac{\mathrm{A}\cdot\rho(t)\,\mathrm{A}\cdot\bar{\rho}(t)}{|Z_l|(k+h)^r}.$$

Note the surprising fact that the discrete measure remains unchanged up to a normalization factor as k varies.

Now we can define the quantity $V_k^G(g)$ which appeared in (1.2) as the number associated to a Riemann surface of genus g and the fusion algebra F_{k-h}^G. Combining (2.2) and (2.5) we obtain

$$V_k^G(g) = \sum_{t \in T_r/W, t^k \in Z_l} \left(\frac{|Z_l|k^r}{\mathrm{A}\cdot\rho(t)\,\mathrm{A}\cdot\bar{\rho}(t)}\right)^{g-1},$$

where T_r is the set of regular elements of T. This can be easily seen to give (1.1) for the case of $G = \mathrm{SU}_2$. Indeed, embedding the maximal torus of SU_2 into $\mathbb{C}$ as the unit circle, we have: $\rho(z) = z$, $Z_l = \pm 1$, $h = 2$ and $A(z) = 1/z$.

Finally, note that since the relations in the fusion algebras given in (2.4) have integer coefficients, and the $\{\chi_\lambda\}$ form an orthonormal basis of F, we see that $\int \alpha^g$ from (2.1) has to be an integer. This proves that $V_k^G(g)$ is an integer for all groups and values of k and g.

Remark 2.2. That this definition of the fusion algebras is equivalent to the standard one via coinvariants of current algebras [22] can be shown to be equivalent to Verlinde's conjecture on the diagonalization of the fusion rules. which gives a formula for the product in F_k^G using the S-matrix. The definition given above is simpler to use for calculations and it gives the correct prescription for non-simply-connected groups (see also [8]).

3. Residue formulas

In this section we study $V_k^G(g)$ as a function of k. We show that $V_k^G(g)$ is a polynomial in k for $G = \mathrm{SL}_3$, and give a simple formula for the coefficients of this polynomial. The generalization of these results to SL_n is straightforward.

Consider the case $G = \mathrm{SL}_2$ first. Again, as at the end of the previous section embed the maximal torus of SL_2 into $\mathbb{C} \subset \mathbb{P}^1$.

Consider the differential form

$$\mu = \frac{dz}{z}\frac{z+z^{-1}}{z-z^{-1}} \text{ on } \mathbb{P}^1.$$

This form has simple poles: at $z = \pm 1$ with residue 1, and at $z = 0, \infty$ with residue -1. Thus if we pull back μ by the k-th power map we obtain a differential form μ_k with poles at the $2k$-th roots of unity and residues $+1$, and simple poles at $z = 0, \infty$ with residue $-k$. It is given by the following formula:

$$\mu_k = k\frac{dz}{z}\frac{z^k+z^{-k}}{z^k-z^{-k}}.$$

Note that μ_k is invariant under multiplication by a $2k$th root of unity and under the Weyl reflection $z \to 1/z$.

Now suppose we have a function $f(z)$, with poles only at $z = \pm 1$, vanishing at 0 and ∞, and invariant under the substitution $z \to z^{-1}$. Then by applying the Residue Theorem to the differential form $f\mu_k$ and using the Weyl symmetry at hand, we have

$$\sum_{j=1}^{k-1} f(\exp(\pi\sqrt{-1}j/k)) = -\operatorname*{Res}_{z=1} \mu_k f(z).$$

Applying this argument to the function

$$f(z) = \left(\frac{2k}{-(z-z^{-1})^2}\right)^{g-1}$$

we obtain the formula

$$V_k^{\mathrm{SL}_2}(g) = (-1)^g (2k)^{g-1} \operatorname*{Res}_{z=1} \frac{k\,dz}{z}\frac{z^k+z^{-k}}{z^k-z^{-k}}\left(\frac{1}{(z-z^{-1})^2}\right)^{g-1}.$$

Now using the invariance of the residue under substitutions we can obtain different formulas for $V_k^{\mathrm{SL}_2}(g)$. For example, the polynomial nature of $V_k^{\mathrm{SL}_2}(g)$ becomes transparent if we perform the substitution $z \to \exp(Ix)$:

$$V_k^{\mathrm{SL}_2}(g) = -(2k)^{g-1} \operatorname*{Res}_{x=0} \frac{k \cot(kx)\, dx}{(2 \sin x)^{2(g-1)}}.$$

It is easy to check using this formula that the degree of $V_k^{\mathrm{SL}_2}(g)$ as a polynomial in k is $3(g-1)$, which, as expected, coincides with the dimension of $\mathfrak{M}_C^{\mathrm{SL}_2}$.

Before we proceed, we need an understanding of higher dimensional residues. The notion that a top dimensional differential form has an invariantly defined number assigned to it, does not carry over to higher dimensions. The correct object in $\mathbb{C}^n$ is Res : $\mathrm{H}^n_{\mathrm{loc}}(\Omega^n, \mathbb{C}^n) \to \mathbb{C}$ mapping from the nth local Čech cohomology group in a neighborhood of 0 with values in holomorphic n-forms to complex numbers. To define this map let ω be a meromorphic n-form defined in a neighborhood of 0 in $\mathbb{C}^n$. Then ω can be represented in the form $dz_1\, dz_2 \ldots dz_n h(z)/f(z)$ where f and h are holomorphic functions. The additional data necessary to represent an element of $\mathrm{H}^n_{\mathrm{loc}}(\Omega^n, \mathbb{C}^n)$ is a splitting of f into the product of n functions $f = a_1 a_2 \ldots a_n$. Such a splitting defines n open sets in a neighborhood of 0: $A_i = \{a_i \neq 0\}$. These define a local Čech cocycle. A detailed explanation of this and an algorithm to calculate the residue can be found in [9, 10].

We will call a differential n-form with such a splitting a *residue form.*

Definition 3.1. A non-trivial residue form ω is called flaglike if a_i only depends on $z_1, \ldots z_i$. This notion depends on choice and the order of the coordinates $z_1, \ldots, z_n$.

Lemma 3.1. *Let ω be a flaglike residue form. Then*

$$\operatorname{Res}(\omega) = \operatorname*{Res}_{z_n} \ldots \operatorname*{Res}_{z_1} \omega.$$

Here Res_{z_i} *is the ordinary 1-dimensional residue, taken assuming all the other variables to be constants.*

The proof is straightforward. Note that the order of the variables is important, while there is some freedom in the way the denominator is split up.

For simplicity we restrict ourselves to the case of SL_3. According to Lemma 2.3 and (2.2) the Verlinde formula can be written as

(3.1)
$$V_k^{\mathrm{SL}_3}(g) = (3k^2)^{g-1} \sum_{i,j,k-i-j>0} (8 \sin(i\pi/k) \sin(j\pi/k) \sin((i+j)\pi/k))^{-2(g-1)}$$

Now we can write down the main result of the paper:

Theorem 3.2.

$$(3.2)\quad V_k^{\mathrm{SL}_3}(g) = (3k^2)^{g-1} \operatorname*{Res}_{Y=1} \operatorname*{Res}_{X=1} \frac{X^k + X^{-k}}{X^k - X^{-k}} \frac{Y^k + Y^{-k}}{Y^k - Y^{-k}} \times$$
$$\times \frac{(-1)^{g-1}}{((X - X^{-1})(Y - Y^{-1})(XY - (XY)^{-1}))^{2(g-1)}} \frac{k^2\, dX\, dY}{XY}.$$

The proof is analogous to the case of SL_2. Denote the residue form in (3.2) by $\omega_k(g)$. Again at the points $p_{ij} = (e^{iI\pi/k}, e^{jI\pi/k})$, with $i, j, k - i - j \geq 1$ the residues of $\omega_k(g)$ reproduce the sum (3.1). However, now it is not immediately obvious that the residue theorem can localize this sum at the point $(1,1)$, since the residue form in (3.2) has non-trivial residues at other points as well. To illustrate the situation consider the matrix M_k whose (i,j)th entry is the residue of $\omega_k(g)$ taken at the point p_{ij} instead of $(1,1)$, where $i, j = 0, \ldots, k-1$.

Example for $g = 2$:

$$M_6 = \begin{bmatrix} 166 & -45 & -29 & -18 & -29 & -45 \\ -45 & 36 & 9 & 9 & 36 & -45 \\ -29 & 9 & 4 & 9 & -29 & 36 \\ -18 & 9 & 9 & -18 & 9 & 9 \\ -29 & 36 & -29 & 9 & 4 & 9 \\ -45 & -45 & 36 & 9 & 9 & 36 \end{bmatrix}.$$

We can apply the Residue Theorem to "each column" by fixing a value of X. By degree count, one can see that $\omega(g)$ has trivial residues at $Y = 0, \infty$ and this implies that the sum of the entries in each column of $M_k(g)$ is 0. Next, note that $M_k(0,i) = M_k(i,0)$, since these residues are *split*, i.e. they have the form $dX\,dY\,X^{-m}Y^{-n}f(X,Y)$, where f is holomorphic at the point where the residue is taken.

Now to prove the Theorem it is sufficient to show that $M_k(j,0) = M_k(j,k-j)$ for every $j > 0$. To see this, note that both residues are simple (first order) in X at $\alpha = \exp(j\pi/k)$. This means that after taking the X-residue, we are left with the form

$$\omega_\alpha = \text{const}\cdot \frac{dY}{Y} \frac{Y^k + Y^{-k}}{Y^k - Y^{-k}} \frac{1}{((Y - Y^{-1})(\alpha Y - \alpha^{-1}Y^{-1}))^{2(g-1)}}.$$

The two numbers we need to compare are the residues of this form at 0 and α respectively. But these two residues clearly coincide since ω_α is invariant under the substitution $Y \to \alpha^{-1}Y^{-1}$. □

The formula for $G = \mathrm{SL}_n$ reads as follows:

(3.3)

$$V_k^G(g) = (-1)^{n-1+(g-1)|\Delta^+|}(nk^{n-1})^{g-1} \operatorname*{Res}_{X_{n-1}=1} \cdots \operatorname*{Res}_{X_1=1} W^{-2(g-1)} \prod_{i=1}^{n-1} \frac{X_i^k + 1}{X_i^k - 1} \cdot \frac{k\, dX_i}{2X_i}$$

where X_i, $i = 1, \ldots, n-1$ are the simple (multiplicative) roots and $W = \prod_{\alpha\in\Delta^+}(\alpha^{\frac{1}{2}} - \alpha^{-\frac{1}{2}})$ is the Weyl measure.

As we pointed out after Lemma 3.1, the ordering of the variables matters when taking the subsequent residues. In the special case of $G = \mathrm{SL}_3$ this ordering does not matter (i.e. $M_k(i,j) = M_k(j,i)$), but for higher rank groups a finer argument is necessary.

Finally, note that similarly to the case of SL_2, (3.3) gives a simple prescription for calculating the coefficients of $V_k^G(g)$ as a polynomial in k, via the exponential substitution. For example, for SL_3 we obtain

$$V_k^{\mathrm{SL}_3}(g) = (3k^2)^{g-1} \operatorname*{Res}_{y=0} \operatorname*{Res}_{x=0} \frac{k^2 cot(kx)\cot(ky)\,dx\,dy}{(8\sin(x)\sin(y)\sin(x+y))^{2(g-1)}}. \tag{3.4}$$

A different generating function was obtained for the case of $G = \mathrm{SL}_3$ by Zagier [26].

4. Multiple ζ-values and intersection numbers of moduli spaces

In this final section we show how (1.2) and (3.2) can be related via the Riemann-Roch formula to Witten's conjectures on the intersection numbers of moduli spaces. Our argument below gives a quick proof of (1.2) for SL_n assuming Witten's formulas. This is a generalization of the work of Thaddeus who considered the case of SL_2 [18].

4.1. Multiple ζ-values and intersection numbers of moduli spaces. Consider the case of SL_2 first. If we want to find the asymptotic behavior of $V_k^{\mathrm{SL}_2}$ for large k, the best way to think about the formula is that it is a discrete approximation to the (divergent) integral $\int_0^1 \sin(\pi x)^{-2(g-1)}\,dx$. To find the leading asymptotics, we can replace $\sin(x)$ by x, and taking the large k limit we obtain: $(k/2)^{g-1}\sum_{j=1}^{\infty}(k/(j\pi))^{2(g-1)} = k^{3(g-1)}\zeta(2(g-1))/(2^{g-1}\pi^{2(g-1)})$. This can be easily proven, and in fact, a generalization of this formula for arbitrary groups appeared in Witten's work [25].

Below we will concentrate on the case of SL_3, however the formulas can be extended to SL_n as well.

If we perform the trick above for SL_3, up to a constant, the leading behavior of the Verlinde formula appears to be

$$V_g^{\mathrm{SL}_3}(k) \sim \text{const}\cdot k^{8(g-1)}/\pi^{6(g-1)} \sum_{i,j=1}^{\infty} (ij(i+j))^{-2(g-1)}.$$

One can write down more general sums, e.g.:

$$S(a,b,c) = \sum_{i,j=1}^{\infty} i^{-a} j^{-b} (i+j)^{-c},$$

closely related to the so-called *multiple zeta values* [27].

It was discovered by Witten that all intersection numbers of moduli spaces are given by combinations of multiple ζ-values [24, 25]. Below we give a couple of useful formulas for them. We restrict ourselves to the case $S(2g, 2g, 2g)$ for simplicity. Similar formulas exist in greater generality.

Lemma 4.1.

$$S(2g,2g,2g) = \frac{1}{2}\int_0^1 \bar{B}_{2g}(x)^3\,dx, \tag{4.1}$$

where $\bar{B}_n(x)$ is a modified nth Bernoulli polynomial, $\bar{B}_n(x) = -(2\pi I)^n B_n(x)/n!$.

$$S(2g,2g,2g) = \frac{1}{3}\operatorname*{Res}_{(0,0)} \cot(x)\cot(y)(xy(x+y))^{-2g}. \tag{4.2}$$

Sketch of Proof: The first formula follows from the definition of the Bernoulli polynomials:

$$\bar{B}_n(x) = \sum_{j\neq 0} e^{2\sqrt{-1}j\pi x}/j^n.$$

Indeed, substituting this into $\frac{1}{2}\int_0^1 B_{2g}(x)^3\,dx$, one obtains $S(2g,2g,2g)$ on the nose; the coefficient $1/2$ is a combinatorial factor.

The proof of the second formula is similar to the proof of Theorem 3.2. One has to apply the Residue Theorem in two steps. That the residue at infinity vanishes follows from the expansion $\cot(x)$:

$$\pi\cot(\pi x) = \sum_{n\in\mathbb{Z}}(x-n)^{-1}. \quad \square$$

4.2. Intersection numbers of the moduli spaces. First we recall some facts about the cohomology of the moduli spaces. We will ignore that the moduli spaces are not smooth in general, and accordingly, we will assume the existence of a universal bundle, Riemann-Roch formula, etc. However, formulas analogous to (3.2) exist for the smooth moduli spaces as well (e.g. when the degree and rank are coprime for SL_n), and all of our statements are rigorous for these cases. Some of the singular moduli spaces (e.g. vector bundles) can be handled using the methods of [3]. We will also ignore certain difficulties which arise for Spin_n, $n>6$, and the exceptional groups, where the ample line bundle exists only for $k=0 \mod l$, for some l, depending on the type of the group. In these cases the Verlinde formula is a polynomial only when restricted to these values. Thus what follows should be perceived as a scheme of a proof, which works as it is in some cases, but requires modification and more work in greater generality.

There is a universal principal G-bundle U over the space $C\times\mathfrak{M}_C^G$, which induces a map $\mathfrak{M}_C^G\to BG$, and consequently a map $s:\mathrm{H}^*(BG)\to\mathrm{H}^*(\mathfrak{M}_C^G)\otimes\mathrm{H}^*(C)$.

Recall that $\mathrm{H}^*(BG)\cong\mathrm{Sym}(\mathfrak{g}^*)^G$, the space of G-invariant polynomial functions on $\mathfrak{g}$. This is a polynomial ring itself in $\mathrm{rank}(G)$ generators and it is isomorphic by restriction to $S^G=\mathrm{Sym}(\mathfrak{t}^*)^W$. For every $\alpha\in\mathrm{H}_i(C)$ and $P\in S^G$ we obtain a cohomology class of $\alpha\cap s(P)\in\mathrm{H}^{2i-j}(\mathfrak{M}_C^G)$, the α-component of $s(P)$. In fact, s induces a map $\bar{s}:\mathrm{H}_*(C)\cap S^G\to\mathrm{H}^*(\mathfrak{M}_C^G)$, where $\mathrm{H}^*(C)\cap S^G$ is the free commutative differential algebra generated by the ring S^G and the differentials of negative degree modeled on $\mathrm{H}_*(C)$. For the case of SL_n and coprime degree and rank it is known that $\bar{s}$ is surjective [2, 14]. In particular, denoting the fundamental class of C by η_C, and the basic invariant scalar product from §2 by P_2 we obtain a class $\omega = \eta_C \cap s(P_2) \in \mathrm{H}^2(\mathfrak{M}_C^G)$, which turns out to be the first Chern class of the line bundle from (1.2). To simplify the notation, below we omit the map s and also α if $\alpha = 1$, when writing down the classes $\mathrm{H}^*(\mathfrak{M}_C^G)$. Thus $1 \cap s(P)$ will be denoted simply by P.

Any power series in the variables $\alpha \cap P$ can be integrated over $\mathfrak{M}_C^G$ and these numbers are called the intersection numbers of the moduli space. Naturally, only the terms of degree $\dim \mathfrak{M}_C^G = \dim(G)(\mathrm{genus}(C) - 1)$ will contribute.

Witten, using non-rigorous methods, gave a complete description of these intersection numbers in the most general case [25]. His formulas are combinations of multiple ζ-values, and are rather difficult to calculate. In this paper, we will focus only on a subset of these intersection numbers, which are of the form $\int_{\mathfrak{M}} \omega^l P$, where $l \in \mathbb{N}$ and P is a not necessarily homogeneous Weyl-symmetric function on $\mathfrak{t}$.

Conjecture 4.2. *For every group G, there exists a residue form Ω^G depending on g, defined in a neighborhood of $0 \in \mathfrak{t}$, the Cartan subalgebra of G, such that*

$$\int_{\mathfrak{M}} e^{\omega} P = \operatorname*{Res}_{\text{at } 0 \in \mathfrak{t}} \Omega^G P, \tag{4.3}$$

For $G = \mathrm{SL}_n$,

$$\Omega = n^{g-1} \operatorname*{Res}_{x_{n-1}=0} \dots \operatorname*{Res}_{x_1=0} \prod_{\alpha \in \Delta^+} L_\alpha^{2(g-1)} \prod_{i=1}^{n-1} \cot(x_i)\, dx_i, \tag{4.4}$$

where the x_i-s are halves of the simple (additive) roots of SL_n, ordered according to the Dynkin diagram.

Let us write down the formula for SL_3 more explicitly and inserting the "grading":

$$\int_{\mathfrak{M}} e^{k\omega} P = (3k^2)^{g-1} \operatorname*{Res}_{y=0} \operatorname*{Res}_{x=0} \frac{k^2 \cot(kx) \cot(ky)\, dx\, dy}{(8xy(x+y))^{2(g-1)}} P \tag{4.5}$$

Remark 4.1. It can be shown that (4.4) is consistent with Witten's formulas. We will not give the proof here, but note that the link between the two types of formulas is given by equalities like (4.2).

At the moment we do not know Ω^G for general G.

Our formulas seem to be related to those given in the works of Jeffrey and Kirwan [11, 12]. □

Finally, we present another evidence for (4.4): the consistency with the Verlinde formula. First we need a few facts about the moduli spaces. Fix a curve C of genus g and a group G. They will be omitted from the notation.

Lemma 4.3.

(1) $c_1(T_{\mathfrak{M}}) = h\omega$,
(2) $p(T_{\mathfrak{M}}) = c(\mathrm{Ad}\, U_z)^{2(g-1)} = \prod_{\alpha \in \Delta}(1+\alpha)^{2(g-1)}$,
(3) $\hat{A}(T_{\mathfrak{M}}) = \prod_{\alpha \in \Delta^+} \left(\frac{\alpha/2}{\sinh(\alpha/2)}\right)^{2(g-1)}$.

Here h is the dual Coxeter number of G, p denotes the total Pontryagin class, c the total Chern class, U_z is the bundle over $\mathfrak{M}$ obtained by restricting the universal principal bundle U to a slice $z \times \mathfrak{M}$ for some $z \in C$ and $\operatorname{Ad} U_z$ is the vector bundle associated to U_z via the adjoint representation of G.

For the proof of the first two statements in some partial cases see [2]. The second statement follows from the Kodaira-Spencer construction. From the second statement we find that the Pontryagin roots of $T_{\mathfrak{M}}$ are the roots of the Lie algebra $\mathfrak{g}$, and this in turn implies the third statement.

Finally, we can put everything together. We will calculate $\dim \mathrm{H}^0(\mathfrak{M}_C^G, \mathcal{L}^k)$. Again, consider $G = \mathrm{SL}_3$ for simplicity. First, the Kodaira vanishing theorem applies to $\mathcal{L}^k$, because the canonical bundle of $\mathfrak{M}$ is negative, (this follows from the Lemma 4.3(1), see also [3]). Thus we can replace the dimension of H^0 by the Euler characteristic, and apply the Riemann-Roch theorem:

$$\dim \mathrm{H}^0(\mathfrak{M}_C^G, \mathcal{L}^k) = \chi(\mathfrak{M}_C^G, \mathcal{L}^k) = \int_{\mathfrak{M}} e^{k\omega}\, \mathrm{Todd}(\mathfrak{M}).$$

According to Lemma 4.3(1), and using the standard shifting trick we can rewrite this integral as

$$\int_{\mathfrak{M}} e^{(k+h)\omega} \hat{A}(\mathfrak{M}).$$

We can calculate this integral using (4.5) and Lemma 4.3(3), and the result is exactly $V_k^G(g)$ according to (3.4). This proves (1.2). $\square$

References

1. M.F. Atiyah, The geometry and physics of knots, Cambridge University Press, 1990.
2. M.F. Atiyah and R. Bott, *Yang-Mills equations over Riemann surfaces*, Phil. Trans. Royal Soc. London **308** (1982) 523-615.
3. A. Bertram, A. Szenes, *Hilbert polynomials of moduli spaces of rank 2 vector bundles II*, Topology **32** (1993) 599-609.
4. R. Bott, *Homogeneous vector bundles*, Ann. of Math. **66** (1957) 203-248.
5. A. Beauville, *Conformal blocks, fusion rules and the Verlinde formula*, preprint, 1994.
6. A. Beauville, Y. Laszlo, *Conformal blocks and generalized theta functions*, preprint, 1993.
7. G. Faltings, *Proof of the Verlinde formula*, preprint.
8. D. Gepner, *Fusion Rings and Geometry*, Comm. Math. Phys. **141** (1991) 381.
9. P. Griffiths, J. Harris, Principles of Algebraic Geometry, Wiley-Interscience, 1978.
10. R. Hartshorne, Residues and Duality, LNM 20, Springer Verlag, 1966.
11. L. Jeffrey, F. Kirwan, *Localization for Nonabelian Group Actions*, preprint, 1993.
12. L. Jeffrey, F. Kirwan, in preparation.
13. V. Kac, Infinite dimensional Lie algebras, 3rd edition, Cambridge University Press, 1990.
14. F. Kirwan, *The cohomology rings of moduli spaces of vector bundles over Riemann surfaces*, preprint.
15. S. Kumar, *Demazure character formula in arbitrary Kac-Moody setting*, Inv. Math. **89** (1987) 395-423.
16. S. Kumar, M.S. Narasimhan, A. Ramanathan, *Infinite Grassmannian and moduli space of G-bundles*, preprint, 1993.
17. O. Mathieu, *Formules de charactères pour les algébres Kac-Moody générales*, Astérisque **159-160** (1988) 1-267.
18. M. Thaddeus, *Conformal field theory and the moduli space of stable bundles*, preprint, 1991.
19. A. Pressley, G. Segal, Loop groups, Clarendon Press, 1986.
20. G. Segal, Two Dimensional Conformal Field Theories and Modular Functors in IXth Intern. Conf. on Math. Physics, eds. B. Simon, A. Truman, I.M. Davies, 1989.
21. A. Szenes, *The Verlinde formulas and moduli spaces of vector bundles*, PhD thesis, Harvard University, 1992.
22. A. Tsushiya, K. Ueno, Y. Yamada, *Conformal field theory on universal family of stable curves with gauge symmetries*, Adv. Stud. in Pure Math. **19** (1989) 459-565.
23. E. Verlinde, *Fusion rules and modular transformations in 2d conformal field theory*, Nucl. Phys. B **300** (1988) 360-376.
24. E. Witten, *On Quantum Gauge Theories in Two Dimensions*, Comm. Math. Phys. **141** (1991) 153.
25. E. Witten, *Two dimensional gauge theories revisited*, preprint, 1992.
26. D. Zagier, *Elementary aspects of the Verlinde formula and the Harder-Narasimhan-Atiyah-Bott formula*, preprint.
27. D. Zagier, *Values of zeta functions and their applications*, preprint.

Massachusetts Institute of Technology, Department of Mathematics, Cambridge, MA 02139

E-mail address: szenes@math.mit.edu

CANONICAL AND ALMOST CANONICAL SPIN POLYNOMIALS OF AN ALGEBRAIC SURFACE

ANDREI TYURIN

Steklov Mathematical Institute

0. INTRODUCTION

Classically, two approaches have been proposed in algebraic surface theory: the *first* uses standard stuff on linear systems, adjunction and singularity theory. As usual, one gets results about some particular class of surfaces. The *second* uses a representation of the surface as a pencil of algebraic curves; that is, as a curve over a function field. This method is very useful for arithmetic applications. If the genus of the fibre is small, it can be used to describe certain classes of surfaces, such as elliptic surfaces and pencils of genus 2 (Xiao Gang), or to obtain some information on 'atomic structure' of surfaces, in Miles Reid's terminology.

Both of these methods use some geometric subobjects of the surface, such as curves and points. On the other hand, there exist also geometric objects lying *over* a surface, such as vector bundles and torsion-free sheaves, which are expressive enough to describe the geometry of the surface itself.

From a technical point of view, rank 2 vector bundles are extremely useful for understanding the underlying smooth structure of an algebraic surface. But I will try to convince you that they are also useful in algebraic geometry. To do this, let me recall two examples of the use of rank 2 vector bundles in the two approaches referred to above. In the first, any base point of a complete linear system determines a rank 2 vector bundle, whose geometry gives good information about this linear system. In the second, using a fibration of a surface one can try (following Miyaoka) to construct fibrewise a stable rank 2 vector bundle with c_1 equal to the canonical class of the surface, and Euler characteristic zero. The existence of such a bundle implies the geographical inequality $K_S^2 \leq 2c_2(S)$ (see for example [T3]). The interplay between geometric 'subobjects' and 'overobjects' is described by the theory of *jumping curves*.

In the present article we propose a new system of notions, constructions and notations for a *third* approach in algebraic surface theory.

CONTENTS

1. JACOBIAN OF A SURFACE

Let S be a nonsingular, compact, nonruled, regular algebraic surface. Then the map $m\colon S \to S_{\min}$ of S to the minimal model $S_{\min}$ of S is uniquely determined. Let $K_{\min} = m^*(K_{S_{\min}}) \in H^2(S,\mathbb{R})$ be the pullback of the canonical class of the minimal model. Then $K_{\min} \in H^2(S,\mathbb{R})$ is contained in the closure of the Kähler cone $K(S) \subset \operatorname{Pic} S \otimes \mathbb{R}$.

Recall that a polarisation H is *almost canonical* (an *ac*-polarisation for short) if the ray $\mathbb{R}^+ \cdot H$ in the projectivisation of the Kähler cone is close to the ray $\mathbb{R}^+ \cdot K_{\min}$ in the sense of the Lobachevski metric (see [T5]). The symbol $M^{ac}(r, c_1, c_2)$ will denote the moduli space of *ac*-slope stable bundles on S of rank r with Chern classes c_1, c_2. By analogy with the Jacobian of an algebraic curve we have proposed (see [T5]) the following definition.

Definition 1. *The Gieseker closure (see [G])*

$$\overline{M^{ac}(2, K_S, c_2(S))} = J(S) \tag{1.1}$$

is called the Jacobian of S.

The Jacobian $J(S)$ contains a distinguished point $T^*S = \Omega S \in J(S)$, the cotangent bundle of S, which is stable by a theorem of Bogomolov. Hence $J(S)$ is always nonempty and $\dim J(S) \geq \operatorname{v.dim} J(S)$ (= the virtual, or expected, dimension of $J(S)) = 4c_2(S) - K_S^2 - 3(p_g + 1)$.

Now it is easy to see from the last formula for virtual dimension of moduli spaces of vector bundles that the virtual dimension of $J(S)$ can be described as follows. By Noether's formula the constant $\mu(S) = (3c_2(S) - K_S^2)/4$ is an integer—called the ***Miyaoka number*** of S—and the expected dimension of the Jacobian of S is

$$\operatorname{v.dim} J(S) = 5\mu(S). \tag{1.2}$$

In particular $\dim J(S) \geq 5\mu(S) \geq 0$, because of the Bogomolov-Miyaoka-Yau inequality. Moreover, one has:

(1) $\mu(S) = 0 \Longrightarrow S = \mathbb{P}^2$, and $J(\mathbb{P}^2) = T^*\mathbb{P}^2$ is a single point.
(2) For a K3 surface $\mu(S) = 18$ is the same number as for the projective plane with 18 points blown up.

(3) If $\sigma\colon \tilde{S} \to S$ is the blow-up a point of S then $\mu(\tilde{S}) = \mu(S) + 1$.
(4) For an unramified cover $\phi : S' \to S$ of degree d, $\mu(S') = d \cdot \mu(S)$.

In fact, it is also useful to consider the following generalisation of our $J(S)$ (see [T3]).

Definition 2. *The Gieseker closure (see [G])*

$$\overline{M^{ac}(2, K_S, c_2(S) + k)} = J_k(S) \tag{1.3}$$

is called the k-th Jacobian of S.

By analogy with $J(S)$ the k-th Jacobian contains a distinguished subscheme:

$$\{\Omega S\}_k = \{F \in J_k(S) | F^{**} = \Omega S\}, \tag{1.4}$$

the subset of torsion-free sheaves which have the cotangent bundle of S as reflexive hull. The structure of $\{\Omega S\}_k$ can be described as a projectivisation of the standard vector bundle on the Hilbert scheme of S and $\dim\{\Omega S\}_k = 3k$.

The virtual (expected) dimension of the kth Jacobian of S is given by

$$\text{v.}\dim J_k(S) = 5\mu(S) + 4k. \tag{1.5}$$

A priori the geometrical dimension of $J(S)$ can be bigger than this, but for $J_k(S)$ with large k the virtual dimension is equal to the geometric dimension (by a Donaldson type theorem). This is one reason for considering the generalisation. Another reason is illustrated by the following example.

Example 1: Blow-up of a point. As usual, any point $s \in S$ defines the so-called Poincaré $\mathbb{P}^1$-bundle

$$\pi : P_s(J_k(S)) \to J_k(S), \tag{1.6}$$

where for any vector bundle $E \in J_k(S)$, the fibre

$$\pi^{-1}(E) = \mathbb{P}E_s$$

is the projectivisation of the fibre of E over s.

Of course, this construction works only over an open subset of the moduli space. We can extend this $\mathbb{P}^1$-bundle to the blow-up along the subvariety

$$J_k(S)_s = \{F | s \in \operatorname{Sing} F\} \subset J_k(S).$$

Now if $\sigma\colon \tilde{S} \to S$ is the blow-up of a point $s \in S$ and ℓ is the exceptional curve on $\tilde{S}$ then it is easy to see that

$$J_k(\tilde{S}) = P_s(J_{k+1}(S)) \tag{1.7}$$

birationally. The birational map $i_k\colon P_s(J_{k+1}(S)) \to J_k(\tilde{S})$ is defined by

$$i_k(E)^* = \ker(\sigma^* E^* \to \mathcal{O}_\ell)^*. \tag{1.8}$$

The inverse map i_k^{-1} is well-defined over an open subvariety

$$J_k(\tilde{S})_0 = \{F \in J_k(\tilde{S}) | F|_\ell = \mathcal{O}_\ell \oplus \mathcal{O}_\ell(-1)\} \tag{1.9}$$

of $J_k(\tilde{S})$, namely

$$i_k^{-1}(F) = R^0\sigma(\ker(F \to \mathcal{O}_\ell(-1))). \tag{1.10}$$

A full filtration of $J_k(\tilde{S})$ is given by subvarieties

$$J_k(\tilde{S})_n = \{F \in J_k(\tilde{S}) | F|_\ell = \mathcal{O}_\ell(n) \oplus \mathcal{O}_\ell(-1-n)\} \tag{1.11}$$

which are the spaces of stable quasi-parabolic bundles for the pair $\ell \in \tilde{S}$, where ℓ is the exceptional curve, described by Kronheimer in [K]. There exists a beautiful description due to Brieskorn of a versal filtration of this type, which predicts the birational modification which we need to make in order to get a regular map.

Hence if $\tilde{S_n}$ is the result of blowing up n points on S then $J(\tilde{S_n})$ is rationally fibred over $J_n(S)$:

$$i_k^{-1} : J(\tilde{S_n}) \to J_n(S) \tag{1.12}$$

with $(\mathbb{P}^1)^n$ as a fibre, and we can specify (at least theoretically) birational modifications to resolve the indeterminacies of this birational correspondence.

The final task is to extend this construction to sheaves with singularities on ℓ.

Now we can define our system of discrete invariants of S. Namely, the standard definition of the slant-product in the algebraic geometric context (see [T4]) defines a polynomial

$$a\gamma_S \in S^{5\mu(S)}(H^2(S,\mathbb{Z}) \oplus H^4(S,\mathbb{Z})). \tag{1.13}$$

(If the Jacobian $J(S)$ has the expected dimension this construction is straightforward, but otherwise we need to use some trick; as for example in [P-T] or in §7 of [T1].) As is usual we will sometimes consider only the polynomials in H^2.

The same construction gives a collection of polynomials

$$a\gamma_k \in S^{5\mu(S)+4k}(H^2(S,\mathbb{Z}) \oplus H^4(S,\mathbb{Z})). \tag{1.14}$$

Definition 3. *The polynomial (1.13) is called the canonical polynomial of S. The polynomials (1.14) are called the almost canonical polynomials of S.*

Continuation of Example 1. The following result is a consequence of the descriptions of the birational maps between $J_k(\tilde{S})$ and $J_{k+1}(S) \times \mathbb{P}^1$.

Proposition 1. *Any canonical or almost canonical polynomial of $\tilde{S}$ is a linear combination of canonical and almost canonical polynomials of S with polynomials in ℓ as coefficients.*

By induction this also holds for the blow-up of n points of S, but as coefficients we need to use polynomials in the exceptional linear forms and in the intersection form on the blown-up lattice.

Remarks. (1) In the differential geometry set-up this proposition was proved by J. Morgan and T. Mrowka (see [M-M]) for the first step of the birational correspondence. We shall not give here any more precise statements and constructions because V. Pidstrigach has proved this in a rather more general situation. Namely, by Morgan's Theorem (see [M]) our almost canonical polynomials coincide with $w_2(S)$-Donaldson polynomials of the underlying smooth 4-manifold of S. Instead of S and $\tilde{S}_n$ one can consider a 4-manifold of the form

$$M_n = M \# N, \tag{1.15}$$

where M is a simply connected 4-manifold with $b_2^+ > 1$, and N is any negative definite 4-manifold with rank $H^2(N, \mathbb{Z}) = n$.

(2) Actually the situation described above holds only in the case when K_S^2 is odd. If K_S^2 is even then a priori there exists a finite set of chambers around $K_{\min}$ in the Kähler cone, and there is a finite set $\{J_n(S)^i\}$ of birationally equivalent Jacobians (for details see [T6], Chap. 1, §2).

(3) The geometric situation of Example 1 (formulas (1.7)-(1.12)) was described in [Q] and [B2] and successfully used by Okonek and Van de Ven in [O-V].

(4) Recently R. Fintushel and R. Stern have presented a beautiful blow-up formula (see [F-S]).

2. Jumping filtration and spin polynomials

Next we shall consider the subset

$$\Theta = \Theta(S) = \{F \in J(S) | h^0(F) \geq 1\} \tag{2.1}$$

of $J(S)$. (We denote this subspace by theta by analogy with Riemann's Theorem in the case of algebraic curves.)

Actually the jumping conditions define a filtration (see [T5]):

$$J(S) \supseteq \Theta \supseteq W^1(S) \supseteq \cdots \supseteq W^r(S) \supseteq \cdots \tag{2.2}$$

If the family of torsion-free sheaves is in 'general position' near F, then the fibre of the normal bundle to $W^r(S)$ at F is given by

$$N_{W^r/J(S)}|_F = \mathrm{Hom}(H^0(F), H^1(F)) \tag{2.3}$$

with $H^0(F) = \mathbb{C}^r$, $H^1(F) = \mathbb{C}^{r-\chi(F)}$ (if $\chi(F) \leq 0$).

Thus the virtual (expected) codimension of Θ is

$$\mathrm{v.codim}\,\Theta = 1 - \chi(T^*S) = 1 + \frac{c_2(S)}{3} + \frac{2\mu(S)}{3} \tag{2.4}$$

(in particular note that Θ is never a divisor) and the virtual dimension of the theta locus is

$$\mathrm{v.dim}\,\Theta = \frac{35c_2(S) - 13K^2}{12} - 1 = 4\mu(S) - (p_g(S) + 1) - 1. \tag{2.5}$$

This integer is nonnegative if and only if the inequality

$$(2 \cdot 696969 \cdots)c_2(S) > K_S^2 \tag{2.6}$$

holds.

Again the standard definition of the slant-product homomorphism in the algebraic geometric context (see [T1]) gives the polynomials

$$a\gamma_\Theta \in S^{\dim\Theta}(H^2(S,\mathbb{Z}) \oplus H^4(S,\mathbb{Z})); \tag{2.7}$$

but here we must be careful with the choice of polarization, as explained in [T6].

Definition 4. *The polynomials (2.7) are called canonical spin polynomials of S.*

Of course, we can do the same with the k-th Jacobian: consider the subset $\Theta_k = \{F | h^0(F) \geq 1\} \subset J_k(S)$. The jumping conditions again define a filtration defined by

$$J_k(S) \supseteq \Theta_k \supseteq W_k^1(S) \supseteq \cdots \supseteq W_k^r(S) \supseteq \cdots$$

(see [T5]). The description of the normal bundle is exactly the same as before, and thus the virtual codimension of Θ_k is given by

$$\mathrm{v.codim}\,\Theta_k = 1 - \chi(T^*S) + k = 1 + \frac{c_2(S)}{3} + \frac{2\mu(S)}{3} + k; \tag{2.8}$$

and the virtual dimension of Θ_k is

$$\begin{aligned}\mathrm{v.dim}\,\Theta_k &= \frac{35c_2(S) - 13K^2}{12} + 3k - 1 \\ &= 4\mu(S) - (p_g(S) + 1) + 3k - 1.\end{aligned} \tag{2.9}$$

The standard definition of the slant-homomorphism gives polynomials

$$a\gamma_{\Theta_k(S)} \in S^{\mathrm{v.dim}\,\Theta_k}(H^2(S,\mathbb{Z}) \oplus H^4(S,\mathbb{Z})). \tag{2.10}$$

Definition 4′. *The polynomials (2.10) are called almost canonical spin polynomials of S.*

Remark. An interesting observation here is the following: $\operatorname{codim} \Theta_k = 1 + \frac{c_2(S)}{3} + \frac{2\mu(S)}{3} + k$ and $\dim\{\Omega S\}_k = 3k$ (see (1.4)) but since our surface S is regular, the intersection $\{\Omega S\}_k \bigcap \Theta_k$ is always empty. This means that the Gieseker compactification is very far from minimal, and Θ_k is very far from ample.

Continuation of Example 1. First of all, if we blow up a point s on S as in (1.6) and (1.7) and consider the restriction of the Poincaré bundle (1.6) to $\Theta_k(S)$:

$$\pi\colon P_s(\Theta_k(S)) \to \Theta_k(S) \tag{2.11}$$

then birationally

$$P_s(\Theta_{k+1}(S)) = \Theta_k(\tilde{S}). \tag{2.12}$$

Indeed, we have the exact sequence

$$0 \to \sigma^*(E) \to i_k(E) \to \mathcal{O}_\ell(-1) \to 0$$

and every section of $i_k(E)$ comes from E. The same is true for a sheaf belonging to each subvariety of the Brieskorn filtration (1.11) and again we can describe the modification precisely.

Repeating the constructions for Jacobians (see(1.6)–(1.11)) in the set-up of theta loci yields the next result.

Proposition 2. *Almost canonical spin polynomials of $\tilde{S}$ are superpositions of almost canonical spin polynomials of S with polynomials in ℓ as coefficients; and the same holds for repeated blow-ups, but then the coefficients must be polynomials in the exceptional linear forms and in the intersection form on the blown-up lattice.*

Here again, a stronger result in the differential geometical set-up was announced by V. Pidstrigach in [P1]. In [P2] he successfully used the gluing constructions for jumping instantons to obtain a proof of the Van de Ven conjecture.

What do we know about general properties of Jacobians and theta loci?

Application of general theorems to our particular case yields the following properties if k is sufficiently large:

(1) $J_k(S)$ has the right dimension (Donaldson [D], Zuo [Z]).
(2) $J_k(S)$ is irreducible (Gieseker-Li [G-L]).
(3) If S is of general type and minimal, then $J_k(S)$ is also of general type (Li).

About Θ_k we know that if it has the right dimension then Θ_k is fibred over the Hilbert scheme of S

$$\Theta_k \to \mathrm{Hilb}^{c_2(S)+k} \tag{2.13}$$

with rational fibres $\mathbb{P}^n$, where

$$n = c_2(S) + k - K_S^2 - p_g(S) - 2. \tag{2.14}$$

One can see that for large k the theta locus loses property (3) even for minimal surfaces of general type.

The important property, which implies (3) for Jacobians is the following (Li):

(4) For a generic holomorphic form $\omega \in H^{2,0}(S)$ one has $a\gamma_k(\omega + \overline{\omega}) > 0$.

In contrast to this property of almost canonical polynomials, for spin polynomials we have:

(5) For any holomorphic form $\omega \in H^{2,0}(S)$ one has

$$a\gamma_{\Theta_k}(\omega + \overline{\omega}) = 0. \tag{2.15}$$

This means that the shape of spin polynomials is simple, as we shall prove in this article. The main result is the following:

Shape Theorem. *There exists on S a finite set of irreducible curves*

$$C_1, \cdots, C_{N(S)} \tag{2.16}$$

subject to the conditions

$$2C_i \cdot K_{min} \leq K_{min}^2, \tag{2.17}$$

such that any almost canonical spin polynomial on H^2 with $k \gg 0$ has the form

$$a\gamma_{\Theta_k} \in S^{\mathrm{v.dim}\,\Theta_k}(q_S, K_S, C_1, ..., C_{N(S)}). \tag{2.18}$$

In other words, $a\gamma_{\Theta_k}$ is a polynomial in the intersection form q_S as a quadratic form, and in the classes of curves (2.16) and the canonical class as linear forms.

As a corollary of this fact, one has the following result:

sV Theorem. *If either $p_g(S) > 0$ and S is not a K3 surface, or S is the Barlow surface, then there exists a proper nontrivial sublattice*

$$sV(S) \subset Pic\ S \subset H^2(S, \mathbb{Z}) \tag{2.19}$$

such that:

(1) $sV(S)$ is invariant with respect to the action of Diff S;

(2) $sV(S)$ contains K_S and all -1-curves:

(3) if S' is any other algebraic surface structure on the underlying differentiable 4-manifold of S then

$$sV(S) = sV(S'). \tag{2.20}$$

In particular, this means that the minimal algebraic subgroup $A(S)$ of the complex orthogonal group $O_{\mathbb{C}}(q_S)$ containing the image of the representation of the diffeomorphism group Diff S is reducible.

Remark. The diffeomorphism invariant sublattice $sV(S)$ is contained in the lattice $\langle K_S, C_1, ..., C_{N(S)}\rangle \subset \text{Pic } S \subset H^2(S, \mathbb{Z})$ generated by the collection of classes (2.16) and the canonical class, but it may be a proper sublattice of $\langle K_S, C_1, ..., C_{N(S)}\rangle$. In fact, direct computations show that on the one hand we can remove all (-2)-rational curves from the system of generators (2.16) (see the end of [T4]), but on the other hand we can't remove the canonical class or the (-1)-rational curves.

From the present point of view, the lattice sV can be called the *smooth Picard lattice*, and our next aim is to prove that:

(1′) $sV(S)$ contains a nontrivial subcone of ample divisors

$$sK(S) = K(S) \cap sV(S) \subset sV(S), \tag{2.21}$$

where $K(S)$ is the Kähler cone of S; and

(2′) if S' is any other algebraic surface structure on the underlying differentiable 4-manifold of S then the intersection

$$sK(S) \cap sK(S') \neq \emptyset. \tag{2.22}$$

Thus, using the 'first approach' of the introduction we could prove that S and S' can be embedded in some projective space simultaneously, and we could try to prove that they are deformation equivalent.

Recently P. Kronheimer and T. Mrowka announced, for a certain class of simply-connected 4-manifolds M, the existence of a finite collection of 2-cohomology classes $K_1, ..., K_p$ invariant with respect to the action of Diff M (see [K-M]). Hence these classes generate a sublattice

$$KM(M) \subset H^2(M, \mathbb{Z}) \tag{2.23}$$

which is invariant under the action of Diff M. R. Brussee has remarked (see [B1]) that if $M = S$ is an algebraic surface then the classes $K_1, ..., K_p$ are algebraic, that is, of Hodge type (1,1). This means that in case $p_g > 0$

the sublattice $KM(S)$ is proper. It would be rather interesting to compare $sV(S)$ with $KM(S)$.

By Theorem 2 of [K-M], the collection of classes $K_1, \cdots, K_p$ defines a piecewise-linear seminorm on $KM(S)$, estimating the minimal genus of a smooth surface realising a cohomology class. On the other hand, the collection of classes (2.16) defines a convex seminorm in $sV(S)$ with generators which are realised as smooth Riemann surfaces of minimal genus. In some sense these seminorms are dual.

Finally, E. Witten has announced recently that the Kronheimer-Mrowka classes $K_1, ..., K_p$ are precisely the classes of irreducible components of a general canonical divisor on the algebraic surface S (see [W]).

In order to prove the shape theorem, we need to use the 'geometric approximation procedure' (GAP for short). First, though, we need to explain the second reason for restricting oneself to the case $c_1 = K_S$, and how the shape theorem implies the sV theorem. The main observation is that our algebraic geometric picture is a slice of a much more general picture in the set-up of Riemannian geometry.

3. SPACES OF JUMPING INSTANTONS

Recall that if the underlying smooth manifold of an algebraic surface S is equipped with a Riemannian metric g then for every $SO(3)$-bundle E of topological type $(2, w_2, p_1)$ the gauge orbit space $\mathcal{B}(E) = \mathcal{A}_h^*(E)/\mathcal{G}$ of irreducible connections contains the subspace $\mathcal{M}^g(2, w_2(S), p_1) \subset \mathcal{B}(E)$ of anti-self-dual connections with respect to the Riemannian metric g, oriented by the choice of the lift to the anticanonical class $-K_S$ of the Stiefel-Whitney class $w_2(S)$, and by an orientation of a maximal positive subspace in $H^2(S, \mathbb{R})$. This space determines via the slant product the homogeneous polynomial

$$\gamma^g(2, K_S, p_1) \in S^d(H^2(S, \mathbb{Z}) \oplus H^4(S, \mathbb{Z})) \tag{3.1}$$

(see [D], [D-K]).

These polynomials behave naturally under diffeomorphisms of S. Namely, if $p_g > 0$ then for any $\sigma \in H^2(S, \mathbb{Z})$, and for any $\phi \in \operatorname{Diff} S$ preserving the orientation of a maximal positive subspace in $H^2(S, \mathbb{R})$, we have

$$\gamma^g(2, K_S, p_1)(\sigma) = \gamma^g(2, K_S, p_1)(\phi(\sigma)). \tag{3.2}$$

This means that some aspects of the shape of the polynomials are invariants of the smooth structure of the 4-manifold.

Now a new system of invariants of the underlying differentiable structure of 4-manifolds was proposed in [P-T], [T1] and [T2] . These so-called spin polynomial invariants will be used to prove the sV theorem.

Recall that if we consider the anticanonical class $-K_S$ as a $Spin^{\mathbb{C}}$-structure on S (see [P-T], [T1] or [T2]) equipped with a Riemannian metric g then for every $U(2)$-bundle E of topological type $(2, c_1, c_2)$ the gauge orbit subspace $\mathcal{M}^g(E) \subset \mathcal{B}(E)$ of anti-self-dual connections contains the subspace:

$$\mathcal{M}_r^{g,-K_S}(E) = \{(a) \in \mathcal{M}^g(E) | \operatorname{rank}\ker D_a^{-K_S,\nabla_0} \geq r\}, \tag{3.3}$$

where $D_a^{-K_S,\nabla_0}$ is the coupled Dirac operator (see [P-T]) with a Hermitian connection ∇_0 on the line bundle with first Chern class $-K_S$ (see [P-T], [T1] or [T2]).

This space determines by slant product the homogeneous polynomial

$$s_r\gamma^{g,-K_S}(2, c_1, c_2) \in S^{d_1}(H^2(S,\mathbb{Z}) \oplus H^4(S,\mathbb{Z})). \tag{3.4}$$

These are the so-called ***spin polynomials*** (see [T1] or [T2]).

Actually the space $\mathcal{M}_r^{g,-K_S}(E)$, and therefore also the polynomial (3.4), depends only on the pair $(-K_S + c_1, p_1)$. This means that if $p_g(S) > 0$ then the polynomial $s_r\gamma^{H,C}(2, -C, c_2)$ is an invariant (up to sign, of course) of the smooth structure of S. Or, more precisely, for any $\sigma \in H^2(S,\mathbb{Z})$ and any $\phi \in \operatorname{Diff} S$, preserving the orientation of a maximal positive subspace in $H^2(S,\mathbb{R})$, we have

$$s_r\gamma^{g,-K_S}(2, K_S, c_2)(\sigma) = s_r\gamma^{g,-K_S}(2, K_S, c_2)(\phi(\sigma)). \tag{3.5}$$

Fortunately, this equality holds even if $p_g(S) = 0$. Namely (see [T4] and [T6]) the polynomials $s_r\gamma^{H,C}(2, -C, c_2)$ are invariant up to sign under diffeomorphisms if $-K_S^2 < 8(r-1)$.

Thus to get differential geometrical information, one must restrict oneself to the equality $c_1 = K_S$ and investigate the geometry of the moduli spaces of stable vector bundles and torsion-free sheaves of topological type $(2, K_S, c_2)$; that is, Θ_k and the almost canonical spin polynomials. Moreover it is easy to see that if

$$s_r\gamma^{g,-K_S}(2, K_S, c_2) \in S^{\mathrm{v.dim}\,\Theta_k}(q_S, K_S, C_1, ..., C_{N(S)}),$$

$$\operatorname{rank}\langle K_S, C_1, ..., C_{N(S)}\rangle \leq \operatorname{rank} H^2(S,\mathbb{Z}) - 2$$

then the sublattice $\langle K_S, C_1, ..., C_{N(S)}\rangle$ contains some diffeomorphism invariant sublattice (which may be $\langle K_S, C_1, ..., C_{N(S)}\rangle$ itself).

Now, if a Riemannian metric g is a Hodge metric g_H, then the Donaldson Identification Theorem says that

$$\mathcal{M}^g(E) = M^H(2, c_1(E), c_2(E)), \tag{3.6}$$

i.e. the moduli space of H-slope stable vector bundles, and making the identification $(a) = E$, we have the following identifications of the kernel and cokernel of the coupled Dirac operator:

$$\ker D_a^{-K_S} = H^0(E) \oplus H^2(E), \tag{3.7}$$

$$\operatorname{coker} D_a^{-K_S} = H^1(E).$$

In our particular case, where $c_1(E) = K_S$, we have a skew symmetric isomorphism $E = E^*(K_S)$ and Serre duality provides an identification

$$H^0(E) = H^2(E)^*,$$

and a nondegenerate symmetric quadratic form

$$q_E \colon H^1(E) \to H^1(E)^*. \tag{3.8}$$

This means that a jumping h^0 implies a jumping h^2 and the theta locus has nontrivial multiple structure; that is, as a subscheme, it has nilpotents and it is not reduced. Fortunately, however, we can describe the scheme-theoretic structure precisely (see [T4] and [T6]). In the simplest case $r = 1$ the normal cone of the theta locus Θ_k at $E \in \Theta_k$ is the light cone of the quadratic form (3.8).

From this and a comparison of the Gieseker and Uhlenbeck compactifications (or, more precisely, a repetition of Morgan's arguments from [M]) we obtain the Identification Theorem.

Proposition 3. *If $p_g(S) > 0$ or $K_S^2 > 0$ then*

$$s_1\gamma^{g_H, -K_S}(2, K_S, c_2) = 2 \cdot a\gamma_{\Theta_{c_2 - c_2(S)}}. \tag{3.9}$$

The proof of this coincidence is contained in [T6] (see also [T4]).

Now, to get the sV theorem as a corollary of the shape theorem, we need to combine this proposition and the property (3.5) with the following fact.

Nondegeneracy Theorem. *If S is a simply connected surface of general type then*

$$a\gamma_{\Theta_k} \neq 0 \qquad \text{for } k \gg 0. \tag{3.10}$$

We will prove this theorem as a partial result of the geometric approximation procedure.

As a corollary of the nondegeneracy theorem we obtain nontriviality of the sV lattice: indeed, if the almost canonical spin polynomial satisfies

$$a\gamma_{\Theta_k} \in S^{\text{v.dim}\,\Theta_k}(q_S), \tag{3.11}$$

then for some (2,0)-form ω one has

$$a\gamma_{\Theta_k}(\omega + \overline{\omega}) \neq 0,$$

but this contradicts (2.15) for large k. Thus the diffeomorphism invariant sublattice in $\langle K_S, C_1, ..., C_{N(S)} \rangle$ is nontrivial.

Now we can explain the remark following proposition 2. In [P1], instead of S and $\tilde{S}$ Pidstrigach considered a 4-manifold of type $M_n = M\#N$ where M is a simply connected 4-manifold with $b_2^+ > 1$, and N is any negative definite 4-manifold with rank $H^2(N, \mathbb{Z}) = n$. He announced the following result:

Proposition. *Any spin polynomial of M_n is a linear combination of spin polynomials of S with polynomials in linear forms and the intersection form on $H^2(N, \mathbb{Z})$.*

We now need to discuss the geometric approximation procedure for describing the shape of an almost canonical spin polynomial. It consists of a number of steps, and we describe them one by one.

4. First step of GAP

First of all, we recall the construction of the geometric approximation $GA\Theta_k^0$ for the Θ locus (see [T1] and [T5]). This variety is a complete intersection in the projectivisation of a certain standard vector bundle on the Hilbert scheme of S and one can compute the polynomial for this family of torsion-free sheaves easily.

Remark. We shall only recall this construction for the H^2 part of the polynomial (see [T1] and [T5]), as the general construction is exactly the same.

Let

$$Z_d \subset S \times \mathrm{Hilb}^d S, \quad \text{with} \quad Z_d \cap (S \times \{\xi\}) = \{\xi\}, \tag{4.1}$$

be the universal subscheme, and

$$\begin{array}{ccc} & Z_d & \\ {}^{\overline{p}_S} \swarrow & & \searrow {}^{\overline{p}_H} \\ S & & \mathrm{Hilb}^d S \end{array} \tag{4.2}$$

be the two projections, induced by the projections p_H and p_S of the direct product $S \times \mathrm{Hilb}^d S$.

For any divisor class $D \in \mathrm{Pic}\, S$ consider the vector bundle

$$\mathcal{E}_D^d = R^0 \overline{p}_H (\overline{p}_S^* \mathcal{O}_S(D)). \tag{4.3}$$

This sheaf is locally free because the canonical homomorphism is surjective. Let H be the divisor class of the Grothendieck sheaf $\mathcal{O}_{\mathbb{P}\mathcal{E}_D^*}(1)$ on the projective bundle $\mathbb{P}\mathcal{E}_D^* \to \mathrm{Hilb}^d S$ associated with $\mathcal{E}_D^d$. Each section s of $\mathcal{E}_D^d$ corresponds to a section $\overline{s}$ of $\mathcal{O}_{\mathbb{P}\mathcal{E}_D^*}(1)$.

The restriction homomorphism $H^0(\mathcal{O}_S(D)) \xrightarrow{res} H^0(\mathcal{E}_D^d)$ can be written as the composite

$$H^0(\mathcal{O}_S(D)) \otimes \mathcal{O}_{\text{Hilb}} \to H^0(\mathcal{E}_D^d) \otimes \mathcal{O}_{\text{Hilb}} \xrightarrow{can} \mathcal{E}_D^d.$$

That is, any section $s \in H^0(\mathcal{O}_S(D))$ defines a section $res(s)$ of the vector bundle $\mathcal{E}_D^d$ and a section $\overline{res(s)}$ of $\mathcal{O}_{\mathbb{P}\mathcal{E}_D^*}(1)$.

If $s_1, \cdots, s_{h^0(\mathcal{O}(D))}$ is a basis of the space of sections $H^0(\mathcal{O}_S(D))$ then the intersection of divisors

$$_{0,0}E_{d,D-K_S} = \bigcap_{i=1}^{h^0(\mathcal{O}(D))} H_{\overline{res(s_i)}} \subset \mathbb{P}\mathcal{E}_D^* \tag{4.4}$$

is the space of all nontrivial extensions of type

$$0 \to \mathcal{O}_S \to E \to J_\xi(D - K_S) \to 0, \tag{4.5}$$

up to $\mathbb{C}^*$, for all $\xi \in \text{Hilb}^d S$ (see [T1]) if $H^1(\mathcal{O}(D)) = 0$. To describe the space of all nontrivial extensions of type (4.5) in general we have to consider the projectivisation

$$P_D = \mathbb{P}(H^1(\mathcal{O}(D)) \otimes \mathcal{O}_{\mathbb{P}\mathcal{E}_D^*} \oplus \mathcal{O}_{\mathbb{P}\mathcal{E}_D^*}(1))$$

of the vector bundle on $\mathbb{P}\mathcal{E}_D^*$ and its restriction to the complete intersection (4.4):

$$_{0,0}E_{d,D-K_S} = P_D|_{\bigcap_{i=1}^{h^0(\mathcal{O}(D))} H_{\overline{res(s_i)}}}. \tag{4.4'}$$

For example,

$$GA\Theta^0_{d-c_2(S)} = \bigcap_{i=1}^{h^0(2K_S)} H_{\overline{res(s_i)}} \subset \mathbb{P}\mathcal{E}^*_{2K_S} \tag{4.5'}$$

is a first geometric approximation to $\Theta_{d-c_2(S)}$ (see [T5]) for a minimal surface S.

It is easy to see that if the hypersurfaces (4.5′) are in general position then

$$\dim GA\Theta^0_k = \text{v.}\dim \Theta_k \tag{4.6}$$

Now $_{0,0}E_{d,D-K_S}$ is the base of the universal family of torsion-free sheaves given on the direct product $S \times {}_{0,0}E_{d,D-K_S}$ as the universal extension

$$0 \to p_1^*\mathcal{O}(H) \to \mathbb{E} \to (id \times \pi)^*\mathcal{I}_Z \otimes p_S^*\mathcal{O}_S(D - K_S) \to 0. \tag{4.7}$$

(For the cocycle of this extension see the diagram (4.25) and (4.28) from [T1].)

By the slant-product, $\mathbb{E}$ defines, as a family of sheaves, a polynomial $ga\gamma_{_{0,0}E_{d,D-K_S}}$: namely, we can consider the (2,2)-Künneth component $\mu_{\mathbb{E}}$ of the class

$$4c_2(\mathbb{E}) - c_1^2(\mathbb{E}) \tag{4.8}$$

as a cohomological correspondence

$$\mu_{\mathbb{E}} : H^2(S,\mathbb{Z}) \to H^2({}_{0,0}E_{d,D-K_S},\mathbb{Z}). \tag{4.9}$$

From (4.7), we have for any $\sigma \in H^2(S,\mathbb{Z})$,

$$\mu_{\mathbb{E}}(\sigma) = 4\pi^*(\widetilde{\sigma}) + 2(\sigma \cdot (D - K_S))H, \tag{4.10}$$

where π is the standard projection of ${}_{0,0}E_{d,D-K_S}$ to Hilb. We need to recall that the universal subscheme (4.2), as an algebraic correspondence, defines the cohomological correspondence

$$\mu_{\mathrm{Hilb}} : H^2(S,\mathbb{Z}) \to H^2(\mathrm{Hilb},\mathbb{Z}), \tag{4.11}$$

$$\widetilde{\sigma} = \mu_{\mathrm{Hilb}}(\sigma).$$

Roughly speaking, if a fundamental 2-cycle σ is given as a smooth oriented surface Σ then the fundamental class $\widetilde{\sigma}$ of $\mu_{\mathrm{Hilb}}(\sigma)$ is given as

$$\widetilde{\sigma} = \mu_{\mathrm{Hilb}}(\sigma) = [\mu_{\mathrm{Hilb}}(\sigma)] = \{\xi \in \mathrm{Hilb}\,|Supp\xi \cap \Sigma \neq \emptyset\}.$$

In other words, $\widetilde{\sigma}$ contains the clusters $\xi = p_1 + \cdots + p_d$ such that at least one point p_i is contained in Σ.

Now the value of our approximation to the spin polynomial at σ is the intersection number

$$ga\gamma_{({}_{0,0}E_{d,D-K_S})}(\sigma) = \left(4\pi^*(\widetilde{\sigma}) + 2(\sigma \cdot (D - K_S))H\right)^{\dim({}_{0,0}E_{d,D-K_S})} H^{h^0(\mathcal{O}(D))} \tag{4.12}$$

of 2-cocycles on $\mathbb{P}\mathcal{E}_D^{d*}$. In particular, the value of our approximation to the almost canonical spin polynomial at σ is the intersection number

$$ga\gamma_{\Theta_k^0}(\sigma) = \left(4\pi^*(\widetilde{\sigma}) + 2(\sigma \cdot K_S)H\right)^{\mathrm{v.dim}\,\Theta_k} H^{h^0(2K_S)}. \tag{4.12$'$}$$

(Here, as usual, $k = d - c_2(S)$.)

Definition 5. *The polynomial (4.12′) is called the geometric approximation of the almost canonical spin polynomial.*

It is easy to see that for $k \geq 2p_g(S) - c_2(S)$ one has

$$\dim GA\Theta_k^0 = \mathrm{v.}\dim \Theta_k$$

and $GA\Theta_k^0$ is irreducible (see [T5]).

Now using (4.12), the projection formula

$$\pi^*(\widetilde{\sigma})^j H^{\dim \mathbb{P}\mathcal{E}-j} = \widetilde{\sigma}^j \pi_* H^{\dim \mathbb{P}\mathcal{E}-j},$$

and the definition of the Segre classes $s_n(\mathcal{E}) = \pi_* H^{rank\mathcal{E}-1+n}$, we have in the final analysis

$$ga\gamma_{({}_{0,0}E_{d,D-K_S})}(\sigma) = 2^\delta((D-K_S)\cdot\sigma)^{\delta-2d} \times \sum_{j=0}^{2d} 2^j \binom{\delta}{j}((D-K_S)\cdot\sigma)^{2d-j} s_{2d-j}(\mathcal{E}_D^d)\widetilde{\sigma}^j, \tag{4.13}$$

where

$$\delta = 3d - 1 - h^0(\mathcal{O}(D)) = \dim {}_{0,0}E_{d,D-K_S}.$$

In particular (see [T5]) we obtain the following result.

Proposition 3. *If $GA\Theta_k^0$ has the expected dimension and $k > K_S^2 + (p_g(S)+1) - c_2(S)$ then*

$$ga\gamma_{\Theta_k}^0(\sigma) = 2^{\mathrm{v.dim}\,\Theta_k}(K_S\cdot\sigma)^{d-K_S^2-p_g-2} \times \sum_{j=0}^{2d} 2^j \binom{\mathrm{v.dim}\,\Theta_k}{j}(K_S\cdot\sigma)^{2d-j} s_{2d-j}(\mathcal{E}_{2K_S}^d)\widetilde{\sigma}^j \tag{4.13'}$$

where $d = c_2(S) + k$.

One can see that if $k \gg 0$ then the geometric approximation of the almost canonical spin polynomial is divisible by the linear form K_S.

To see the shape of our approximation of the spin polynomial it is enough to remark that $ga\gamma^0_{{}_{0,0}E_{d,D-K_S}}$ is a superposition of the standard polynomials

$$p_k^j(D)(\sigma) = s_{2c_2(S)+2k-j}(\mathcal{E}_D^{c_2(S)+k})\widetilde{\sigma}^j,$$

and to use the following result.

Ellingsrud's Proposition. $p_k^j(D) \in S^j(q_S, D)$. *In other words, these polynomials are polynomials in the intersection form q_S as a quadratic form, and D as a linear form.*

Hence, $ga\gamma_{\Theta_k}$ are polynomials in the intersection form q_S as a quadratic form and K_S as a linear form. Now one can see the shape of the polynomials (4.13) and of the first approximation of the almost canonical spin polynomial:

$$ga\gamma_{({}_{0,0}E_{d,D-K_S})} \in S^j(q_S, K_S, D), \tag{4.14}$$

$$ga\gamma_{\Theta_k}^0 \in S^j(q_S, K_S).$$

To get the almost canonical spin polynomials we have to apply a chain of elementary transformations to these families of sheaves and polynomials. In the next section we shall describe the centres of these elementary transformations for the case of spin canonical polynomials.

5. Beginning of the correcting procedure: crosses

A nonruled algebraic surface S carries a finite set of nonempty complete linear systems

$$\{|C_i|\}_{i=0,1,\dots,N} \qquad \text{such that } \ 2C_i \cdot K_{\min} \leq K^2_{\min}. \tag{5.1}$$

The set (5.1) admits the following order:

$$|0|; \quad \{E_i\}_{i=1,\dots,N_{-1}}; \quad \{E'_j\}_{j=1,\dots,N_{-2}}; \tag{5.2}$$

where E_i is a (-1)-curve; E'_j is a (-2)-curve of the minimal model. It is easy to see that these curves are determined by the equality $C_i \cdot K_{\min} = 0$.

Moreover, the positive semigroup $P_{-2,-1}$ generated by the finite collection of all exceptional curves acts naturally on the set (5.1). Up to this action we have the subset of the algebraic canonical walls (see [T5]),

$$\{|C_n|\}^w_{n=0,1,\dots,N_w}, \quad C_n^2 < 0; \tag{5.2$'$}$$

and the finite set

$$\{|C_m|\}^p_{m=1,\dots,N_p}, \quad C_m^2 \geq 0 \tag{5.2$''$}$$

of inverse images of curves of degree $< \frac{1}{2}N \cdot K^2_{\min}$ on the canonical model of S.

If a sheaf F has a regular section (that is, its zero set does not contain a curve), then F is contained in $GA\Theta^0_k$. Hence a priori we have lost in this family the sheaves with no regular sections. By the stability condition, a curve from the zero set of a section is contained in the collection (5.1). Hence for every class of (5.1) we need to consider the full collection of nontrivial extensions of the type

$$0 \to \mathcal{O}_S(C_i) \to F \to J_{\xi_i}(K_S - C_i) \to 0.$$

At this point it is very convenient to introduce some special notation. Namely, let

$${}_{d_1,D_1}E_{d_2,D_2} = \{0 \to J_{\xi_1}(D_1) \to F \to J_{\xi_2}(D_2) \to 0\}/\mathbb{C}^* \tag{5.3}$$

be the set of all nontrivial extensions for all clusters $\xi_i \in \mathrm{Hilb}^{d_i}$, with $i = 1, 2$. This set has the projectivisation of the vector space $\mathrm{Ext}^1(J_{\xi_2}(D_2), J_{\xi_1}(D_1))$ as fibre over (ξ_1, ξ_2). In particular (see (4.5)),

$$GA\Theta^0_k = {}_{0,0}E_{c_2(S)+k,K_S},$$

and as the geometric approximation of Θ we have to consider the union

$$GA\Theta_k = \bigcup_{C_i^2 \geq 0} {}_{0,C_i}E_{c_2(S)+k-C_i\cdot K_S+C_i^2,K_S-C_i}.$$

It is easy to see that each variety of this union is a component, that is, its dimension is $\geq \mathrm{v.dim}\,\Theta_k$. It is natural to denote these components by

$$\begin{aligned} GA\Theta_k^{\bar{C}_i} &= {}_{0,\bar{C}_i}E_{c_2(S)+k-\bar{C}_i\cdot K_S+\bar{C}_i^2,K_S-\bar{C}_i} \\ &= {}_{0,0}E_{c_2(S)+k-\bar{C}_i\cdot K_S+\bar{C}_i^2,K_S-2\bar{C}_i}, \end{aligned}$$

and for every component the polynomial is of the type (4.13). This polynomial is

$$ga\gamma_{\Theta_k}^{\bar{C}_i} \in S^j(q_S, K_S, C_i).$$

Now we need to remove from every component the subset containing nonstable bundles, and for this we need to introduce a second notation: let

$${}^{d_1',D_1'}_{d_1,D_1}CR^{d_2',D_2'}_{d_2,D_2} = \left({}_{d_1,D_1}E_{d_2,D_2}\right) \cap \left({}_{d_1',D_1'}E_{d_2',D_2'}\right) \tag{5.4}$$

be the set of all diagrams of type

$$\begin{array}{ccccc} & & 0 & & \\ & & \uparrow & & \\ & & J_{\eta_2}(D_2') & & \\ & \phi' \nearrow & \uparrow & & \\ 0 \to J_{\xi_1}(D_1) & \to & F & \to & J_{\xi_2}(D_2) \to 0 \\ & & \uparrow & \phi \nearrow & \\ & & J_{\eta_1}(D_1') & & \\ & & \uparrow & & \\ & & 0 & & \end{array} \tag{5.5}$$

(called a *cross* for short), where $\xi_i \in \mathrm{Hilb}^{d_i} S$ and $\eta_j \in \mathrm{Hilb}^{d_j'} S$. Obviously a cross can be extended to a diagram

$$\begin{array}{ccccccc}
 & 0 & & 0 & & 0 & \\
 & \uparrow & & \uparrow & & \uparrow & \\
0 \to & J_{\xi_1}(D_1) & \to & J_{\eta_2}(D'_2) & \to & \mathcal{O}_C(\varepsilon) \to 0 & \\
 & \uparrow & \phi' \nearrow & \uparrow & & \uparrow & \\
0 \to & J_{\xi_1}(D_1) & \to & F & \to & J_{\xi_2}(D_2) \to 0 & \\
 & \uparrow & & \uparrow & \phi \nearrow & \uparrow & \\
 & 0 & \to & J_{\eta_1}(D'_1) & \to & J_{\eta_1}(D'_1) \to 0 & \\
 & & & \uparrow & & \uparrow & \\
 & & & 0 & & 0 &
\end{array} \tag{5.6}$$

where C is an effective curve such that

$$C = D_2 - D'_1 = D'_2 - D_1 \tag{5.7}$$

as a divisor class. This effective curve is the zero locus of the reflexive hull of the homomorphisms

$$\begin{aligned}
\phi &: J_{\eta_1}(D'_1) \to J_{\xi_2}(D_2) \\
\phi' &: J_{\xi_1}(D_1) \to J_{\eta_2}(D'_2).
\end{aligned} \tag{5.8}$$

Moreover, we have the cycle μ which is a subcycle of C determined by the diagram

$$\begin{array}{cccccccc}
 & 0 & & 0 & & 0 & & \\
 & \downarrow & & \downarrow & & \downarrow & & \\
0 \to & J_{\eta_1}(D'_1) & \to & J_{\xi_2}(D_2) & \to & \mathcal{O}_C(\varepsilon) & \to 0 & \\
 & \downarrow & & \downarrow & & \downarrow & & \\
0 \to & \mathcal{O}_S(D'_1) & \to & \mathcal{O}_S(D_2) & \to & \mathcal{O}_C(D_2) & \to 0 & \\
 & \downarrow & & \downarrow & & \downarrow & & \\
0 \to & \mathcal{O}_{\eta_1}(D'_1) & \to & \mathcal{O}_{\xi_2}(D_2) & \to & \mathcal{O}_\mu(D_2) & \to 0 & \\
 & \downarrow & & \downarrow & & \downarrow & & \\
 & 0 & & 0 & & 0 & &
\end{array} \tag{5.9}$$

This means that every cycle can be represented as a sum:

$$\begin{aligned} \eta_i &= \eta_i' + \eta_i^C \\ \xi_j &= \xi_j' + \xi_j^C \end{aligned} \tag{5.10}$$

where η_i^C and ξ_j^C are the maximal subcycles contained in C. Then

$$\begin{aligned} \eta_1' = \xi_2'; \qquad \xi_1' &= \eta_2' \\ \varepsilon = C \cdot D_2 + \eta_1^C - \xi_2^C &= C \cdot D_2' + \xi_1^C - \eta_2^C \\ \mu = \xi_2^C - \eta_1^C &= \eta_2^C - \xi_1^C \end{aligned} \tag{5.11}$$

as divisors classes on C.

In the particular case when we need to describe nonstable bundles belonging to components of $GA\Theta_k$, we have

$$D_1 = C_i, \quad D_2 = K_S - C_i, \quad D_1' = K_S - C_i - C, \quad D_2' = C_i + C \tag{5.12}$$

but D_1' must destabilise our sheaf F, and so we have

$$2(K_S - C_i - C) \cdot K_{\min} \geq K_{\min}^2.$$

But this means that the effective curve $C_i + C$ is contained in our collection (5.1) (and of course C is contained in it).

Now for the clusters we have the relations:

$$\begin{aligned} \xi_1 \subset C_i &\implies \xi_1^C \subset C \cdot C_i \\ \eta_1' = \xi_2', &\qquad \xi_2^C = \eta_1^C + \mu \end{aligned} \tag{5.13}$$

as divisors on C; and

$$\xi_1' = \eta_2', \qquad \eta_2^C = \xi_1^C + \mu.$$

So to describe the nonstable locus of $GA\Theta_k$ we need to use crosses of type

$${}^{d_1', K_S - C_i - C_j}_{0, C_i} CR^{d_2', C_i + C_j}_{d_2, K_S - C_i}, \tag{5.14}$$

where C_i and C_j are the classes from (5.1) such that $C_i + C_j$ is also a class from (5.1), and where

$$d_2 = d_1' + d_2' + (K_S - C_j) \cdot C_j - 2C_i \cdot C_j. \tag{5.15}$$

This space is fibred over $\mathrm{Hilb}^{d_1'} S \times |C_j|$:

$$\pi_1 \times c : {}^{d_1', K_S - C_i - C_j}_{0, C_i} CR^{d_2', C_i + C_j}_{d_2, K_S - C_i} \to \mathrm{Hilb}^{d_1'} S \times |C_j|, \tag{5.16}$$

and for the fibre of this fibration over $(\eta, C) \in \mathrm{Hilb}^{d_1'} S \times |C_j|$ we shall write

$$CR^{d_2', C_i + C}_{d_2, K_S - C_i}, \tag{5.17}$$

$$CR^{d_2', C_i + C_j}_{d_2, K_S - C_i} = \bigcup_{C \in |C_j|} CR^{d_2', C_i + C}_{d_2, K_S - C_i}.$$

This space is fibred over $\mathrm{Hilb}^{d_2'} C \times \mathrm{Hilb}^{d_2 - d_1'} C$:

$$p_1 \times p_2 : CR^{d_2', C_i + C}_{d_2, K_S - C_i} \to \mathrm{Hilb}^{d_2'} C \times \mathrm{Hilb}^{d_2 - d_1'} C \tag{5.18}$$

where p_1 sends a cross to η_2 and p_2 sends a cross to ξ_2^C.

Proposition 4. *The map of CR to the direct product of all targets of arrows is an isomorphism.*

Proof. First of all, see [T1] and [T5] for $\operatorname{Hilb}^d C$, the subscheme of clusters contained in C. Namely, the section $s \in H^0(\mathcal{O}_S(C))$ which cuts out C defines a section $\overline{s}$ of the standard vector bundle $\mathcal{E}_{C}$ on $\operatorname{Hilb} S$ (see (4.3)) whose zero-scheme

$$(\overline{s})_0 = \operatorname{Hilb} C \qquad (5.19)$$

is the subvariety of $\operatorname{Hilb} S$ containing the clusters lying on C.

Now if $|C|$ is a linear system, write

$$\operatorname{Hilb}|C| = \bigcup_{C' \in |C|} \operatorname{Hilb} C' \qquad (5.20)$$

for the union of all clusters lying on curves $C' \in |C|$. As a cohomology class,

$$[\operatorname{Hilb} C] = c_{top}(\mathcal{E}_{C}) \qquad (5.21)$$

and

$$[\operatorname{Hilb}|C|] = c_m(\mathcal{E}_{C}) \qquad (5.22)$$

where

$$m = \operatorname{rank} \mathcal{E}_{C} - h^0(\mathcal{O}_S(C)) - 1.$$

It is enough to prove proposition 4 for the particular case when $\eta_1 = 0$, i.e. for the fibre of π_1 (5.16). We need to recall that any homomorphism

$$\phi_\xi : \mathcal{O}_S(K_S - C_i - C_j) \to J_{\xi_2}(K_S - C_i)$$

can be lifted to a homomorphism $\mathcal{O}_S(K_S - C_i - C_j) \to F$ in the horizontal short exact sequence (5.5), if and only if the element $e \in \operatorname{Ext}^1(J_{\xi_2}, \mathcal{O}_S(2C_i - K_S))$ which defines the extension (5.5) is in the kernel of the homomorphism

$$\operatorname{Ext}^1(J_{\xi_2}, \mathcal{O}_S(2C_i - K_S)) \xrightarrow{\phi^1} \operatorname{Ext}^1(\mathcal{O}_S(-C_j), \mathcal{O}_S(2C_i - K_S))$$

induced by ϕ. But this kernel is

$$\ker \phi^1 = \operatorname{Ext}^1(\mathcal{O}_{C}(-\xi_2), \mathcal{O}_S(2C_i - K_S)) \qquad (5.23)$$

and by Serre duality

$$\begin{aligned} \operatorname{Ext}^1(\mathcal{O}_{C}(-\xi_2), \mathcal{O}_S(2C_i - K_S))^* &= \operatorname{Ext}^1(\mathcal{O}_S(2C_i - K_S), \mathcal{O}_{C}(K_S|_{C} - \xi_2)) \\ &= H^1(\mathcal{O}_{C}(2(K_S)|_{C} - 2C_i \cdot C - \xi_2)) \\ &= H^0(\mathcal{O}_{C}((C)^2 + 2C_i \cdot C + (\xi_2 - K_S|_{C}))^*. \end{aligned}$$

Thus

$$\pi^{-1}(\xi_2) = |(C)^2 - K_S \cdot C + 2C_i \cdot C + \xi_2|. \tag{5.24}$$

It is easy to see that in our cross the horizontal extension is given by the element

$$\eta_2 \in |(C)^2 - K_S \cdot C + 2C_i \cdot C + \xi_2|, \tag{5.25}$$

which proves proposition 4.

The map of a component (5.14) to $GA\Theta_k^{\overline{C}_i}$ induces a map

$$e\colon \operatorname{Hilb}^{d_1'} S \times \operatorname{Hilb}^{d_2-d_1'} C \to \operatorname{Hilb}^{d_1+d_2} S. \tag{5.26}$$

This is the restriction of the Ellingsrud map (see [P-T]) to $\operatorname{Hilb} C$ and it is easy to prove that in this case the map is regular. Hence we can describe the space of crosses and its normal bundle as a subscheme of a component of $GA\Theta_k$ over $\operatorname{Hilb}^{d_1'} S$ purely in terms of the geometry of the curve C or of the complete linear systems $|C_j|$ (see [T5]). Namely, for a curve C, there exists a universal subscheme, with its two projections

$$\begin{array}{ccc} & Z \subset C \times \operatorname{Hilb} C & \\ \overline{p}_{\overline{C}} \swarrow & & \searrow^{\overline{p}_H} \\ C & & \operatorname{Hilb} C. \end{array} \tag{5.27}$$

(See for example [ACGH], or Example 14.4.17 of [F].) Any divisor class $D \in \operatorname{Pic} C$ induces a standard rank d vector bundle

$$\mathcal{E}_D^d = R^0 \overline{p}_H(\overline{p}_{\overline{C}}^* \mathcal{O}_{\overline{C}}(D)) \tag{5.28}$$

on $\operatorname{Hilb}^d C$, with fibre $H^0(\mathcal{O}_\eta(D))$ over $\eta \in \operatorname{Hilb}^d C$.

The standard vector bundle on $\operatorname{Hilb} S$ (4.3) can be restricted to $\operatorname{Hilb} C$ if $C \subset S$ and for $D \in \operatorname{Pic} S$ we have the equality

$$\mathcal{E}_{D\cdot \overline{C}}^d = \mathcal{E}_D^d|_{\operatorname{Hilb} \overline{C}}. \tag{5.29}$$

Now $CR_{d_2,K_S-\overline{C}_i}^{d_2',\overline{C}_i+\overline{C}}$ over $\operatorname{Hilb}^{d_2-d_1'} C$ (see (5.18)) admits the same description as $GA\Theta_k^0$ over $\operatorname{Hilb} S$. Namely $CR_{d_2,K_S-\overline{C}_i}^{d_2',\overline{C}_i+\overline{C}}$ is the basc subscheme of the complete linear system on

$$\mathbb{P}\mathcal{E}_{2K_S-2\overline{C}_i}^{c_2+k}|_{\operatorname{Hilb} \overline{C}} = \mathbb{P}(\mathcal{E}_{(2K_S-2\overline{C}_i)\cdot \overline{C}}^{c_2+k}) \otimes \mathcal{O}_{\operatorname{Hilb} \overline{C}}),$$

given by the rational map

$$\mathbb{P}(\mathcal{E}_{(2K_S-2\overline{C}_i)\cdot \overline{C}}^{c_2+k})^* \to \mathbb{P}(H^0(\mathcal{O}_{\overline{C}}(2K_S - 2C_i))^*. \tag{5.30}$$

Hence, if the restriction homomorphism

$$r : H^0(\mathcal{O}_S(2K_S - 2C_i)) \to H^0(\mathcal{O}_{C}(2K_S - 2C_i)) \qquad (5.31)$$

is an epimorphism then

$$CR^{d'_2, C_i + C}_{d_2, K_S - C_i} = \pi^{-1}(\operatorname{Hilb} C) \cap GA\Theta^{C_i}_k \qquad (5.32)$$

$$N_{CR^{d'_2, C_i+C}_{d_2,K_S-C_i} \subset GA\Theta_k} = \pi^* N_{\operatorname{Hilb} C \subset \operatorname{Hilb} S} = \pi^* \mathcal{E}_C|_{\operatorname{Hilb} C} \qquad (5.33)$$

$$CR^{d'_2, C_i + C_j}_{d_2, K_S - C_i} = \pi^{-1}(\operatorname{Hilb} |C_j|) \cap GA\Theta_k,$$

$$N_{CR^{d'_2, C_i+C_j}_{d_2,K_S-C_i} \subset GA\Theta_k} = \pi^* N_{\operatorname{Hilb} |C| \subset \operatorname{Hilb} S} \qquad (5.34)$$

and so on (see [T5]). Now to describe the partial modification of $GA\Theta_k$ along $CR^{d'_2, C_i+C_j}_{d_2,K_S-C_i}$, we use the second projection p_1 of (5.18).

6. Partial modification and regularisation

Roughly speaking, the partial modification of the component $GA\Theta_k$ along the component $CR^{d'_2, C_i+C_j}_{d_2,K_S-C_i}$ is the blow-up of this subvariety in $GA\Theta_k$, an 'elementary transformation' of the lift of the universal family $\mathbb{E}$ (4.7), followed by a recomputation of the cohomological correspondence (4.10) and of the new polynomial (or more precisely, of correction terms to it).

The geometric description of the blow-up of a component of CR was given in section 5 over $\operatorname{Hilb}^{d'_1} S$, and this information is enough to prove the shape theorem. First, let

$$\sigma_{0,j} : \widetilde{\mathbb{P}\mathcal{E}^*_{2K_S-2C_i}} \to \mathbb{P}\mathcal{E}^*_{2K_S-2C_i} \qquad (6.1)$$

be the blow-up of $CR^{d'_2, C_i+C_j}_{d_2,K_S-C_i}$ in $\mathbb{P}\mathcal{E}^*_{2K_S-2C_i}$. Then the blow-up in $GA\Theta^{C_i}_k$ of $CR^{d'_2, C_i+C_j}_{d_2,K_S-C_i}$ is the intersection

$$\widetilde{GA\Theta^{C_i}_k} = \bigcap_{i=1}^{h^0(\mathcal{O}_S(2K_S-2C_i))} \widetilde{H}_{\overline{res(s_i)}} \subset \widetilde{\mathbb{P}\mathcal{E}}^*_{2K_S-2C_i}, \qquad (6.2)$$

where $\widetilde{H}_{\overline{res(s_i)}}$ is the geometric inverse image of the hypersurface $H_{\overline{res(s_i)}}$ under (6.1). Denote the exceptional divisor of this blow up by $\widetilde{CR}$. Then the projection p_1 of (5.18) defines a projection

$$\operatorname{id} \times p_1 : S \times \widetilde{CR} \to S \times \operatorname{Hilb}^{d'_2} S. \qquad (6.3)$$

Let Z' be the inverse image of the universal subscheme Z (3.1) under this morphism. Then on the divisor $S \times \widetilde{CR}$ of the variety $S \times \widetilde{GA\Theta_k}$, we have the torsion-free sheaf $(\mathrm{id} \times (\sigma_{0,j} \cdot p_1))^*(\mathcal{I}_{Z'} \otimes p_S^* \mathcal{O}_S(C_i + C_j))$ and the epimorphism of lifting of the restriction of the universal extension (4.7):

$$(\mathrm{id} \times \sigma_{0,j})^* \mathbb{E}|_{S \times \widetilde{CR}} \to (\mathrm{id} \times (\sigma_{0,j} \cdot p_1))^*(\mathcal{I}_{Z'} \otimes p_S^* \mathcal{O}_S(C_i + C_j)) \to 0 \quad (6.4)$$

(see the diagram (5.5)).

Remark. Actually if the linear system $|C_j|$ has positive dimension, then we must twist the last sheaf by $\mathcal{O}_{|C_j|}(1)$. It is easy to see that the cohomological correspondence remains the same after such twisting.

We can consider the epimorphism in (6.4) as an epimorphism of lifting of the universal sheaf $\mathbb{E}$ to a torsion sheaf on $S \times \widetilde{GA\Theta_k}$ which determines an exact sequence:

$$0 \to \mathbb{E}' \to (\mathrm{id} \times (\sigma_{0,j} \cdot p_1))^* \mathbb{E} \to (\mathrm{id} \times (\sigma_{0,j} \cdot p_1))^*(\mathcal{I}_{Z'} \otimes p_S^* \mathcal{O}_S(C_i + C_j)) \to 0. \quad (6.5)$$

Its kernel $\mathbb{E}'$ is the 'elementary transformation' which we need and we thus realise $\widetilde{GA\Theta_k}$ as the base of a family of bundles and torsion-free sheaves. It is easy to see that its restriction to $S \times \widetilde{CR}$ is contained in $GA\Theta_k$:

$$\mathbb{E}'|_{S \times \widetilde{CR}} \subset GA\Theta_k^{C_i + C_j}. \quad (6.6)$$

Continuing this procedure step by step, ordering the system of curves by the divisibility property, we obtain as the last step of this procedure the family

$$\mathbb{E}^{st} \to S \times \widetilde{GA\Theta}^{C_j}_{k,st}, \quad (6.7)$$

which does not contain nonstable sheaves, and $\widetilde{GA\Theta}^{C_j}_{k,st}$ is the intersection of type (6.2) in the result $\widetilde{\mathbb{P}\mathcal{E}}^*$ of a finite chain of blow-ups of the projectivisation of the standard vector bundle on Hilb. For the description of the exact algorithm of elementary transformations of the universal sheaves we use the notion of δ-stable pair, where δ is some linear polynomial, introduced by Huybrechts and Lehn in [H-L]. This description is contained in [T6].

Now before recomputing the new cohomological correspondence, we must compute the fundamental class of the right moduli space if $\dim GA\Theta_k^{C_i}$ is greater than its virtual dimension. For this we need to use the regularisation argument described in the last section of [T1]. Precisely the same arguments as in Section 7 of [T1] give the following result.

Proposition 5. *The fundamental class of the right moduli space in* $\widetilde{GA\Theta}^{C_j}_{k,st}$ *is the class*

$$[M] = \bigcap_{m=1}^{C_i^2} \widetilde{H}_i, \quad (6.8)$$

where $\widetilde{H}$ is the geometric inverse image of the class H (see (4.4)) with respect to the chain of blow-ups.

Now we are ready to recompute the polynomial provided by this component step by step.

It follows from the exact sequence (6.5) that

$$\mu_{\mathbb{E}'}(\sigma) = 4\widetilde{\sigma} + 2((K_S - C_i)\cdot\sigma)H_m - 2((K_S - C_i + 2C_j)\cdot\sigma)\widetilde{CR}, \qquad (6.9)$$

and we have a new polynomial

$$\begin{aligned} ga_j\gamma_{\Theta_k^{C_i}}(\sigma) = \big(4\pi^*(\widetilde{\sigma}) &+ 2(\sigma\cdot(K_S - C_i))H \\ &- 2((K_S - C_i + 2C_j)\cdot\sigma)\widetilde{CR}\big)^{\mathrm{v.dim}\,\Theta_k - \bar{C}_i^2} \\ &\times (H - \widetilde{CR})^{h^0(\mathcal{O}_S(2K_S - 2\bar{C}_i)) + \bar{C}_i^2}, \end{aligned} \qquad (6.10)$$

and the binomial decomposition gives

$$ga_j\gamma_{\Theta_k^{C_i}}(\sigma) = ga\gamma_{\Theta_k^{C_i}}(\sigma) + \widetilde{CR}\times\Sigma$$

where

$$\begin{aligned} \Sigma = \sum_{p=0}^{h^0+\bar{C}_i^2}\ \sum_{q=1}^{\mathrm{v.dim}\,\Theta_k - \bar{C}_i^2} &\binom{h^0 + C_i^2}{p}\binom{\mathrm{v.dim}\,\Theta_k - C_i^2}{q}(-1)^{p+q}2^q \\ &\times \Big(((K_S - C_i + 2C_j)\cdot\sigma)^q\,\widetilde{CR}^{p+q-1}\,H^{h^0+\bar{C}_i^2-p} \\ &\times \big(4\pi^*(\widetilde{\sigma}) + 2(\sigma\cdot(K_S - C_i))H\big)^{\mathrm{v.dim}\,\Theta_k - \bar{C}_i^2 - q}\Big), \end{aligned} \qquad (6.11)$$

where $h^0 = h^0(\mathcal{O}_S(2K_S - 2C_i))$.

Now we can compute the correction term to the polynomial as the intersection of divisors on $\left({}^{d'_1,K_S-\bar{C}_i-\bar{C}_j}_{0,\bar{C}_i}CR^{d'_2,\bar{C}_i+\bar{C}_j}_{d_2,K_S-\bar{C}_i}\right)$; for this we need to lift all divisors from the formula (6.11) to $\left({}^{d'_1,K_S-\bar{C}_i-\bar{C}_j}_{0,\bar{C}_i}CR^{d'_2,\bar{C}_i+\bar{C}_j}_{d_2,K_S-\bar{C}_i}\right)$. Let H' be the Grothendieck divisor in $\mathrm{Pic}\,\mathbb{P}\mathcal{E}_{\bar{C}^2}, C \in |C_j|$. Then $\widetilde{CR}^2 = -H'$ and

$$\widetilde{\sigma}\,\mathrm{Hilb}\,C = (C_j\cdot\sigma)H_{Schw} \qquad (6.12)$$

is the Schwarzenberger divisor (see [ACGH]).

Now we compute the correction sum in (6.11) on the exceptional divisor over

$${}^{d'_1,K_S-\bar{C}_i-\bar{C}_j}_{0,\bar{C}_i}CR^{d'_2,\bar{C}_i+\bar{C}_j}_{d_2,K_S-\bar{C}_i}$$

as the relative product described in (5.18), (5.32)-(5.34). Fibrewise this computation coincides with the computation in [T5] (formulas (4.39)-(4.42)). Using (6.12) we get the correcting sum in the form

$$\sum a_{nmk}(\sigma \cdot \sigma)^n (K_S \cdot \sigma)^m (C_j \cdot \sigma)^k (C_i \cdot \sigma)^{\text{v.dim}\,\Theta_k - 2n - m - k}, \tag{6.13}$$

where the constants a_{nmk} arise from the products of Segre classes of the standard vector bundles on Hilb S and Hilb C for all curves from (5.1) and the binomial coefficients.

Repeating these arguments for all possible combinations of the finite chains $C' \cup C'' \cup \cdots \cup C^{(n)}$ of curves from (5.1) whose union is contained in (5.1) and all possible combinations of degrees of clusters gives the proof of the shape theorem (2.18).

REFERENCES

[ACGH] E. Arbarello, M. Cornalba, P. A. Griffiths, J. Harris, *Geometry of Algebraic Curves* (1985), Springer-Verlag, New York.

[B1] R. Brussee, *A remark on the Kronheimer-Mrowka classes of algebraic surfaces*, preprint (1993).

[B2] R. Brussee, *Stable rank 2 vector bundles on blow-up surfaces*, Dissertation, Leiden (1991).

[Bo] F. Bogomolov, *On stability of vector bundles on surfaces and curves.*, submitted to Matem. Sbornik.

[D] S. Donaldson, *Polynomial invariants for smooth 4-manifolds*, Topology **29** (1990), 257–315.

[D-K] S. K. Donaldson and P. B. Kronheimer, *The Geometry of Four-Manifolds*, OUP, 1990.

[F] W. Fulton, *Intersection Theory*, Springer-Verlag, New York, 1987.

[F-S] R. Fintushel and R. J. Stern, *The blow-up formula for Donaldson invariants*, Duke preprint 94 05 002 (May 1994).

[G] D. Gieseker, *On the moduli of vector bundles on an algebraic surface*, Ann. of Math. **106** (1977), 45–60.

[G-L] D. Gieseker and J. Li, *Irreducibility of moduli of rank two vector bundles*, submitted to J. Diff. Geom..

[H-L] D. Huybrechts and M. Lehn, *Stable pairs on curves and surfaces*, J. Alg. Geom., to appear.

[K] P. B. Kronheimer, *The genus-minimizing property of algebraic curves*, Bull. Amer. Math. Soc. (1994).

[K-M] P. B. Kronheimer and T. S. Mrowka, *Recurrence relations and asymptotics for four-manifold invariants*, Bull. Amer. Math. Soc. **30** (1994), 215–221.

[M] J. Morgan, *Comparison of the Donaldson polynomial invariants with their algebro-geometric analogues*, in print.

[M-M] J. W. Morgan and T. Mrowka, *A note on Donaldson polynomial invariants*, International Mathematics Reserch Notices **10** (1992), 223–230.

[O-V] C. Okonek and A. Van de Van, *Stable bundles and differentiable structures on certain elliptic surfaces*, Invent. Math. **86** (1986), 357–370.

[P1] V. Pidstrigach, *Cutting along 3-sphere for spin polynomial invariants*, preprint, Göttingen (1994).

[P2] V. Pidstrigach, *Some glueing formulas for spin polynomials and the Van de Ven conjecture*, Izv. RAN (to appear) 54 (1994), 28.

[P-T] V. Pidstrigach and A. Tyurin, *Invariants of the smooth structures of an algebraic surface arising from the Dirac operator*, Izv. AN SSSR, Ser. Math. **52:2** (1992), 279–371 (Russian); English translation (1994).

[Q] Z. B. Qin, *Stable rank 2 sheaves on blown-up surface*, preprint, McMaster (1991).

[T1] A. Tyurin, *The spin polynomial invariants of the smooth structures of algebraic surfaces*, Iz. AN SSSR **57:2** (1993), 279–371 (Russian); English translation in Mathematica Gottingensis, Heft 6 (1993).

[T2] A. Tyurin, *The simple method of distinguishing the underlying differentiable structures of algebraic surfaces*, Mathematica Gottingensis, Sonderforschungsbereichs Geometry and Analysis **Heft 25** (1992), 1–24.

[T3] A. Tyurin, *The geometry of 1-special components of moduli spaces of torsion-free sheaves on algebraic surfaces*, Proceedings of the Conference 'Complex Algebraic Varieties' Bayreuth 1990, Lecture Notes Math. 1507, Springer-Verlag, New york, 1991, pp. 166–175.

[T4] A. Tyurin, *Spin canonical invariants of 4-manifolds and algebraic surfaces*, preprint, Warwick (1994).

[T5] A. Tyurin, *On almost canonical polynomials of algebraic surface*, Proceedings of the Conference 'Algebraic Geometry', Yaroslavl 1992, Aspects of Mathematics, 1993, pp. 250–275.

[T6] A. Tyurin, *The canonical spin polynomial of algebraic surfaces I*, in print in Iz. AN SSSR (1994).

[W] E. Witten, *Supersymmetric Yang-Mills theory on a four-manifold*, preprint IASSNS-HEP-94 (1994).

[Z] K. Zuo, *Generic smoothness of the moduli spaces of rank two stable vector bundles over algebraic surfaces*, Math. Z. **207** (1991), 629–643.

Algebra Section, Steklov Math Institute, Ul. Vavilova 42, Moscow 117966 GSP–1, Russia

E-mail address: Tyurin@top.mian.su *or* Tyurin@CFGauss.Uni-Math.Gwdg.De *or* Tyurin@Maths.Warwick.Ac.UK

ON CONFORMAL FIELD THEORY

KENJI UENO

Department of Mathematics, Faculty of Science, Kyoto University, Kyoto, 606-01 Japan

Introduction

Belavin-Polyakov-Zamodolochikov ([BPZ]) initiated conformal field theory as a certain limit of the theory of the two-dimensional lattice model. This theory has a deep relationship with string theory and a rich mathematical structure. It is a two-dimensional quantum field theory invariant under conformal transformations; in fact, as we shall see below, it is invariant under a much bigger group of transformations, and this gives a relationship with the moduli space of algebraic curves ([FS], [EO]).

A typical example of conformal field theory is abelian conformal field theory, the theory of free fermions over a compact Riemann surface. For a mathematically rigorous treatment of abelian conformal field theory we refer the reader to [KNTY]. This theory has a deep relationship with various fields of mathematics, such as the moduli theory of algebraic curves, KP hierarchy, theta functions, complex cobordism rings and formal groups ([KNTY], [KSU2], [KSU3]).

For non-abelian conformal field theory the first mathematically rigorous treatment was given by Tsuchiya-Kanie ([TK]), who constructed the theory over $\mathbf{P}^1$. Later Tsuchiya-Ueno-Yamada ([TUY]) generalized this to algebraic curves of arbitrary genus.

Let us explain briefly the main ideas of conformal field theory. It can be decomposed into two parts, holomorphic and anti-holomorphic, and in the following we shall only consider the holomorphic theory. This is often called chiral conformal field theory by the physicists.

Let ξ be a complex coordinate. An infinitesimal conformal transformation is expressed in the form

$$\xi \mapsto \xi + \epsilon f(\xi) \tag{1}$$

where ϵ is the dual number $x \mod (x^2)$ in $\mathbf{C}[x]/(x^2)$, and $f(\xi)\frac{d}{d\xi}$ is a local holomorphic vector field. Let us choose special vector fields

$$\ell_n = \xi^{n+1}\frac{d}{d\xi}, \quad n = -1, 0, 1, 2, \ldots, n, \ldots . \tag{2}$$

These vector fields then satisfy the following commutation relations:

$$[\ell_m, \ell_n] = (m-n)\ell_{m+n}.$$

Thus, $\{\ell_n\}_{n\geq -1}$ forms an infinite dimensional Lie algebra.

Note that $\{\ell_{-1}, \ell_0, \ell_1\}$ forms a Lie algebra isomorphic to $\mathfrak{sl}(2, \mathbf{C})$. This is the Lie algebra of linear fractional transformations of the Riemann sphere $\mathbf{P}^1$. Conformal field theory will thus be invariant under such transformations.

The most important feature of conformal field theory is the condition imposed by physicists that the theory should be invariant not only under the transformations (1), but also under the same transformations where $f(\xi)\frac{d}{d\xi}$ is a local meromorphic vector field ([BPZ]). This is, of course, no longer an infinitesimal conformal transformation; however, we may interpret such a transformation as an infinitesimal change of complex structure. In this way conformal field theory is related to the moduli theory of algebraic curves. We need to generalize the Lie algebra $\{\ell_n\}$ by adding negative powers of ξ in (2).

As we said at the beginning conformal field theory is a kind of quantum field theory. This means that it has infinitely many degrees of freedom and divergence appears, so we need to renormalize the theory. Renormalization is very simple in our case. It can be done by using the normal ordering of operators, which we shall explain below. Renormalization induces a central extension of the infinite Lie algebra $\{\ell_n\}$, called the Virasoro algebra.

The Virasoro algebra $\{L_n\}$ has the following commutation relations:

$$[L_n, L_m] = (n-m)L_{n+m} + \frac{c}{12}(n^3 - n)\cdot id.$$

The number c is called the central charge. Then, conformal invariance of the theory is expressed by the commutation relation

$$[L_n, X(z)] = z^n(z\frac{d}{dz} + n + 1)X(z).$$

This means that the field $X(z)$ behaves like a meromorphic one-form on a Riemann surface. (A more precise formulation will be given in Chapter I §1.2 (c) below.)

Conformal field theory can be formulated not only over a compact Riemann surface but also over a semi-stable algebraic curve, that is, a complete curve with at worst ordinary double points as singularities. These singular curves correspond to points on the boundary of the moduli space of smooth curves. Precisely speaking, we need to consider pointed stable curves.

Let C be a semi-stable curve of genus g. To formulate conformal field theory we also need a set Λ of labels which describe basic particles (or 'primary fields', as physicists call them) on the curve C. There is a distinguished label $0 \in \Lambda$ which corresponds to the state without particles. Also we assume that the set Λ has an involution

$$\dagger : \Lambda \rightarrow \Lambda,$$

where $\lambda^\dagger$ describes the anti-particle of the particle corresponding to $\lambda \in \Lambda$.

Usually the set Λ is related to certain representations of the infinite dimensional Lie algebra or more general algebras describing symmetry of the theory. In conformal field theory with gauge symmetry (which for simplicity we will call non-abelian conformal field theory) the Lie algebra is an affine Lie algebra associated with a simple Lie algebra. In the following we shall also give a sketch of an abelian conformal field theory, which is a generalization of the usual abelian conformal field described in [KNTY]. In this case we take as our algebra the one generated by vertex operators and an affine Lie algebra attached to the one-dimensional abelian Lie algebra.

Let us choose N smooth points $Q_1, Q_2, \ldots, Q_N$ on the curve C with local coordinates ξ_j with center Q_j. Assign an element λ_j of the set Λ to each point Q_j. We also fix a local coordinate ξ_j with centre Q_j.

Conformal field theory is a theory which now attaches a certain finite dimensional vector space—the 'conformal block'—to the data

$$\mathfrak{X} = (C; Q_1, Q_2, \ldots, Q_N; \xi_1, \xi_2, \ldots, \xi_N)$$

and $\vec{\lambda} = (\lambda_1, \lambda_2, \ldots, \lambda_N) \in \Lambda^{\oplus N}$. The conformal block $\mathcal{V}^\dagger_{\vec{\lambda}}(\mathfrak{X})$ satisfies the following conditions.

1) Let P be a smooth point on the curve C with coordinate η. Put

$$\widetilde{\mathfrak{X}} = (C; Q_1, Q_2, \ldots, Q_N, P; \xi_1, \xi_2, \ldots, \xi_N, \eta).$$

Then, there is a canonical isomorphism

$$\mathcal{V}^\dagger_{\vec{\lambda},0}(\widetilde{\mathfrak{X}}) \simeq \mathcal{V}^\dagger_{\vec{\lambda}}(\mathfrak{X}).$$

2) Let $\nu : \widetilde{C} \rightarrow C$ be a normalization of the curve C and P_+ and P_- be the inverse image of a double point of C, with local coordinates η_+, η_-, respectively. Then, there exists a canonical isomorphism

$$\bigoplus_{\mu \in \Lambda} \mathcal{V}^\dagger_{\vec{\lambda},\mu\mu^\dagger}(\widetilde{\mathfrak{X}}) \simeq \mathcal{V}^\dagger_{\vec{\lambda}}(\mathfrak{X})$$

where we put

$$\widetilde{\mathfrak{X}} = (\widetilde{C}; Q_1, Q_2, \ldots, Q_N, P_+, P_-; \zeta_1, \eta_2, \ldots, \zeta_N, \eta_+, \eta_-).$$

3) $\dim \mathcal{V}^\dagger_{\vec{\lambda}}(\mathfrak{X})$ depends only on the genus of the curve C, the number N of the points Q_j and $\vec{\lambda}$.

4) The collection $\bigoplus_{[\mathfrak{X}]} \mathcal{V}^\dagger_{\vec{\lambda}}(\mathfrak{X})$ over the moduli space of N-pointed curves forms a vector bundle and it carries a projectively flat connection with regular singularities along the boundary consisting of the locus of N-pointed singular curves.

The conformal block is also called the 'space of vacua' ([TUY]) or 'modular functor' ([S]). The above property 2) is often called factorization of conformal blocks and physically it is used to construct conformal field theory. The property 2) gives a way to calculate the dimensions of conformal blocks by reducing to the case of three-pointed $\mathbf{P}^1$. This is a combinatorial problem and not easy to calculate. The best way is to use the Verlinde formula, first conjectured by Verlinde ([Ve]) and proved in ([MS1]).

The above property 4) is a vague statement. The conformal block $\mathcal{V}^\dagger_{\vec{\lambda}}(\mathfrak{X})$ depends on the choice of local coordinates ξ_j at the point Q_j and one can show that it only depends on the first-order infinitesimal neighbourhood of each of the points. Therefore, we need to take as our moduli space that of N-pointed stable curves together with first-order infinitesimal neighbourhoods.

In the present notes we shall briefly explain a method of constructing the conformal block for conformal field theory with gauge symmetries with simple Lie algebra as gauge group, and for abelian conformal field theory. We shall follow [TK] and [TUY]: this method is often called operator formalism. It is a primitive but direct way to construct the theory by following the physicists' description. It also gives a mathematically rigorous interpretation of the physical theory. There are other approaches ([H], [BL], [BF], [Fe], [KNR], [Ku]), which give descriptions of the conformal blocks different from ours. Finally, it is an important problem to generalize the integral representation of conformal blocks of N-pointed $\mathbf{P}^1$ given in [BF] and [Ku] to those of N-pointed curves of genus $g \geq 2$.

The details of our arguments on non-abelian conformal field theory can be found in [U].

Table of contents

Chapter I Conformal Field Theory with Gauge Symmetries

§1.1 Affine Lie algebras and integrable highest weight modules

In this section we recall basic facts on integrable highest weight representations of affine Lie algebras. Then we shall define the energy momentum tensor which plays an important role in conformal field theory. For the details of integrable highest weight representations of affine Lie algebras we refer the reader to Kac's book [Ka].

a) Affine Lie algebras

Let $\mathfrak{g}$ be a simple Lie algebra over the complex numbers $\mathbf{C}$ and $\mathfrak{h}$ its Cartan subalgebra. By Δ we denote the root system of $(\mathfrak{g}, \mathfrak{h})$. We have the root space decomposition

$$\mathfrak{g} = \mathfrak{h} \oplus \sum_{\alpha \in \Delta} \mathfrak{g}_\alpha.$$

Let $\mathfrak{h}^*_{\mathbf{R}}$ be the linear span of Δ over $\mathbf{R}$. Fix a lexicographic ordering of $\mathfrak{h}^*_{\mathbf{R}}$ once for all. This gives the decomposition $\Delta = \Delta_+ \sqcup \Delta_-$ of the root system into the positive roots and the negative roots. Put $\mathfrak{h}^* = \mathfrak{h}^*_{\mathbf{R}} \otimes_{\mathbf{R}} \mathbf{C}$, the linear span of Δ over $\mathbf{C}$. Let $(\ , \ )$ be a constant multiple of the Cartan-Killing form of the simple Lie algebra $\mathfrak{g}$. For each element of $\lambda \in \mathfrak{h}^*$, there exists a unique element $H_\lambda \in \mathfrak{h}^*$ such that

$$\lambda(H) = (H_\lambda, H)$$

for all $H \in \mathfrak{h}$. For $\alpha \in \Delta$, H_α is called the *root vector* corresponding to the root α.

On $\mathfrak{h}^*$ we introduce an inner product by

(1.1-1) $$(\lambda, \mu) = (H_\lambda, H_\mu).$$

Let us normalize the inner product (,) and call it the normalized Cartan-Killing form. Let θ be the highest (or longest) root, that is,

$$\theta \in \Delta_+, \quad \theta + \alpha_i \notin \Delta$$

for all simple roots α_i. For example, the highest root of the simple Lie algebra of type A_ℓ is given by

$$\theta = \sum_{i=1}^{\ell} \alpha_i.$$

The normalized Cartan-Killing form is given by the condition

(1.1-2) $$(\theta, \theta) = 2.$$

Note that the Cartan-Killing form has the following property.

(1.1-3) $$([X, Y], Z) + (Y, [X, Z]) = 0.$$

By $\mathbf{C}[[\xi]]$ and $\mathbf{C}((\xi))$ we mean the ring of formal power series in ξ and the field of formal Laurent power series in ξ, respectively.

Definition 1.1.1. The affine Lie algebra $\widehat{\mathfrak{g}}$ over $\mathbf{C}((\xi))$ associated with $\mathfrak{g}$ is defined to be

$$\widehat{\mathfrak{g}} = \mathfrak{g} \otimes \mathbf{C}((\xi)) \oplus \mathbf{C}c$$

where c is an element of the center of $\widehat{\mathfrak{g}}$ and the Lie algebra structure is given by

(1.1-4) $$[X \otimes f(\xi), Y \otimes g(\xi)] = [X, Y] \otimes f(\xi)g(\xi) + c \cdot (X, Y) \operatorname*{Res}_{\xi=0}(g(\xi)df(\xi))$$

for

$$X, Y \in \mathfrak{g}, \quad f(\xi), g(\xi) \in \mathbf{C}((\xi)).$$

Put

(1.1-5) $$\widehat{\mathfrak{g}}_+ = \mathfrak{g} \otimes \mathbf{C}[[\xi]]\xi, \quad \widehat{\mathfrak{g}}_- = \mathfrak{g} \otimes \mathbf{C}[\xi^{-1}]\xi^{-1}.$$

We regard $\widehat{\mathfrak{g}}_+$ and $\widehat{\mathfrak{g}}_-$ as Lie subalgebras of $\widehat{\mathfrak{g}}$. Also in the following we often identify $\mathfrak{g}$ and $\mathfrak{g} \otimes 1$ so that we may regard $\mathfrak{g}$ as a Lie subalgebra of $\widehat{\mathfrak{g}}$. We have a decomposition

(1.1-6) $$\widehat{\mathfrak{g}} = \widehat{\mathfrak{g}}_+ \oplus \mathfrak{g} \oplus \mathbf{C}c \oplus \widehat{\mathfrak{g}}_-.$$

In what follows we use the following notation freely.

$$X(n) := X \otimes \xi^n, \quad X \in \mathfrak{g}, \quad n \in \mathbf{Z}$$
$$X = X(0) = X \otimes 1$$

Remark 1.1.2. Note that usually the affine Lie algebra is defined by using $\mathbf{C}[\xi, \xi^{-1}]$. Namely, put

$$\mathfrak{g}_{aff} := \mathfrak{g} \otimes \mathbf{C}[\xi, \xi^{-1}] \oplus \mathbf{C} \cdot c$$

with commutation relation (1.1-4). Put

$$\mathfrak{g}_{+} = \mathfrak{g} \otimes \mathbf{C}[\xi]\xi.$$

This is a Lie subalgebra and we have

$$\mathfrak{g}_{aff} = \mathfrak{g}_{+} \oplus \mathfrak{g} \oplus \mathbf{C} \cdot c \oplus \widehat{\mathfrak{g}}_{-}.$$

Moreover, the Lie algebra $\widehat{\mathfrak{g}}_{+}$ is contained in the ξ-adic completion of $\mathfrak{g}_{+}$.

b) Integrable highest weight modules

Let us recall briefly the representation theory of a simple Lie algebra $\mathfrak{g}$. An irreducible left $\mathfrak{g}$-module V_λ is called a highest weight module with highest weight λ, if there exists a non-zero vector $e \in V_\lambda$ (called a highest weight vector) such that

$$He = \lambda(H)e, \quad Xe = 0 \quad \text{for all } X \in \mathfrak{g}_\alpha, \quad \alpha \in \Delta_{+}.$$

It is well-known that a finite dimensional irreducible left $\mathfrak{g}$-module is a highest weight module and two irreducible left $\mathfrak{g}$-modules are isomorphic if and only if they have the same highest weight. A weight $\lambda \in \mathfrak{h}^*_{\mathbf{R}}$ is called an integral weight, if

$$2(\lambda, \alpha)/(\alpha, \alpha) \in \mathbf{Z}$$

for any $\alpha \in \Delta$. A weight $\lambda \in \mathfrak{h}^*_{\mathbf{R}}$ is called a dominant weight, if

$$w(\lambda) \leq \lambda$$

for any element w of the Weyl group W of $\mathfrak{g}$. By P_{+} we denote the set of dominant integral weights of $\mathfrak{g}$. A weight λ is the highest weight of an irreducible left $\mathfrak{g}$-module if and only if $\lambda \in P_{+}$.

Let us fix a positive integer ℓ (called the *level*) and put

$$P_\ell = \{\lambda \in P_{+} \mid 0 \leq (\theta, \lambda) \leq \ell\}.$$

Let us introduce the involution

$$\begin{aligned} \dagger : P_\ell &\to P_\ell \\ \lambda &\mapsto \lambda^\dagger \end{aligned}$$

by

$$\lambda^{\dagger} = -w(\lambda)$$

where w is the longest element of the Weyl group of the simple Lie algebra $\mathfrak{g}$. Note that $\lambda^{\dagger}$ is also characterized by the fact that $-\lambda^{\dagger}$ is the lowest weight of the $\mathfrak{g}$-module V_{λ}.

For each element $\lambda \in P_{\ell}$ the Verma module $\mathcal{M}_{\lambda}$ is defined as follows. Put

$$\widehat{\mathfrak{p}}_{+} := \widehat{\mathfrak{g}}_{+} \oplus \mathfrak{g} \oplus \mathbf{C} \cdot c.$$

Then $\widehat{\mathfrak{p}}_{+}$ is a Lie subalgebra of $\widehat{\mathfrak{g}}$. Let V_{λ} is the irreducible left $\mathfrak{g}$-module of highest weight λ. The action of $\widehat{\mathfrak{p}_{+}}$ on V_{λ} is defined as

$$\begin{aligned} cv &= \ell v \quad \text{for all } v \in V_{\lambda} \\ av &= 0 \quad \text{for all } a \in \widehat{\mathfrak{g}}_{+} \text{ and } v \in V_{\lambda} \end{aligned}$$

Put

(1.1-8) $$\mathcal{M}_{\lambda} := U(\widehat{\mathfrak{g}}) \otimes_{\widehat{\mathfrak{p}}_{+}} V_{\lambda}.$$

Then $\mathcal{M}_{\lambda}$ is a left $\widehat{\mathfrak{g}}$-module and is called a *Verma module*. The Verma module $\mathcal{M}_{\lambda}$ is not irreducible and contains the maximal proper submodule $\mathcal{J}_{\lambda}$. The quotient module $\mathcal{H}_{\lambda} := \mathcal{M}_{\lambda}/\mathcal{J}_{\lambda}$ has the following properties.

Theorem 1.1.3. *For each $\lambda \in P_{\ell}$, the left $\widehat{\mathfrak{g}}$-module $\mathcal{H}_{\lambda}$ is the unique left $\widehat{\mathfrak{g}}$-module (called the* integrable highest weight $\widehat{\mathfrak{g}}$-module*) satisfying the following properties.*

(1) *$V_{\lambda} = \{\, |v\rangle \in \mathcal{H}_{\lambda} \mid \widehat{\mathfrak{g}}_{+}|v\rangle = 0 \,\}$ is the irreducible left $\mathfrak{g}$-module with highest weight λ.*
(2) *The central element c acts on $\mathcal{H}_{\lambda}$ as $\ell \cdot \mathrm{id}$.*
(3) *$\mathcal{H}_{\lambda}$ is generated by V_{λ} over $\widehat{\mathfrak{g}}_{-}$ with only one relation*

(1.1-9) $$(X_{\theta} \otimes \xi^{-1})^{\ell-(\theta,\lambda)+1}|\lambda\rangle = 0$$

where $X_{\theta} \in \mathfrak{g}$ is the element corresponding to the maximal root θ and $|\lambda\rangle \in V_{\lambda}$ is the highest weight vector.

The theorem says that the maximal proper submodule $\mathcal{J}_{\lambda}$ is given by

(1.1-10) $$\mathcal{J}_{\lambda} = U(\widehat{\mathfrak{p}}_{-})|J_{\lambda}\rangle$$

where we put

(1.1-11) $$|J_{\lambda}\rangle = (X_{\theta} \otimes \xi^{-1})^{\ell-(\theta,\lambda)+1}|\lambda\rangle.$$

For the details see Kac [Ka, (10.4.6)].

Similarly we can define the integrable lowest weight right $\hat{\mathfrak{g}}$-module $\mathcal{H}_\lambda^\dagger$ which will be discussed in below.

c) Energy momentum tensor and Virasoro algebra

Next we shall define the energy momentum tensor and the Virasoro algebra. For that purpose we introduce the following notation.

$$X(z) = \sum_{n \in \mathbf{Z}} X(n) z^{-n-1}$$

where z is a variable. The normal ordering $\circ\atop\circ$ $\circ\atop\circ$ is defined by

$$ {}^\circ_\circ X(n)Y(m) {}^\circ_\circ = \begin{cases} X(n)Y(m), & n < m, \\ \frac{1}{2}(X(n)Y(m) + Y(m)X(n)) & n = m, \\ Y(m)X(n) & n > m. \end{cases}$$

Note that by (1.1-4), if $n > m$ and $X = Y$, we have

$$ {}^\circ_\circ X(n)X(m) {}^\circ_\circ = X(n)X(m) - n\delta_{n+m,0}(X, Y) \cdot c. \tag{1.1-12}$$

Definition 1.1.4. The *energy-momentum tensor* $T(z)$ of level ℓ is defined by

$$T(z) = \frac{1}{2(g^* + \ell)} \sum_{a=1}^{\dim \mathfrak{g}} {}^\circ_\circ J^a(z) J^a(z) {}^\circ_\circ$$

where $\{J^1, J^2, \dots\}$ is an orthonormal basis of $\mathfrak{g}$ with respect to the Cartan-Killing form $(\ ,\ )$ and g^* is the dual Coxeter number of $\mathfrak{g}$.

Put

$$L_n = \frac{1}{2(g^* + \ell)} \sum_{m \in \mathbf{Z}} \sum_{a=1}^{\dim \mathfrak{g}} {}^\circ_\circ J^a(m) J^a(n-m) {}^\circ_\circ . \tag{1.1-13}$$

Then we have the expansion

$$T(z) = \sum_{n \in \mathbf{Z}} L_n z^{-n-2}.$$

The operator L_n is called the Virasoro operator which acts on $\mathcal{H}_\lambda$. By (1.1-12), if $n \neq 0$, in the definition of L_n, we need not use the normal ordering, that is, we have

$$L_n = \frac{1}{2(g^* + \ell)} \sum_{m \in \mathbf{Z}} \sum_{a=1}^{\dim \mathfrak{g}} J^a(m) J^a(n-m).$$

For $n = 0$ we need the normal ordering to define L_0. The operator

$$\frac{1}{2(g^* + \ell)} \sum_{m \in \mathbf{Z}} \sum_{a=1}^{\dim \mathfrak{g}} J^a(m) J^a(-m)$$

cannot operate on $\mathcal{H}_\lambda$. The operator L_0 is a generalization of the Casimir element of the simple Lie algebra $\mathfrak{g}$

$$\Omega = \sum_{a=1}^{\dim \mathfrak{g}} J^a J^a$$

to the affine lie algebra $\hat{\mathfrak{g}}$. The Casimir element Ω belongs to the center of $U(\mathfrak{g})$, hence acts on V_λ by

$$2(g^* + \ell)\Delta_\lambda \cdot \mathrm{id}$$

where

(1.1-14.) $$\Delta_\lambda = \frac{(\lambda, \lambda) + 2(\lambda, \rho)}{2(g^* + \ell)}, \quad \rho = \frac{1}{2} \sum_{\alpha \in \Delta_+} \alpha$$

The following lemma can be proved by direct calculations. It plays an important role in the theory.

Lemma 1.1.5. *The set $\{L_n\}$ forms a Virasoro algebra and we have*

$$[L_n, X(m)] = -mX(n+m), \quad \text{for } X \in \mathfrak{g}$$
$$[L_n, L_m] = (n-m)L_{n+m} + \frac{c_v}{12}(n^3 - n)\delta_{n+m,0}$$

where

$$c_v = \frac{\ell \dim \mathfrak{g}}{g^* + \ell}$$

is the central charge of the Virasoro algebra.

Corollary 1.1.6.

$$[L_n, X(z)] = z^n (z\frac{d}{dz} + n + 1)X(z).$$

For $X \in \mathfrak{g}$, $f = f(z) \in \mathbf{C}((z))$ and $\underline{\ell} = \ell(z)\frac{d}{dz} \in \mathbf{C}((z))\frac{d}{dz}$ we use the following notation.

$$X[f] = \operatorname*{Res}_{z=0}(X(z)f(z)dz)$$
$$T[\underline{\ell}] = \operatorname*{Res}_{z=0}(T(z)\ell(z)dz).$$

In particular, we have

(1.1-15) $$L_0 = T[\xi\frac{d}{d\xi}].$$

Lemma 1.1.7. *$X[f]$ and $T[\underline{\ell}]$ act on $\mathcal{H}_\lambda$ and we have*

$$
\begin{aligned}
X[f] &= X \otimes f(\xi), \\
[T[\underline{\ell}], X[f]] &= -X[\underline{\ell}(f)], \\
[T[\underline{\ell}], T[\underline{m}]] &= -T[[\underline{\ell}, \underline{m}]] + \frac{c_v}{12}\operatorname*{Res}_{z=0}(\ell''' m dz).
\end{aligned}
\tag{1.1-16}
$$

Remark 1.1.8. In the last formula of (1.1-16) we can use other expressions based on the following equalities.

$$
\operatorname*{Res}_{z=0}(\ell'''(z)m(z)dz) = -\operatorname*{Res}_{z=0}(m'''(z)\ell(z)dz) = \frac{1}{2}\operatorname*{Res}_{z=0}\left(\begin{vmatrix} \ell'(z)m'(z) \\ \ell''(z)m''(z) \end{vmatrix} dz\right).
$$

d) Filtration on $\mathcal{H}_\lambda$ and right $\widehat{\mathfrak{g}}$-modules

Next Let us introduce the filtration $\{F_\bullet\}$ on $\mathcal{H}_\lambda$. For that purpose first define the subspace $\mathcal{H}_\lambda(d)$ of $\mathcal{H}_\lambda$ for a non-negative integer d by

$$
\mathcal{H}_\lambda(d) = \{\, |v\rangle \in \mathcal{H}_\lambda \mid \; L_0|v\rangle = (d+\Delta_\lambda)|v\rangle \,\}
\tag{1.1-17}
$$

where Δ_λ is defined in (1.1-14). Note that by (1.1-13), on V_λ the operator L_0 acts as

$$
L_0|v\rangle = \frac{1}{2(g^*+\ell)}\sum_{a=1}^{\dim \mathfrak{g}} J^a J^a |v\rangle = \frac{1}{2(g^*+\ell)}\Omega|v\rangle = \Delta_\lambda|v\rangle
$$

for $|v\rangle \in \mathcal{H}_\lambda$. For a positive integer m and $|v\rangle \in V_\lambda$, we have

$$
\begin{aligned}
L_0 X(-m)|v\rangle &= L_0 X(-m)|v\rangle + mX(-m)|v\rangle \\
&= (\Delta_v + m)X(-m)|v\rangle.
\end{aligned}
$$

Hence, we have $X(-m)|v\rangle \in F_m\mathcal{H}_\lambda$.

Similarly, for positive integers $m_1, \dots, m_k$, we have

$$
\begin{aligned}
&L_0 X_1(M_1)\cdots X_k(-m_k)|v\rangle \\
&\qquad = (\Delta_\lambda + m_1 + \cdots + m_k)X_1(M_1)\cdots X_k(-m_k)|v\rangle, \quad |v\rangle \in V_\lambda.
\end{aligned}
$$

¿From this, it is easy to show that $\mathcal{H}_\lambda(d)$ is a finite dimensional vector space and we have

$$
\mathcal{H}_\lambda = \bigoplus_{d=0}^{\infty} \mathcal{H}_\lambda(d).
$$

For a negative integer $-d$ we put

$$\mathcal{H}_\lambda(-d) = \{0\}.$$

Now we define the filtration $\{F_p\mathcal{H}_\lambda\}$ by

$$F_p\mathcal{H}_\lambda = \sum_{d=0}^{p} \mathcal{H}_\lambda(d). \tag{1.1-18}$$

Note that all the filtrations defined above are the increasing ones.

Put

$$\mathcal{H}_\lambda^\dagger(d) = \mathrm{Hom}_{\mathbf{C}}(\mathcal{H}_\lambda(d), \mathbf{C}). \tag{1.1-19}$$

Then the dual space $\mathcal{H}_\lambda^\dagger$ of $\mathcal{H}_\lambda$ is defined to be

$$\mathcal{H}_\lambda^\dagger = \mathrm{Hom}_{\mathbf{C}}(\mathcal{H}_\lambda, \mathbf{C}) = \prod_{d=0}^{\infty} \mathcal{H}_\lambda^\dagger(d). \tag{1.1-20}$$

By our definition $\mathcal{H}_\lambda^\dagger$ is a right $\widehat{\mathfrak{g}}$-module. A decreasing filtration $\{F^p\mathcal{H}_\lambda^\dagger\}$ is defined by

$$F^p\mathcal{H}_\lambda^\dagger = \prod_{d \geq p} \mathcal{H}_\lambda^\dagger(d). \tag{1.1-21}$$

There is a canonical perfect bilinear pairing

$$\langle\ |\ \rangle : \mathcal{H}_\lambda^\dagger \times \mathcal{H}_\lambda \longrightarrow \mathbf{C}, \tag{1.1-22}$$

which satisfies the following equality for each $a \in \widehat{\mathfrak{g}}$.

$$\langle u|av\rangle = \langle ua|v\rangle, \quad \text{for all } \langle u| \in \mathcal{H}_\lambda^\dagger \text{ and } |v\rangle \in \mathcal{H}_\lambda.$$

Note that the filtrations $\{F_p\}$ and $\{F^p\}$ define a uniform topology on $\mathcal{H}_\lambda$ and $\mathcal{H}_\lambda^\dagger$, respectively. With respect to this topology $\mathcal{H}_\lambda^\dagger$ is complete and is the integrable highest weight right $\widehat{\mathfrak{g}}$-module with the lowest weight λ. Put

$$V_\lambda^\dagger = \{\, \langle v| \in \mathcal{H}_\lambda^\dagger \mid \langle v|\widehat{\mathfrak{g}}_- = 0 \,\}.$$

It is easy to show that $V_\lambda^\dagger = \mathcal{H}_\lambda^\dagger(0)$ and $V_\lambda^\dagger$ is the irreducible right $\mathfrak{g}$-module with lowest weight λ. The integrable highest weight right $\widehat{\mathfrak{g}}$-module with lowest weight λ is generated by $V_\lambda^\dagger$ over $\widehat{\mathfrak{g}}_+$ with only one relation

$$\langle\lambda|(X_{-\theta} \otimes \xi)^{\ell-(\theta,\lambda)+1} = 0.$$

Lemma 1.1.9.

$$\begin{aligned} X(m)\mathcal{H}_\lambda(d) &\subset \mathcal{H}_\lambda(d-m) \\ L_m\mathcal{H}_\lambda(d) &\subset \mathcal{H}_\lambda(d-m) \\ \mathcal{H}_\lambda^\dagger(d)X(m) &\subset \mathcal{H}_\lambda^\dagger(d+m) \\ \mathcal{H}_\lambda^\dagger(d)L_m &\subset \mathcal{H}_\lambda^\dagger(d+m). \end{aligned}$$

In the similar manner we can introduce a filtration on the Verma module $\mathcal{M}_\lambda$.

Proposition 1.1.10. *For a root vector $X_\alpha \in \mathfrak{g}_\alpha$ of the simple Lie algebra $\mathfrak{g}$ and any element $f(\xi) \in \mathbf{C}((\xi))$ the actions of $X_\alpha \otimes f(\xi)$ on $\mathcal{H}_\lambda$ and $\mathcal{H}_\lambda^\dagger$ are locally nilpotent. That is, for elements $|\phi\rangle \in \mathcal{H}_\lambda$ and $\langle\psi| \in \mathcal{H}_\lambda^\dagger$ there exists a positive integer m such that for each $n \geq m$ we have*

$$(X_\alpha \otimes f(\xi))^n |\phi\rangle = 0 \quad \langle\psi|(X_\alpha \otimes f(\xi))^n = 0.$$

In particular

$$\sum \frac{1}{n!}(X_\alpha \otimes f(\xi))^n$$

acts on $\mathcal{H}_\lambda$ and $\mathcal{H}_\lambda^\dagger$.

Now let us introduce the left $\widehat{\mathfrak{g}}$-module structure on $\mathcal{H}_\lambda^\dagger$ by

$$X(n)\langle\Phi| := -\langle\Phi|X(-n). \tag{1.1-23}$$

It is easy to check that this indeed defines the left $\widehat{\mathfrak{g}}$-module structure on $\mathcal{H}_\lambda^\dagger$.

Now we give the relationship between the *left* $\widehat{\mathfrak{g}}$-modules $\mathcal{H}_\lambda^\dagger$ and $\mathcal{H}_{\lambda^\dagger}$.

Lemma 1.1.11. *There exists a bilinear pairing*

$$(\ \ |\ \ ) : \mathcal{H}_\lambda \times \mathcal{H}_{\lambda^\dagger} \to \mathbf{C}$$

unique up to a constant multiple such that we have

$$(X(n)u|v) + (u|X(-n)v) = 0$$

for any $X \in \mathfrak{g}$, $n \in \mathbf{Z}$, $|u\rangle \in \mathcal{H}_\lambda$, $|v\rangle \in \mathcal{H}_{\lambda^\dagger}$ and $(\ \ |\ \ )$ is zero on $\mathcal{H}_\lambda(d) \times \mathcal{H}_{\lambda^\dagger}(d')$, if $d \neq d'$.

Proof. Since $V_\lambda \otimes V_{\lambda^\dagger}$, considered as a $\mathfrak{g}$-module by the diagonal action, contains only one-dimensional trivial $\mathfrak{g}$-module $\mathbf{C}|0_{\lambda,\lambda^\dagger}\rangle$, we have a bilinear form $(\ \ |\ \ ) \in \mathrm{Hom}_{\mathfrak{g}}(V_\lambda \otimes V_{\lambda^\dagger}, \mathbf{C})$ unique up to the constant multiple. Assume that we have a bilinear form $(\ \ |\ \ ) \in \mathrm{Hom}(F_p\mathcal{H}_\lambda \otimes F_p\mathcal{H}_{\lambda^\dagger}, \mathbf{C})$ with desired properties. For an element

$$X(-m)|u\rangle \in F_{p+1}\mathcal{H}_\lambda, \quad |u\rangle \in F_p\mathcal{H}_\lambda, \quad m > 0$$

and an element $|v\rangle \in F_{p+1}\mathcal{H}_{\lambda^\dagger}$ define

$$(X(-m)u|v) = -(u|X(m)v). \tag{1.1-24}$$

Note that since $X(m)|v\rangle \in F_{p+1-m}\mathcal{H}_{\lambda^\dagger}$, the right hand side is defined already. It is easy to show that in this way we can define the bilinear form $(\ \ |\ \ )$ satisfying the conditions of Lemma 1.1.11. QED.

Corollary 1.1.12. *There is a canonical left $\mathfrak{g}$-module isomorphism*

$$\mathcal{H}_\lambda^\dagger \simeq \widehat{\mathcal{H}}_{\lambda^\dagger}$$

where $\widehat{\mathcal{H}}_{\lambda^\dagger}$ is the completion of $\mathcal{H}_{\lambda^\dagger}$ with respect to the filtration $\{F_p\}$.

§1.2 Conformal blocks attached to N-pointed stable curves

a) N-pointed stable curves

Definition 1.2.1. Data $\mathfrak{X} = (C; Q_1, Q_2, \dots, Q_N)$ consisting of a curve C and points $Q_1, \dots, Q_N$ on C are called an *N-pointed stable curve*, if the following conditions are satisfied.

(1) The curve C is a reduced connected complete algebraic curve defined over the complex numbers $\mathbf{C}$. The singularities of the curve C are at worst ordinary double points. That is, C is a semi-stable curve.

(2) $Q_1, Q_2, \dots, Q_N$ are non-singular points of the curve C.

(3) If an irreducible component C_i is a projective line (i.e. Riemann sphere) $\mathbf{P}^1$ (resp. a rational curve with one double point, resp. an elliptic curve), the sum of the number of intersection points of C_i and other components and the number of Q_j's on C_i is at least three (resp. one).

(4) $\dim_{\mathbf{C}} H^1(C, \mathcal{O}_C) = g$.

Note that the above condition (3) is equivalent to saying that $\mathrm{Aut}(\mathfrak{X})$ is a finite group so that $\mathfrak{X}$ has no infinitesimal automorphisms. In the following we often add the following condition (Q) for an N-pointed stable curve $\mathfrak{X}$.

(Q) Each component C_i contains at least one Q_j.

The meaning of the condition (Q) will be clarified in the following Lemma 1.2.5 and Lemma 1.2.6.

Definition 1.2.2. Let C be a curve and Q a non-singular point on C. An *formal neighbourhood* s of C at the point Q is a $\mathbf{C}$-algebra isomorphism

$$s : \widehat{\mathcal{O}}_{C,Q} = \varinjlim \mathcal{O}_{C,Q}/\mathfrak{m}_Q^{n+1} \simeq \mathbf{C}[[\xi]]/(\xi^{n+1}) \tag{1.2-1}$$

where $\mathfrak{m}_Q$ is the maximal ideal of $\mathcal{O}_{C,Q}$ consisting of germs of holomorphic functions vanishing at Q.

Definition 1.2.3. Data

$$\mathfrak{X} = (C; Q_1, Q_2, \dots, Q_N; s_1, s_2, \dots, s_N)$$

are called an *N-pointed stable curve of genus g with formal neighbourhoods*, if

(1) $(C; Q_1, Q_2, \dots, Q_N)$ is an N-pointed stable curve of genus g.
(2) s_j is a formal neighbourhood of C at Q_j.

Let C be a semi-stable curve, that is, C is a reduced curve with at most ordinary double points and proper over $\mathbf{C}$. Let Ω^1_C be a sheaf of Kähler differentials of the curve C and ω_C be the dualizing sheaf of the curve C. Near a singular point P, the curve C is analytically isomorphic to the variety defined by

$$xy = 0.$$

By these coordinates the sheaf Ω^1_C is expressed as

$$\Omega^1_C = (\mathcal{O}_C dx + \mathcal{O}_C dy)/\mathcal{O}_C(xdy + ydx). \tag{1.2-2}$$

On the other hand, near the singular point P the dualizing sheaf ω_C is an invertible sheaf generated by the differential ζ given by dx/x outside $x = 0$ and $-dy/y$ outside $y = 0$. Moreover, outside singular points of the curve C, the sheaves Ω^1_C and ω_C coincide. Thus, we have

$$\Omega^1_C = \mathfrak{m}\omega_C \tag{1.2-3}$$

where $\mathfrak{m}$ is the defining ideal sheaf of the singular points of C. Hence, we have the following exact sequence.

$$0 \to \Omega^1_C \to \omega_C \to \omega_C \otimes \mathcal{O}_{C_{Sing}} \to 0.$$

Let $\nu : \widetilde{C} \to C$ be the normalization of the curve C. We let $\{P_1, \dots, P_q\}$ be the set of double points of the curve C and for each double point P_i, put $\nu^{-1}(P_i) = \{P_{i,+}, P_{i,-}\}$. Then, we have the following exact sequence.

$$0 \to \omega_C \to \nu_* \omega_{\widetilde{C}}(\sum_i^k (P_{i,+} + P_{i,-})) \xrightarrow{r} \bigoplus_i^q \mathbf{C} \to 0 \tag{1.2-4}$$

where at each double point P_i, the mapping r is given by

$$\mathrm{res}_{P_{i,+}}(\tau) - \mathrm{res}_{P_{i,-}}(\tau).$$

This means that a local holomorphic section of the dualizing sheaf ω_C is regarded as a local meromorphic section of one-form on $\widetilde{C}$ which has a pole of order one at $P_{i,+}$ and $P_{i,-}$ such that the sum of the residues is zero and holomorphic outside $P_{i,\pm}$'s. In the following we shall often use this interpretation. The following lemma is an easy consequence of this interpretation.

Lemma 1.2.4. *Let τ be a meromorphic section of the dualizing sheaf ω_C holomorphic at the double points. Then the sum of the residues of τ is zero.*

Lemma 1.2.5. *Assume that an N-pointed stable curve*

$$\mathfrak{X} = (C; Q_1, Q_2, \ldots, Q_N; s_1, s_2, \ldots, s_N)$$

with formal neighbourhoods satisfies the condition (Q). By t_j we denote the Laurent expansions at Q_j with respect to a formal parameter $\xi_j = s_j^{-1}(\xi)$. Then, the following homomorphisms are injective.

(1.2-5)

$$t = \oplus t_j : H^0(C, \mathcal{O}(*\sum_{j=1}^{N} Q_j)) \longrightarrow \bigoplus_{j=1}^{N} \mathbf{C}((\xi_j))$$

(1.2-6)

$$t = \oplus t_j : H^0(C, \omega_C(*\sum_{j=1}^{N} Q_j)) \longrightarrow \bigoplus_{j=1}^{N} \mathbf{C}((\xi_j))d\xi_j$$

where ω_C is the dualizing sheaf of the curve C.

By this Lemma $H^0(C, \mathcal{O}(*\sum_{j=1}^{N} Q_j))$ (resp. $H^0(C, \omega_C(*\sum_{j=1}^{N} Q_j)))$ can be regarded as a subspace of $\oplus_{j=1}^{N} \mathbf{C}((\xi_j))$ (resp. $\oplus_{j=1}^{N} \mathbf{C}((\xi_j))d\xi_j$). There is the residue pairing

(1.2-7)

$$\begin{array}{ccc} \bigoplus_{j=1}^{N} \mathbf{C}((\xi_j)) \times \bigoplus_{j=1}^{N} \mathbf{C}((\xi_j))d\xi_j & \rightarrow & \mathbf{C} \\ ((f(\xi_1), .., f(\xi_N), g(\xi_1)d\xi_1, .., g(\xi_N)d\xi_N) & \mapsto & \sum_{j=1}^{N} \operatorname{Res}_{\xi_j=0}(f(\xi_j)g(\xi_j)d\xi_j). \end{array}$$

The following Lemma is well-known and plays an important role in our theory.

Lemma 1.2.6. *Under the residue pairing (1.2-7) the vector space*

$$H^0(C, \mathcal{O}(*\sum_{j=1}^{N} Q_j))$$

and the vector space

$$H^0(C, \omega_C(*\sum_{j=1}^{N} Q_j))$$

are the annihilators to each other.

b) Conformal blocks

For an N-pointed stable curve $\mathfrak{X} = (C; Q_1, Q_2, \ldots, Q_N; \eta_1, \eta_2, \ldots, \eta_N)$ with formal neighbourhoods let us define the conformal block (the space of vacua in [TUY]) $\mathcal{V}^{\dagger}_{\vec{\lambda}}(\mathfrak{X})$ and its dual space, the space of covacua $\mathcal{V}_{\vec{\lambda}}(\mathfrak{X})$ where $\vec{\lambda} = (\lambda_1, \lambda_2, \ldots, \lambda_N)$, $\lambda_j \in P_\ell$. For that purpose we first define a generalized affine Lie algebra $\widehat{\mathfrak{g}}_N$.

Definition 1.2.7. Let $\mathfrak{g}$ be a simple Lie algebra over the complex numbers. A Lie algebra $\widehat{\mathfrak{g}}_N$ is defined as

$$\widehat{\mathfrak{g}}_N = \bigoplus_{j=1}^{N} \mathfrak{g} \otimes_{\mathbf{C}} \mathbf{C}((\xi_j)) \oplus \mathbf{C}c$$

with the following commutation relations.

$$(1.2\text{-}8)\quad [(X_j \otimes f_j), (Y_j \otimes g_j)] = ([X_j, Y_j] \otimes f_j g_j) + c \sum_{j=1}^{N} (X_j, Y_j) \operatorname*{Res}_{\xi_j=0}(g_j df_j)$$

where (a_j) means $(a_1, a_2, \dots, a_N)$ and c belongs to the center of $\widehat{\mathfrak{g}}_N$.

We also put

$$(1.2\text{-}9)\qquad \widehat{\mathfrak{g}}(\mathfrak{X}) = \mathfrak{g} \otimes_{\mathbf{C}} H^0(C, \mathcal{O}_C(*\sum_{j=1}^{N} Q_j)).$$

By Lemma 1.2.5 we have a natural embedding

$$t : H^0(C, \mathcal{O}_C(*\sum_{j=1}^{N} Q_j)) \hookrightarrow \bigoplus_{j=1}^{N} \mathbf{C}((\xi_j)).$$

We regard $H^0(C, \mathcal{O}_C(*\sum_{j=1}^{N} Q_j))$ as a subspace of $\bigoplus_{j=1}^{N} \mathbf{C}((\xi_j))$. By Lemma 1.2.4 we have the following lemma.

Lemma 1.2.8. *$\widehat{\mathfrak{g}}(\mathfrak{X})$ is a Lie subalgebra of $\widehat{\mathfrak{g}}_N$.*

Let us fix a non-negative integer ℓ. For each $\vec{\lambda} = (\lambda_1, \dots, \lambda_N) \in (P_\ell)^N$, a left $\widehat{\mathfrak{g}}_N$-module $\mathcal{H}_{\vec{\lambda}}$ and a right $\widehat{\mathfrak{g}}_N$-module $\mathcal{H}^\dagger_{\vec{\lambda}}$ are defined by

$$\mathcal{H}_{\vec{\lambda}} = \mathcal{H}_{\lambda_1} \otimes_{\mathbf{C}} \cdots \otimes_{\mathbf{C}} \mathcal{H}_{\lambda_N}$$
$$\mathcal{H}^\dagger_{\vec{\lambda}} = \mathcal{H}^\dagger_{\lambda_1} \widehat{\otimes}_{\mathbf{C}} \cdots \widehat{\otimes}_{\mathbf{C}} \mathcal{H}^\dagger_{\lambda_N}.$$

For each element $X_j \in \mathfrak{g}$, $f(\xi_j) \in \mathbf{C}((\xi_j))$, the action ρ_j of $X_j[f_j]$ on $\mathcal{H}_{\vec{\lambda}}$ is given by

(1.2-10)

$$\rho_j(X_j[f_j])|v_1 \otimes \cdots \otimes v_N\rangle = |v_1 \otimes \cdots \otimes v_{j-1} \otimes (X_j[f_j])v_j \otimes v_{j+1} \otimes \cdots v_N\rangle$$

where

$$|v_1 \otimes \cdots \otimes v_N\rangle$$

means

$$|v_1\rangle \otimes \cdots \otimes |v_N\rangle, \quad |v_j\rangle \in \mathcal{H}_{\lambda_j}.$$

The left $\widehat{\mathfrak{g}}_N$-action is given by

$$(1.2\text{-}11)\quad (X_1 \otimes f_1, \ldots, X_N \otimes f_N)|v_1 \otimes \cdots v_N\rangle = \sum_{j=1}^{N} \rho_j(X_j[f_j])|v_1 \otimes \cdots v_N\rangle.$$

Similarly, the right $\widehat{\mathfrak{g}}_N$-action on $\mathcal{H}^\dagger_{\vec{\lambda}}$ is defined by

$$(1.2\text{-}12)\quad \langle u_1 \otimes \cdots u_N|(X_1 \otimes f_1, \ldots, X_N \otimes f_N) = \sum_{j=1}^{N} \langle u_1 \otimes \cdots u_N|\rho_j(X_j[f_j]).$$

As a Lie subalgebra, $\widehat{\mathfrak{g}}(\mathfrak{X})$ operates on $\mathcal{H}_{\vec{\lambda}}$ and $\mathcal{H}^\dagger_{\vec{\lambda}}$ as

(1.2-13)

$$(X \otimes f)|v_1 \otimes \cdots \otimes v_N\rangle = \sum_{j=1}^{N} \rho_j(X \otimes t_j(f))|v_1 \otimes \cdots v_N\rangle$$

(1.2-14)

$$\langle u_1 \otimes \cdots \otimes u_N|(X \otimes f) = \sum_{j=1}^{N} \langle u_1 \otimes \cdots \otimes u_N|\rho_j(X \otimes t_j(f)).$$

The pairing $\langle \quad | \quad \rangle$ introduced in (1.1-22) induces a perfect bilinear pairing

$$(1.2\text{-}15)\qquad \langle \quad | \quad \rangle : \mathcal{H}^\dagger_{\vec{\lambda}} \times \mathcal{H}_{\vec{\lambda}} \to \mathbf{C}$$

$$(\langle u_1 \otimes \ldots \otimes u_N|, |v_1 \otimes \ldots \otimes u_N\rangle) \to \langle u_1|v_1\rangle\langle u_2|v_2\rangle \cdots \langle u_N|v_N\rangle$$

which is $\widehat{\mathfrak{g}}_N$-invariant.

$$\langle \Psi(X_j \otimes f_j)|\Phi\rangle = \langle \Psi|(X_j \otimes f_j)\Phi\rangle.$$

Now we are ready to define the conformal block attached to $\mathfrak{X}$.

Definition 1.2.9. Assume that $\mathfrak{X}$ enjoys the property (Q) in §1.1. Put

$$(1.2\text{-}16)\qquad \mathcal{V}_{\vec{\lambda}}(\mathfrak{X}) = \mathcal{H}_{\vec{\lambda}}/\widehat{\mathfrak{g}}(\mathfrak{X})\mathcal{H}_{\vec{\lambda}}.$$

The vector space $\mathcal{V}_{\vec{\lambda}}(\mathfrak{X})$ is called the *space of covacua* attached to $\mathfrak{X}$. The *conformal block* (or *space of vacua*) attached to $\mathfrak{X}$ is defined as

$$(1.2\text{-}17)\qquad \mathcal{V}^\dagger_{\vec{\lambda}}(\mathfrak{X}) = \mathrm{Hom}_{\mathbf{C}}(\mathcal{V}_{\vec{\lambda}}(\mathfrak{X}), \mathbf{C}).$$

In case $\mathfrak{X}$ does not satisfy the property (Q), we use Theorem 1.2.14 below to define the conformal block.

From the definition (1.2-17) the following lemma follows easily.

Lemma 1.2.10.

$$\mathcal{V}^{\dagger}_{\vec{\lambda}}(\mathfrak{X}) = \{ \langle\Psi| \in \mathcal{H}^{\dagger}_{\vec{\lambda}} \mid \langle\Psi|\, \widehat{\mathfrak{g}}(\mathfrak{X}) = 0 \}. \tag{1.2-18}$$

Corollary 1.2.11. *Let* $\mathfrak{X}_1 = (C_1; Q_1, \dots, Q_M; s_1, \dots, s_M)$ *and* $\mathfrak{X}_2 = (C_2, Q_{M+1}, \dots, Q_N; s_{M+1}, \dots, s_N)$ *be* M*-pointed and* $(N-M)$*-pointed stable curves with formal neighbourhoods, respectively. Let*

$$\mathfrak{X} = (C_1 \sqcup C_2; Q_1, \dots, Q_N; s_1, \dots, s_N)$$

be the N*-pointed stable curve with formal neighbourhoods obtained from* $\mathfrak{X}_1$ *and* $\mathfrak{X}_2$*. Then, we have a canonical isomorphism*

$$\mathcal{V}^{\dagger}_{\vec{\lambda}_1,\vec{\lambda}_2}(\mathfrak{X}) \simeq \mathcal{V}^{\dagger}_{\vec{\lambda}_1}(\mathfrak{X}_1) \otimes \mathcal{V}^{\dagger}_{\vec{\lambda}_2}(\mathfrak{X}_2).$$

Let us study the conformal blocks when the underlying curve is the Riemann sphere $\mathbf{P}^1$. We regard $\mathbf{P}^1$ as $\mathbf{C} \cup \{\infty\}$ and let z be a global coordinate of $\mathbf{C}$. For a positive integer $N \geq 1$ let us choose N-points $z_1, z_2, \dots z_N$ of $\mathbf{P}^1$ and put

$$\xi_j = \begin{cases} z - z_j & \text{if } z_j \neq \infty \\ 1/z & \text{if } z_j = \infty. \end{cases}$$

Then $\mathfrak{X} = (\mathbf{P}^1; z_1, \dots, z_N; \xi_1, \dots, \xi_N)$ is an N-pointed curve with formal coordinates. Choose $\vec{\lambda} = (\lambda_1, \dots \lambda_N) \in (P_\ell)^N$ and put

$$V_{\vec{\lambda}} = V_{\lambda_1} \otimes_{\mathbf{C}} \cdots \otimes_{\mathbf{C}} V_{\lambda_N}.$$

We have the following result.

Proposition 1.2.12. *The natural restriction mapping*

$$\mathrm{Hom}_{\mathbf{C}}(\mathcal{H}_{\vec{\lambda}}, \mathbf{C}) \to \mathrm{Hom}_{\mathbf{C}}(V_{\vec{\lambda}}, \mathbf{C}).$$

induces an injective homomorphism

$$j : \mathcal{V}^{\dagger}_{\vec{\lambda}}(\mathfrak{X}) \hookrightarrow \mathrm{Hom}_{\mathfrak{g}}(V_{\vec{\lambda}}, \mathbf{C}).$$

The proposition implies that the space of vacua attached to the Riemann sphere $\mathbf{P}^1$ is finite dimensional. Gabber's theorem [Ga] implies this is true in general. (For details see [TUY] or [U].)

Theorem 1.2.13. $\mathcal{V}_{\vec{\lambda}}(\mathfrak{X})$ *and* $\mathcal{V}^{\dagger}_{\vec{\lambda}}(\mathfrak{X})$ *are finite-dimensional vector space.*

In §1.3 we shall show that $\dim_{\mathbf{C}} \mathcal{V}^{\dagger}_{\vec{\lambda}}(\mathfrak{X})$ depends only on the genus $g(C)$ of the curve C and $\vec{\lambda}$.

Next we shall discuss important properties of conformal blocks. For $\mathfrak{X} = (C; Q_1, \ldots, Q_N; \eta_1, \ldots, \eta_N)$ let P be a non-singular point of the curve C and t a formal parameter of C at P. Put

$$\widetilde{\mathfrak{X}} = (C; Q_1, \ldots, Q_N, Q_{N+1}; \eta_1, \ldots, \eta_N, \eta_{N+1})$$

where $Q_{N+1} = P$ and $\eta_{N+1} = \eta$.

In the following we fix a highest weight vector $|0\rangle$ of the integrable left $\widehat{\mathfrak{g}}$-module $\mathcal{H}_0$. Since there is a canonical inclusion

$$\begin{aligned} \mathcal{H}_{\vec{\lambda}} &\longrightarrow \mathcal{H}_{\vec{\lambda}} \otimes \mathcal{H}_0 \\ |v\rangle &\longrightarrow |v\rangle \otimes |0\rangle \end{aligned}$$

we have a canonical surjection

$$\widehat{\iota}^* \; : \; \mathcal{H}^\dagger_{\vec{\lambda}} \widehat{\otimes} \mathcal{H}^\dagger_0 \longrightarrow \mathcal{H}^\dagger_{\vec{\lambda}} \, .$$

Theorem 1.2.14. *The canonical surjection $\widehat{\iota}^*$ induces a canonical isomorphism*

$$\mathcal{V}^\dagger_{\vec{\lambda},0}(\widetilde{\mathfrak{X}}) \simeq \mathcal{V}^\dagger_{\vec{\lambda}}(\mathfrak{X}).$$

Corollary 1.2.15. *There is a canonical isomorphism*

$$\mathcal{V}_{\vec{\lambda}}(\mathfrak{X}) \simeq \mathcal{V}_{\vec{\lambda},0}(\widetilde{\mathfrak{X}})$$

Next let us consider a singular curve. For an N-pointed stable curve $\mathfrak{X} = (C; Q_1, \ldots, Q_N; \eta_1, \ldots, \eta_N)$ with formal neighbourhoods, assume that the curve C has a double point P. Let $\nu : \widetilde{C} \to C$ be the normalization at the point P. Put $\nu^{-1}(P) = \{P', P''\}$. Furthermore we introduce formal neighbourhoods η' and η'' at P' and P'', respectively.

Proposition 1.2.16. *Under the above notation, for an N-pointed stable curve $\mathfrak{X} = (C; Q_1, \ldots, Q_N; \eta_1, \ldots, \eta_N)$ with formal neighbourhoods, put*

$$\widetilde{\mathfrak{X}} = (\widetilde{C}; P', P'', Q_1, \ldots, Q_N; \eta', \eta'', \eta_1, \ldots, \eta_N).$$

Then there is a canonical isomorphism

$$\bigoplus_{\mu \in P_l} \mathcal{V}^\dagger_{\mu, \mu^\dagger, \vec{\lambda}}(\widetilde{\mathfrak{X}}) \xrightarrow{\sim} \mathcal{V}^\dagger_{\vec{\lambda}}(\mathfrak{X}).$$

c) Correlation functions

Let $\mathfrak{X} = (C; Q_1, \ldots, Q_N; \eta_1, \ldots, \eta_N)$ be an N-pointed stable curve with formal neighbourhoods. Put

$$\omega_j = \sum_{n \in \mathbf{Z}} \langle \Psi | \rho_j(X(n)) | u \rangle \xi_j^{-n-1} d\xi_j \, , \quad j = 1, 2, \ldots, N + M$$

where $\xi_j = \eta_j^{-1}(\xi)$. For an element $f \in H^0(*\sum_{j=1}^{N+M} Q_j))$ let $f_j(\xi_j) = \sum a_n^{(j)} \xi_j^n$ be the formal Laurent expansion of f at the point Q_j by the formal parameter $\xi_j = \eta_j^{-1}(\xi)$. Hence $t(f) = (f_1(\xi_1), \dots, f_{N+M}(\xi_{N+M}))$. Then we have

$$\sum_{j=1}^{N+M} \operatorname*{Res}_{\xi_j=0}(f_j(\xi_j)\omega_j) = \sum_{j=1}^{N+M} \sum_{n \in \mathbf{Z}} \langle \Psi | \rho_j(X(n)) | u \rangle a_n^{(j)}$$
$$= \langle \Psi | X \otimes t(f) | u \rangle = 0$$

since $\langle \Psi | X \otimes t(f) = 0$ by our assumption. Therefore, by Lemma 1.2.6 there exists an element $\omega \in H^0(C, \omega_C(*\sum_{j=1}^{N+M} Q_j))$ with

$$t(\omega) = (\omega_1, \dots, \omega_{N+M}).$$

The meromorphic form ω is written as

$$\langle \Psi | X(z) | u \rangle dz$$

and called the *correlation function* or *one point function* of the current $X(z)$.

More generally we can show the following result.

Theorem 1.2.17. *Fix $\langle \Psi | \in \mathcal{V}_{\vec{\lambda}}^{\dagger}(\mathfrak{X})$. For each non-negative integer M the data*

$$X_1, X_2, \dots, X_M \in \mathfrak{g}, \quad |\Phi\rangle \in \mathcal{H}_{\vec{\lambda}}$$

define an element

$$F = \langle \Psi | X_1(z_1) X_2(z_2) \dots X_M(z_M) | \Phi \rangle dz_1 dz_2 \dots dz_M$$

of

$$H^0\Big(C^M, \omega_{C^M}(\sum_{1 \le i < j \le M} *\Delta_{ij} + \sum_{i=1}^{M} \sum_{j=1}^{N} *\pi_i^{-1}(Q_j)\Big),$$

where $\Delta_{ij} = \{(P_1, \dots, P_N) | P_i = P_j\}$ is the (i,j) diagonal. The meromorphic form has the following properties.

0) *For $M = 0$, $F = <\Psi|\Phi>$ is the canonical pairing induced by the pairing (1.1-22).*

1) *F is linear with respect to $|\Phi\rangle$ and multi-linear with respect to X_i's.*

2) *For any permutation $\sigma \in \mathfrak{S}_M$, we have*

$$F = \langle \Psi | X_{\sigma(1)}(z_{\sigma(1)}) X_{\sigma(2)}(z_{\sigma(2)}) \cdots X_{\sigma(M)}(z_{\sigma(M)}) | \Phi \rangle dz_1 dz_2 \dots dz_M .$$

For example, for a transposition $(i, i+1)$ we have

$$F = \langle \Psi | X_1(z_1) \cdots X_{i-1}(z_{i-1}) X_{i+1}(z_{i+1}) X_i(z_i)$$
$$X_{i+2}(z_{i+2}) \dots X_M(z_M) | \Phi \rangle dz_1 dz_2 \dots dz_M .$$

3) *For $k = 1, \ldots, N$ and $\xi_k = \eta_k^{-1}(\xi)$, if ξ_k is a holomorphic coordinate, then we have the equality*

$$\operatorname*{Res}_{\xi_k=0}(\xi_k^n\langle\Psi|X(\xi_k)X_1(z_1)X_2(z_2)\ldots X_M(z_M)|\Phi\rangle d\xi_k)dz_1\cdots dz_M$$
$$= \langle\Psi|X_1(z_1)X_2(z_2)\cdots X_M(z_M)|\rho_k(X(n))\Phi\rangle dz_1\cdots dz_M.$$

In other words, we have an expansion

$$\langle\Psi|X(\xi_k)X_1(z_1)X_2(z_2)\ldots X_M(z_M)|\Phi\rangle d\xi_k dz_1\cdots dz_M$$
$$= \sum_{n\in\mathbf{Z}}\langle\Psi|X_1(z_1)X_2(z_2)\cdots X_M(z_M)|\rho_k(X(n))\Phi\rangle\xi_k^{-n-1}d\xi_k dz_1\cdots dz_M$$

4) *For a local holomorphic coordinate z at a nonsingular point P of the curve C, we have the following equality.*

$$\langle\Psi|X(z)Y(w)X_1(z_1)X_2(z_2)\cdots X_M(z_M)|\Phi\rangle$$
$$= \Big\{\frac{\ell\cdot(X,Y)}{(z-w)^2}\langle\Psi|X_1(z_1)X_2(z_2)\cdots X_M(z_M)|\Phi\rangle$$
$$+\frac{1}{z-w}\langle\Psi|[X,Y](w)X_1(z_1)X_2(z_2)\cdots X_M(z_M)|\Phi\rangle$$
$$+\text{regular at } z=w\ \Big\}\,dz_1\cdots dz_M\,.$$

5) *For a local holomorphic coordinate z at Q_i and for $|v\rangle \in V_{\vec{\lambda}} = V_{\lambda_1}\otimes\cdots\otimes V_{\lambda_N} \subset \mathcal{H}_{\vec{\lambda}}$, we have an equality*

$$\langle\Psi|X(w)X_1(z_1)X_2(z_2)\cdots X_M(z_M)|v\rangle$$
$$= \Big\{\frac{1}{z-z(Q_i)}\langle\Psi|X_1(z_1)X_2(z_2)\cdots X_M(z_M)|\rho_i(X)v\rangle$$
$$+\text{ regular at } z=Q_i\Big\}\,dz_1\cdots dz_M\,.$$

The result 4) is often expressed as

$$X(z)Y(w) \sim \frac{\ell\cdot(X,Y)}{(z-w)^2} + \frac{[X,Y](w)}{z-w}$$

and called *operator product expansion*.

For energy momentum tensor we have similar correlation functions.

Theorem 1.2.18.

1) *Put*

$$\begin{aligned}&\langle\Psi|T(z)X_1(z_1)X_2(z_2)\dots X_M(z_M)|\Phi\rangle dz^2dz_1dz_2\cdots dz_M\\&=\frac{1}{2(g^*+l)}\lim_{w\to z}\left\{\sum_{a=1}^{\dim\mathfrak{g}}\langle\Psi|J^a(z)J^a(w)\right.\\&\qquad X_1(z_1)X_2(z_2)\dots X_M(z_M)|\Phi\rangle dzdwdz_1dz_2\cdots dz_M\\&\qquad\left.-\frac{\ell\dim\mathfrak{g}}{(z-w)^2}\langle\Psi|X_1(z_1)X_2(z_2)\dots X_M(z_M)|\Phi\rangle dzdwdz_1dz_2\cdots dz_M\right\}.\end{aligned}$$

Then, this is well-defined and for $k=1,\dots,N$, *we have*

$$\begin{aligned}&\operatorname*{Res}_{\xi_k=0}(\xi_k^{n+1}\langle\Psi|T(\xi_k)X_1(z_1)X_2(z_2)\dots X_M(z_M)|\Phi\rangle d\xi_k)dz_1dz_2\cdots dz_M\\&\qquad=\langle\Psi|X_1(z_1)X_2(z_2)\dots X_M(z_M)|\rho_k(L_n)\Phi\rangle dz_1dz_2\cdots dz_M\end{aligned}$$

where $\{J^1,\dots,J^{\dim\mathfrak{g}}\}$ *is an orthonormal basis of the Lie algebra* $\mathfrak{g}$. *Thus we have an expansion*

$$\begin{aligned}&\langle\Psi|T(\xi_k)X_1(z_1)X_2(z_2)\dots X_M(z_M)|\Phi\rangle d\xi_k^2)dz_1dz_2\cdots dz_M\\&=\sum_{n\in\mathbf{Z}}\langle\Psi|X_1(z_1)X_2(z_2)\dots X_M(z_M)|\rho_k(L_n)\Phi\rangle\xi_k^{-n-2}d\xi_k^2dz_1dz_2\cdots dz_M.\end{aligned}$$

2) *For a holomorphic coordinate transformation* $w=w(z)$ *we have*

$$\begin{aligned}&\langle\Psi|T(w)X_1(z_1)X_2(z_2)\dots X_M(z_M)|\Phi\rangle dw^2dz_1dz_2\cdots dz_M\\&\quad=\langle\Psi|T(z)X_1(z_1)X_2(z_2)\dots X_M(z_M)|\Phi\rangle dz^2dz_1dz_2\cdots dz_M\\&\quad-\frac{c_v}{12}\{w(z);z\}\langle\Psi|X_1(z_1)X_2(z_2)\dots X_M(z_M)|\Phi\rangle dz^2dz_1dz_2\cdots dz_M\end{aligned}$$

where $\{w(z);z\}$ *is the Schwarzian derivative.*

d) $\mathbf{P}^1$ and elliptic curves

In this subsection we shall discuss correlation functions of current and the energy momentum tensor in case of $\mathbf{P}^1$ and elliptic curves.

As above we regard $\mathbf{P}^1$ as $\mathbf{C}\cup\{\infty\}$ and let z be a global coordinate of $\mathbf{C}$. For a positive integer $N\geq 1$ let us choose N-points $z_1,z_2,\dots z_N$ of $\mathbf{P}^1$ and put

$$\xi_j=\begin{cases}z-z_j & \text{if } z_j\neq\infty\\ 1/z & \text{if } z_j=\infty.\end{cases}$$

Let us consider N-pointed $\mathbf{P}^1$ $\mathfrak{X} = (\mathbf{P}^1; z_1, \dots, z_N; \xi_1, \dots, \xi_N)$. formal coordinates. Since there are no holomorphic one-forms on $\mathbf{P}^1$, for $X \in \mathfrak{g}$ and $|v\rangle \in V_{\vec{\lambda}}$ by Theorem 1.2.17, 5), we have

$$\langle\Psi|X(z)|v\rangle dz = \sum_{j=1}^{N} \frac{1}{z - z_j} \langle\Psi|\rho_j(X)v\rangle dz. \tag{1.2-19}$$

Note that if one of z_j, say z_1 is the point at infinity ∞, then in (1.2-19) the term

$$\frac{1}{z - z_j} \langle\Psi|\rho_1(X)v\rangle dz$$

disappears. This is because the residue at the point at infinity ∞ of the form

$$\sum_{j=2}^{N} \frac{1}{z - z_j} \langle\Psi|\rho_j(X)v\rangle dz$$

is

$$-\sum_{j=2}^{N} \langle\Psi|\rho_j(X)v\rangle = \langle\Psi|\rho_1(X)v\rangle$$

by the gauge condition.

For the correlation function of energy momentum tensor, first consider $\langle\Psi|X(z)X(w)|v\rangle dzdw$ for $X \in \mathfrak{g}$. By Theorem 1.2.18 we have

(1.2-20)

$$\begin{aligned} &\langle\Psi|X(z)X(w)|v\rangle dzdw \\ &= \frac{\ell \cdot (X,X)}{(z-w)^2} \langle\Psi|v\rangle dzdw + \sum_{j=1}^{N} \frac{1}{z - z_j} \langle\Psi|X(w)|\rho_j(X)v\rangle dzdw \end{aligned}$$

since there are non holomorphic two-forms on $\mathbf{P}^1 \times \mathbf{P}^1$. By (1.2-19) the equality (1.2-20) is rewritten in the form

(1.2-21)

$$\begin{aligned} \langle\Psi|X(z)X(w)|v\rangle dzdw &= \frac{\ell \cdot (X,X)}{(z-w)^2} \langle\Psi|v\rangle dzdw \\ &+ \sum_{j=1}^{N} \sum_{k=1}^{N} \frac{1}{(z - z_j)(w - z_k)} \langle\Psi|\rho_j(X)\rho_k(X)v\rangle dzdw. \end{aligned}$$

Here, again, if $z_1 = \infty$, then the terms containing z_1 should be omitted. By (1.2-21) and Theorem 1.2.18, for a correlation function of the energy momentum tensor we have the following expression.

(1.2-22)

$$\langle\Psi|T(z)|v\rangle(dz)^2$$

$$= \frac{1}{2(g^*+\ell)}\sum_{j,k}\frac{1}{(z-z_j)(z-z_k)}\cdot\sum_{a=1}^{\dim \mathfrak{g}}\langle\Psi|\rho_j(J^a)\rho_k(J^a)|v\rangle(dz)^2$$

for $\langle\Psi| \in \mathcal{V}^\dagger_{\vec{\lambda}}(\mathfrak{X})$ and $|v\rangle \in V_{\vec{\lambda}}$. Put

$$\Omega_{jk} = \sum_{a=1}^{\dim \mathfrak{g}} \rho_j(J^a)\rho_k(J^a) \tag{1.2-23}$$

and

$$\langle\psi| = j(\langle\Psi|) \in \mathrm{Hom}_{\mathfrak{g}}(V_{\vec{\lambda}}, \mathbf{C}).$$

Then we have the following expression which will be used below.

$$\langle\Psi|T(z)|v\rangle(dz)^2 = \frac{1}{2(g^*+\ell)}\sum_{j=1}^{N}\sum_{k=1}^{N}\frac{\langle\psi|\Omega_{jk}|v\rangle}{(z-z_j)(z-z_k)}(dz)^2. \tag{1.2-24}$$

If $z_1 = \infty$ this formula should read

$$\langle\Psi|T(z)|v\rangle(dz)^2 = \frac{1}{2(g^*+\ell)}\sum_{j=2}^{N}\sum_{k=2}^{N}\frac{\langle\psi|\Omega_{jk}|v\rangle}{(z-z_j)(z-z_k)}(dz)^2. \tag{1.2-25}$$

Next let us consider correlation functions for one pointed elliptic curves. Let E be an elliptic curve with period matrix $(1,\tau)$ with $\tau \in H$, where H is the upper half plane:

$$E = \mathbf{C}/(1,\tau).$$

Let us consider a one-pointed curve $\mathfrak{X} = (\,E\,;\,[0]\,;\,z\,)$ of genus 1 with formal coordinate where $[0]$ is the origin of the elliptic curve E and z is a global coordinate of $\mathbf{C}$. The conformal block $\mathcal{V}^\dagger_\lambda(\mathfrak{X})$ is given by the conditions

$$\langle\Psi|X\otimes\wp^{(n)}(z) = 0 \quad n = 0,1,\ldots$$
$$\langle\Psi|X\otimes 1 = 0$$

where $\wp(z)$ is the Weierstrass $\wp$ function. For an element $\langle\Psi| \in \mathcal{V}^\dagger_\lambda(\mathfrak{X})$, an element $X \in \mathfrak{g}$ and an element $|v\rangle \in V_\lambda$, the one form $\langle\Psi|X(z)|v\rangle dz$ has the expansion

$$\begin{aligned}\langle\Psi|X(z)|v\rangle dz &= \sum_{n\in\mathbf{Z}}\langle\Psi|X(n)v\rangle z^{-n-1}dz\\ &= \sum_{n=0}^{\infty}\langle\Psi|X(-n)v\rangle z^{n-1}dz.\end{aligned}$$

Since there is no meromorphic one form on the elliptic curve E which has a pole of order one at the origin and holomorphic outside the origin, the form $\langle\Psi|X(z)|v\rangle dz$ is holomorphic. Therefore, we have

$$\langle\Psi|X(z)|v\rangle dz = \langle\Psi|X(-1)|v\rangle dz \tag{1.2-26}$$

and if $n \neq 1$ we have

$$\langle\Psi|X(-n)|v\rangle = 0. \tag{1.2-27}$$

By using Theorem 1.2.17 we have

$$\langle\Psi|X(z)X(w)|v\rangle dzdw = \langle\Psi|X(-1)Y(-1)v\rangle dzdw$$

for $\langle\Psi| \in \mathcal{V}^\dagger_\lambda(\mathfrak{X})$, $X, \in \mathfrak{g}$ and an element $|v\rangle \in V_\lambda$. Hence, by Theorem 1.2.18 we also have the following result. For $|v\rangle \in V_\lambda$ we have

$$\langle\Psi|T(z)|v\rangle dz^2 = \frac{1}{2(g^* + \ell)} \sum_{a=1}^{\dim \mathfrak{g}} \langle\Psi|J^a(-1)J^a(-1)|v\rangle dz^2.$$

Again by Theorem 1.2.18, 1) we also have

$$\begin{aligned}\langle\Psi|T(z)|v\rangle dz^2 &= \sum_{n\in\mathbf{Z}} \langle\Psi|L_n v\rangle z^{-n-2} dz^2 \\ &= \langle\Psi|L_0 v\rangle \wp(z) dz^2 + \langle\Psi|L_{-2}v\rangle dz^2 + \cdots \\ &= \Delta_\lambda \langle\Psi|v\rangle \wp(z) dz^2 + \langle\Psi|L_{-2}v\rangle dz^2 + \cdots \\ &= \langle\Psi|L_{-2}v\rangle dz^2 \qquad (1.2\text{-}28)\end{aligned}$$

In particular

$$\langle\Psi|L_{-1}v\rangle = 0.$$

§1.3 Sheaf of conformal blocks

In this section we shall define the sheaf of conformal blocks attached to a family of N-pointed stable curves with formal coordinates and show that it is coherent and locally free.

a) Sheaf of conformal blocks

Definition 1.3.1. Data $(\pi : Y \to B; s_1, s_2, \ldots, s_N; \eta_1, \eta_2, \ldots, \eta_N)$ are called a (holomorphic) family of N-pointed stable curves of genus g with formal neighbourhoods, if the following conditions are satisfied.

(1) Y and B are connected complex manifolds, $\pi : Y \to B$ is a proper flat holomorphic map and $s_1, s_2, \dots, s_N$ are holomorphic sections of π.

(2) For each point $b \in B$ the data $(Y_b := \pi^{-1}(b); s_1(b), s_2(b), \dots, s_N(b))$ is an N-pointed stable curve of genus g.

(3) η_j is an $\mathcal{O}_B$-algebra isomorphism

$$\eta_j : \widehat{\mathcal{O}}_{/s_j} = \varprojlim_{n \to \infty} \mathcal{O}_Y / I_j^n \simeq \mathcal{O}_B[[\xi]],$$

where I_j is the defining ideal of $s_j(B)$ in Y.

Definition 1.3.2. A family $\mathfrak{X} = (\pi : \mathcal{C} \to \mathcal{B}; s_1, s_2, \dots, s_N)$ of N-pointed stable curves of genus g is called to be *versal* (resp. *universal*) at a point $b \in \mathcal{B}$, if for any deformation $\mathfrak{Y} = (\pi : X \to Y; s_1, \dots, s_N)$ of $\pi^{-1}(b) = (C; Q_1, \dots, Q_N)$ with prescribed point $y \in Y$ there exists a holomorphic mapping (resp. unique holomorphic mapping) f from a neighbourhood of y in Y to $\mathcal{B}^{(n)}$ such that the pullback $f^*\mathfrak{X}$ is isomorphic to $\mathfrak{Y}$ in a neighbourhood of y in Y and that df is uniquely determined at the base point. If the family is versal (resp. universal) at each point of $\mathcal{B}$, the family $\mathfrak{X}$ is called a versal (resp. universal) family.

Let $\mathfrak{F} = (\pi : \mathcal{C} \to \mathcal{B}; s_1, \dots, s_N; \eta_1, \dots, \eta_N)$ be a family of N-pointed stable curves of genus g with formal coordinates. We assume that $\mathcal{B}$ is a finite dimensional complex manifold and that each fibre of the family $\mathfrak{F}$ satisfies the condition (Q) in §1.2, but we do *not* assume that the family is connected. The main purpose of the present section is to define the sheaf of conformal blocks $\mathcal{V}^{\dagger}_{\vec{\lambda}}(\mathfrak{F})$ attached to the family $\mathfrak{F}$ and show that it is a coherent $\mathcal{O}_{\mathcal{B}}$-module.

Definition 1.3.3. The sheaf $\widehat{\mathfrak{g}}_N(\mathcal{B})$ of affine Lie algebra over $\mathcal{B}$ is a sheaf of $\mathcal{O}_{\mathcal{B}}$-module

$$\widehat{\mathfrak{g}}_N(\mathcal{B}) = \mathfrak{g} \otimes_{\mathbf{C}} \left(\bigoplus_{j=1}^{N} \mathcal{O}_{\mathcal{B}}((\xi_j))\right) \oplus \mathcal{O}_{\mathcal{B}} \cdot c$$

with the following commutation relation, which is $\mathcal{O}_{\mathcal{B}}$-bilinear.

$$\begin{aligned}
&[(X_1 \otimes f_1, \dots, X_N \otimes f_N), (Y_1 \otimes g_1, \dots, Y_N \otimes g_N)] \\
&\qquad = ([X_1, Y_1] \otimes (f_1 g_1), \dots, [X_N, Y_N] \otimes (f_N g_N)) \\
&\qquad\qquad \oplus c \cdot \sum_{j=1}^{N} (X_j, Y_j) \operatorname*{Res}_{\xi_j = 0} (g_j df_j)
\end{aligned}$$

$$c \in \text{Center}$$

where

$$X_j, Y_j \in \mathfrak{g}, \quad f_j, g_j \in \mathcal{O}_{\mathcal{B}}((\xi_j)).$$

Put

$$(1.3\text{-}1) \qquad \widehat{\mathfrak{g}}(\mathfrak{F}) = \mathfrak{g} \otimes_{\mathbf{C}} \pi_*(\mathcal{O}_{\mathcal{C}}(*S))$$

where we define

$$S = \sum_{j=1}^{N} s_j(\mathcal{B})$$
$$\pi_*(\mathcal{O}_{\mathcal{C}}(*S)) = \varinjlim_{k} \pi_*(\mathcal{O}_{\mathcal{C}}(kS)).$$

There is a sheaf version of homomorphism defined in (1.2-5), by using the formal neighbourhoods η_j.

$$\tilde{t} : \pi_*(\mathcal{O}_{\mathcal{B}}(*S)) \to \bigoplus_{j=1}^{N} \mathcal{O}_{\mathcal{B}}((\xi_j))$$

and we may regard $\widehat{\mathfrak{g}}(\mathfrak{F})$ as a Lie subalgebra of $\widehat{\mathfrak{g}}_N(\mathcal{B})$.

Fix a non-negative integer ℓ. For any $\vec{\lambda} = (\lambda_1, \ldots, \lambda_N) \in (P_\ell)^N$, put

$$(1.3\text{-}2) \qquad \mathcal{H}_{\vec{\lambda}}(\mathcal{B}) = \mathcal{O}_{\mathcal{B}} \otimes_{\mathbf{C}} \mathcal{H}_{\vec{\lambda}},$$

$$(1.3\text{-}3) \qquad \mathcal{H}^{\dagger}_{\vec{\lambda}}(\mathcal{B}) = \underline{\mathrm{Hom}}_{\mathcal{O}_{\mathcal{B}}}(\mathcal{H}_{\vec{\lambda}}(\mathcal{B}), \mathcal{O}_{\mathcal{B}}) = \mathcal{O}_{\mathcal{B}} \otimes_{\mathbf{C}} \mathcal{H}^{\dagger}_{\vec{\lambda}}.$$

The pairing (1.1-22) induces an $\mathcal{O}_{\mathcal{B}}$-bilinear pairing

$$(1.3\text{-}4) \qquad \langle \ | \ \rangle : \mathcal{H}^{\dagger}_{\vec{\lambda}}(\mathcal{B}) \times \mathcal{H}_{\vec{\lambda}}(\mathcal{B}) \to \mathcal{O}_{\mathcal{B}}.$$

The sheaf of affine Lie algebra $\widehat{\mathfrak{g}}_N(\mathcal{B})$ acts on $\mathcal{H}_{\vec{\lambda}}(\mathcal{B})$ and $\mathcal{H}^{\dagger}_{\vec{\lambda}}(\mathcal{B})$ by

(1.3-5)
$$((X_1 \otimes \sum_{n\in\mathbf{Z}} a_n^{(1)} \xi_1{}^n), \ldots, (X_N \otimes \sum_{n\in\mathbf{Z}} a_n^{(N)} \xi_N{}^n))(F \otimes |\Psi\rangle)$$
$$= \sum_{j=1}^{N} \sum_{n\in\mathbf{Z}} (a_n^{(j)} F) \otimes \rho_j(X_j(n))|\Psi\rangle$$

The action of $\widehat{\mathfrak{g}}_N(\mathcal{B})$ on $\mathcal{H}^{\dagger}_{\vec{\lambda}}(\mathcal{B})$ is the dual action of $\mathcal{H}_{\vec{\lambda}}(\mathcal{B})$, that is,

$$(1.3\text{-}6) \qquad \langle \Psi a | \Phi \rangle = \langle \Psi | a \Phi \rangle \quad \text{for any } a \in \widehat{\mathfrak{g}}_N.$$

Definition 1.3.4. Put

$$\mathcal{V}_{\vec{\lambda}}(\mathfrak{F}) = \mathcal{H}_{\vec{\lambda}}(\mathcal{B})/\widehat{\mathfrak{g}}(\mathfrak{F})\mathcal{H}_{\vec{\lambda}}(\mathcal{B})$$
$$\mathcal{V}^{\dagger}_{\vec{\lambda}}(\mathfrak{F}) = \underline{\mathrm{Hom}}_{\mathcal{O}_{\mathcal{B}}}(\mathcal{V}_{\vec{\lambda}}(\mathcal{B}), \mathcal{O}_{\mathcal{B}}).$$

These are sheaves of $\mathcal{O}_{\mathcal{B}}$-modules on $\mathcal{B}$. The sheaf $\mathcal{V}^{\dagger}_{\vec{\lambda}}(\mathfrak{F})$ is called the *sheaf of conformal blocks* attached to the family $\mathfrak{F}$ and $\mathcal{V}_{\vec{\lambda}}(\mathfrak{F})$ is called the *sheaf of covacua*. Note that we have

$$\mathcal{V}^{\dagger}_{\vec{\lambda}}(\mathfrak{F}) = \{ \langle\Psi| \in \mathcal{H}^{\dagger}_{\vec{\lambda}}(\mathfrak{F}) \mid \langle\Psi|a = 0 \quad \text{for any} \quad a \in \widehat{\mathfrak{g}}(\mathfrak{F}) \}.$$

The pairing (1.3-4) induces an $\mathcal{O}_{\mathcal{B}}$-bilinear pairing

$$\langle \mid \rangle : \mathcal{V}^{\dagger}_{\vec{\lambda}}(\mathfrak{F}) \times \mathcal{V}_{\vec{\lambda}}(\mathfrak{F}) \to \mathcal{O}_{\mathcal{B}}. \tag{1.3-7}$$

Lemma 1.3.5. *For a point $s \in \mathcal{B}$ put*

$$\mathfrak{F}_s := (\pi^{-1}(s); s_1(s), \ldots, s_N(s); \eta_1|_{\pi^{-1}(s)}, \ldots, \eta_N|_{\pi^{-1}(s)})$$
$$\mathbf{C}_s := \mathcal{O}_{\mathcal{B},s}/\mathfrak{m}_s$$

where $\mathfrak{m}_s$ is the maximal ideal of the stalk $\mathcal{O}_{\mathcal{B},s}$. Then, we have the following canonical isomorphisms.

$$\mathbf{C}_s \otimes_{\mathcal{O}_{\mathcal{B}}} \mathcal{H}_{\vec{\lambda}}(\mathcal{B}) \simeq \mathcal{H}_{\vec{\lambda}}$$
$$\mathbf{C}_s \otimes_{\mathcal{O}_{\mathcal{B}}} \widehat{\mathfrak{g}}_N(\mathcal{B}) \simeq \widehat{\mathfrak{g}}_N$$
$$\widehat{\mathfrak{g}}(\mathfrak{F}) \otimes_{\mathcal{O}_{\mathcal{B}}} \mathbf{C}_s \simeq \widehat{\mathfrak{g}}(\mathfrak{F}_s)$$
$$\mathcal{V}_{\vec{\lambda}}(\mathfrak{F}) \otimes_{\mathcal{O}_{\mathcal{B}}} \mathbf{C}_s \simeq \mathcal{V}_{\vec{\lambda}}(\mathfrak{F}_s)$$
$$\mathbf{C}_s \otimes_{\mathcal{O}_{\mathcal{B}}} \mathcal{H}^{\dagger}_{\vec{\lambda}}(\mathcal{B}) \simeq \mathcal{H}^{\dagger}_{\vec{\lambda}}.$$

More generally, for a holomorphic mapping $f : Y \to \mathcal{B}$ we let $\mathfrak{F}_Y$ be the pull-back of the family $\mathfrak{F}$ by the morphism f. Then, we have the following canonical isomorphisms.

$$\mathcal{O}_Y \otimes_{\mathcal{O}_{\mathcal{B}}} \mathcal{H}_{\vec{\lambda}}(\mathcal{B}) \simeq \mathcal{H}_{\vec{\lambda}}(Y)$$
$$\widehat{\mathfrak{g}}_N(\mathcal{B}) \otimes_{\mathcal{O}_{\mathcal{B}}} \mathcal{O}_Y \simeq \widehat{\mathfrak{g}}_N(Y)$$
$$\widehat{\mathfrak{g}}(\mathfrak{F}) \otimes_{\mathcal{O}_{\mathcal{B}}} \mathcal{O}_Y \simeq \widehat{\mathfrak{g}}(\mathfrak{F}_Y)$$
$$\mathcal{O}_Y \otimes_{\mathcal{O}_{\mathcal{B}}} \mathcal{V}_{\vec{\lambda}}(\mathfrak{F}) \simeq \mathcal{V}_{\vec{\lambda}}(\mathfrak{F}_Y)$$
$$\mathcal{O}_Y \otimes_{\mathcal{O}_{\mathcal{B}}} \mathcal{H}^{\dagger}_{\vec{\lambda}}(\mathcal{B}) \simeq \mathcal{H}^{\dagger}_{\vec{\lambda}}(Y).$$

Moreover, the actions of $\widehat{\mathfrak{g}}_N(\mathcal{B})$ and $\widehat{\mathfrak{g}}_N(Y)$ on $\mathcal{H}_{\vec{\lambda}}(\mathcal{B})$, $\mathcal{H}^{\dagger}_{\vec{\lambda}}(\mathcal{B})$, $\mathcal{H}_{\vec{\lambda}}(Y)$ and $\mathcal{H}^{\dagger}_{\vec{\lambda}}(Y)$ defined in (1.3-5) and (1.3-6) and the action of $\widehat{\mathfrak{g}}_N$ on $\mathcal{H}_{\vec{\lambda}}$ are compatible with respect to the above canonical isomorphisms.

Note that a priori it is not clear that the natural mapping

$$\mathcal{V}^{\dagger}_{\vec{\lambda}}(\mathfrak{F}) \otimes_{\mathcal{O}_{\mathcal{B}}} \mathbf{C}_s \to \mathcal{V}^{\dagger}_{\vec{\lambda}}(\mathfrak{F}_s)$$

is isomorphic. This is the case if the sheaf $\mathcal{V}_{\vec{\lambda}}(\mathfrak{F})$ is locally free. We shall show later this fact.

For our family, Theorem 1.2.11 takes the following form.

Theorem 1.3.6. *The sheaves $\mathcal{V}_{\vec{\lambda}}(\mathfrak{F})$ and $\mathcal{V}^{\dagger}_{\vec{\lambda}}(\mathfrak{F})$ are coherent $\mathcal{O}_{\mathcal{B}}$-modules.*

b) Local freeness of sheaf of conformal blocks

In this subsection we shall prove local freeness of the sheaf of conformal blocks $\mathcal{V}_{\vec{\lambda}}(\mathfrak{F})$ over the locus $\mathcal{B}\setminus D$ of smooth curves.(For the notation see below.) For that purpose first we introduce a certain $\mathcal{O}_{\mathcal{B}}$-submodule $\mathcal{L}(\mathfrak{F})$ of $\bigoplus_{j=1}^{N}\mathcal{O}_{\mathcal{B}}((\xi_j^{-1}))\frac{d}{d\xi_j}$ and an action of $\mathcal{L}(\mathfrak{F})$ on the sheaves of conformal blocks and covacua as first order twisted differential operators. In the next section this action will be used to define a projectively flat connection on the sheaf of conformal blocks.

Let $\mathfrak{F}^{(0)} = (\pi : \mathcal{C}\to\mathcal{B}; s_1,\dots,s_N)$ be a *versal family* of N-pointed stable curves of genus g. We let Σ be the locus of double points of the fibres of $\mathfrak{F}^{(0)}$ and D be $\pi(\Sigma)$. Note that Σ is a no-singular submanifold of codimension two in $\mathcal{C}$ and D is a divisor in $\mathcal{B}$ whose irreducible components D_i, $i=1,2,\dots,k$ are non-singular. Assume that formal coordinates

$$\eta_j : \widehat{\mathcal{O}}_{\mathcal{C}/s_j(\mathcal{B})} \simeq \mathcal{O}_{\mathcal{B}}[[\,\xi\,]], \qquad j=1,2,\dots,N$$

are given. For simplicity, in the following we assume that $\eta_j^{-1}(\xi)$ is holomorphic in a neighbourhood of $s_j(\mathcal{B})$. The general case can be treated by approximating formal coordinates by holomorphic ones. We use the following notation freely.

$$S_j = s_j(\mathcal{B}), \qquad S=\sum_{j=1}^{N} S_j, \qquad \xi_j=\eta_j^{-1}(\xi).$$

Now $\mathfrak{F}=(\pi:\mathcal{C}\to\mathcal{B}; s_1,\dots,s_N;\eta_1,\dots,\eta_N)$ is a family of N-pointed stable curves of genus g with formal neighbourhoods. For each $\vec{\lambda}\in(P_\ell)^{\oplus N}$ we can define the sheaf of conformal blocks $\mathcal{V}^{\dagger}_{\vec{\lambda}}(\mathfrak{F})$ and the sheaf of covacua $\mathcal{V}_{\vec{\lambda}}(\mathfrak{F})$. These are a subsheaf of $\mathcal{H}^{\dagger}_{\vec{\lambda}}(\mathfrak{B})$ and a quotient sheaf of $\mathcal{H}_{\vec{\lambda}}(\mathfrak{B})$, respectively.

First recall that we have the following exact sequence of $\mathcal{O}_{\mathcal{B}}$-modules.

$$(1.3\text{-}8)\qquad 0\to\pi_*(\Theta_{\mathcal{C}/\mathcal{B}}(*S))\xrightarrow{b}\bigoplus_{j=1}^{N}\mathcal{O}_{\mathcal{B}}[\,\xi_j^{-1}\,]\frac{d}{d\xi_j}\xrightarrow{\vartheta} R^1\pi_*\Theta_{\mathcal{C}/\mathcal{B}}(-S)\to 0$$

More precisely, for any positive integer $m\geq 4g-3$ we have the following exact sequence.

$$(1.3\text{-}9)\quad 0\to\pi_*(\Theta_{\mathcal{C}/\mathcal{B}}(mS))\xrightarrow{b_m}\bigoplus_{j=1}^{N}\bigoplus_{k=0}^{m}\mathcal{O}_{\mathcal{B}}\xi_j^{-k}\frac{d}{d\xi_j}\xrightarrow{\vartheta_m} R^1\pi_*\Theta_{\mathcal{C}/\mathcal{B}}(-S)\to 0$$

Note that the mappings b and b_m correspond to the Laurent expansions with respect to ξ_j up to the zero-th order. To define the first order twisted differential operators acting on the sheaves of conformal blocks and covacua, we need to modify the exact sequence (1.3-8) in the following way.

There is an exact sequence

$$0 \to \Theta_{C/B} \to \Theta_C \stackrel{d\pi}{\to} \pi^*\Theta_B \to 0$$

where $\Theta_{C/B}$ is a sheaf of vector fields tangent to the fibres of π. Put

$$\Theta'_{C,\pi} = d\pi^{-1}(\pi^{-1}\Theta_B(-\log D)).$$

Hence, $\Theta'_{C,\pi}$ is a sheaf of vector field on C tangent along Σ whose vertical components are constant along the fibres of π. That is, in a neighbourhood of a smooth point of a fibre $\Theta'_{C,\pi}$ consists of germs of holomorphic vector fields of the form

$$a(z,u)\frac{\partial}{\partial z} + \sum_{i=1}^{m} b_i(u)u_i\frac{\partial}{\partial u_i} + \sum_{i=m+1}^{n} b_i(u)\frac{\partial}{\partial u_i}$$

where $(z, u_1, \dots, u_n)$ is a system of local coordinates such that the mapping π is expressed as the projection

$$\pi(z, u_1, \dots, u_n) = (u_1, \dots, u_n)$$

and $\pi(\Sigma) = D$ is given by the equation

$$u_1 \cdot u_2 \cdots u_m = 0.$$

More generally, we can define a sheaf $\Theta'_C(mS)_\pi$ as the one consisting of germs of meromorphic vector fields of the form

$$A(z,u)\frac{\partial}{\partial z} + \sum_{i=1}^{m} B_i(u)u_i\frac{\partial}{\partial u_i} + \sum_{i=m+1}^{n} B_i(u)\frac{\partial}{\partial u_i}$$

where $A(z,u)$ has the poles of order at most m along S. Now we have an exact sequence

$$\text{(1.3-10)} \qquad 0 \to \Theta_{C/B}(mS) \to \Theta'_C(mS)_\pi \stackrel{d\pi}{\to} \pi^{-1}\Theta_B(-\log D) \to 0.$$

Note that $\Theta'_C(mS)_\pi$ has a structure of a sheaf of Lie algebras by the usual bracket operation of vector fields and the above exact sequence is the one as sheaves of Lie algebras.

For $m > \frac{1}{N}(2g-2)$, by (1.3-10) we have an exact sequence of $\mathcal{O}_{\mathcal{B}}$-modules.

$$0 \to \pi_*\Theta_{\mathcal{C}/\mathcal{B}}(mS) \to \pi_*\Theta'_{\mathcal{C}}(mS)_\pi \stackrel{d\pi}{\to} \Theta_{\mathcal{B}}(-\log D) \to 0 \tag{1.3-11}$$

which is also an exact sequence of sheave of Lie algebras. Taking $m \to \infty$ we obtain the exact sequence

$$0 \to \pi_*\Theta_{\mathcal{C}/\mathcal{B}}(*S) \to \pi_*\Theta'_{\mathcal{C}}(*S)_\pi \stackrel{d\pi}{\to} \Theta_{\mathcal{B}}(-\log D) \to 0. \tag{1.3-12}$$

The exact sequences (1.3-8) and (1.3-12)are related by the following commutative diagram.

$$\begin{array}{ccccccccc}
0 \to & \pi_*\Theta_{\mathcal{C}/\mathcal{B}}(*S) & \to & \pi_*\Theta'_{\mathcal{C}}(*S)_\pi & \stackrel{d\pi}{\to} & \Theta_{\mathcal{B}}(-\log D) & \to 0 \\
 & \| & & \downarrow p & & \downarrow \rho & \\
0 \to & \pi_*\Theta_{\mathcal{C}/\mathcal{B}}(*S) & \to & \bigoplus_{j=1}^N \mathcal{O}_{\mathcal{B}}[\,\xi_j^{-1}]\frac{d}{d\xi_j} & \stackrel{\vartheta}{\to} & R^1\pi_*\Theta_{\mathcal{C}/\mathcal{B}}(-S) & \to 0
\end{array}$$

where ρ is the Kodaira-Spencer mapping of the family $\mathfrak{F}^{(0)}$ and p is given by taking the non-positive part of the $\frac{d}{d\xi_j}$ part of the Laurent expansions of the vector fields in $\pi_*\Theta_{\mathcal{C}}(mS)_\pi$ at $s_j(\mathcal{B})$. Since our family $\mathfrak{F}^{(0)}$ is versal, the Kodaira-Spencer mapping ρ is an isomorphism of $\mathcal{O}_{\mathcal{B}}$-modules. Therefore, p is isomorphic. Let

$$\tilde{p} : \pi_*\Theta'_{\mathcal{C}}(*S)_\pi \to \bigoplus_{j=1}^{N} \mathcal{O}_{\mathcal{B}}((\,\xi_j\,))\frac{d}{d\xi_j}$$

be the natural lift of the homomorphism p given by taking the $\frac{d}{d\xi_j}$ part of the Laurent expansions at $s_j(\mathcal{B})$. Since p is isomorphic, $\tilde{p}$ is injective. Put

$$\mathcal{L}(\mathfrak{F}) := \tilde{p}(\pi_*\Theta'_{\mathcal{C}}(*S)_\pi).$$

Then, we have the following exact sequence.

$$0 \to \pi_*\Theta_{\mathcal{C}/\mathcal{B}}(*S) \to \mathcal{L}(\mathfrak{F}) \stackrel{\theta}{\to} \Theta_{\mathcal{B}}(-\log D) \to 0 \tag{1.3-13}$$

of $\mathcal{O}_{\mathcal{B}}$-modules. The Lie bracket $[\ \ ,\ \ ]_d$ of $\mathcal{L}(\mathfrak{F})$ is obtained by that of $\pi_*\Theta'_{\mathcal{C}}(*S)_\pi)$ by the mapping p. Thus, for $\vec{\ell}, \vec{m} \in \mathcal{L}(\mathfrak{F})$ we have

$$[\vec{\ell}, \vec{m}]_d = [\vec{\ell}, \vec{m}]_0 + \theta(\vec{\ell})(\vec{m}) - \theta(\vec{m})(\vec{\ell}) \tag{1.3-14}$$

where $[\ ,\ ]_0$ is the usual bracket of formal vector fields and the action of $\theta(\vec{\ell})$ on

$$\vec{m} = (m_1 \frac{d}{d\xi_1}, \dots, m_N \frac{d}{d\xi_N})$$

is defined by

$$(\theta(\vec{\ell})(m_1)\frac{d}{d\xi_1}, \dots, \theta(\vec{\ell})(m_N)\frac{d}{d\xi_N}).$$

Then, the exact sequence (1.3-13) is also that of sheaves of Lie algebras.

Let us define an action of $\mathcal{L}(\mathfrak{F})$ on $\mathcal{H}_{\vec{\lambda}}(\mathcal{B})$.

Definition 1.3.7. *For $\vec{\ell} = (\underline{l}_1, \dots, \underline{l}_N) \in \mathcal{L}(\mathfrak{F})$, the action $D(\vec{\ell})$ on $\mathcal{H}_{\vec{\lambda}}(\mathcal{B})$ is defined by*

$$D(\vec{\ell})(F \otimes |\Phi\rangle) = \theta(\vec{\ell})(F) \otimes |\Phi\rangle - F \cdot (\sum_{j=1}^{N} \rho_j(T[\underline{l}_j])|\Phi\rangle \tag{1.3-15}$$

where

$$F \in \mathcal{O}_{\mathcal{B}}, \quad |\Phi\rangle \in \mathcal{H}_{\vec{\lambda}}.$$

The following proposition is an easy consequence of the definition and (1.1-10).

Proposition 1.3.8. *The action $D(\vec{\ell})$ of $\vec{\ell} \in \mathcal{L}(\mathfrak{F})$ on $\mathcal{H}_{\vec{\lambda}}(\mathcal{B})$ defined above has the following properties.*

1) *For any $f \in \mathcal{O}_{\mathcal{B}}$ we have*

$$D(f\vec{\ell}) = fD(\vec{\ell}).$$

2) *For $\vec{\ell}, \vec{m} \in \mathcal{L}(\mathfrak{F})$ we have*

$$[D(\vec{\ell}), D(\vec{m})] = D([\vec{\ell}, \vec{m}]_d) + \frac{c_v}{12}\sum_{j=1}^{N} \operatorname*{Res}_{\xi_j=0}\left(\frac{d^3\ell_j}{d\xi_j^3} m_j d\xi_j\right) \cdot id.$$

3) *For $f \in \mathcal{O}_{\mathcal{B}}$ and $|\phi\rangle \in \mathcal{H}_{\vec{\lambda}}(\mathcal{B})$ we have*

$$D(\vec{\ell})(f|\phi\rangle) = (\theta(\vec{\ell})(f))|\phi\rangle + fD(\vec{\ell})|\phi\rangle.$$

Namely, $D(\vec{\ell})$ is a first order differential operator, if $\theta(\vec{\ell}) \neq 0$.

We define the dual action of $\mathcal{L}(\mathfrak{F})$ on $\mathcal{H}^{\dagger}_{\vec{\lambda}}(\mathcal{B})$ by

$$D(\vec{\ell})(F \otimes \langle\Psi|) = (\theta(\vec{\ell})F) \otimes \langle\Psi| + \sum_{j=1}^{N} F \cdot \langle\Psi|\rho_j(T[\underline{l}_j]). \tag{1.3-16}$$

where

$$F \in \mathcal{O}_{\mathcal{B}}, \qquad \langle \Psi | \in \mathcal{H}^{\dagger}_{\vec{\lambda}}(\mathcal{B}).$$

Then, for any $|\widetilde{\Phi}\rangle \in \mathcal{H}_{\vec{\lambda}}(\mathcal{B})$ and $\langle \widetilde{\Psi}| \in \mathcal{H}^{\dagger}_{\vec{\lambda}}(\mathcal{B})$, we have

$$\{D(\vec{\ell})\langle\widetilde{\Psi}|\}|\widetilde{\Phi}\rangle + \langle\widetilde{\Psi}|\{D(\vec{\ell})|\widetilde{\Phi}\rangle\} = \theta(\vec{\ell})\langle\widetilde{\Psi}|\widetilde{\Phi}\rangle. \tag{1.3-17}$$

This agrees with the usual definition of the dual connection. See also Proposition 1.3.14 and (1.3-21) below.

Now we shall show that the operator $D(\vec{\ell})$ acts on $\mathcal{V}_{\vec{\lambda}}(\mathfrak{F})$.

Proposition 1.3.9. *For any $\vec{\ell} \in \mathcal{L}(\mathfrak{F})$ we have*

$$D(\vec{\ell})(\widehat{\mathfrak{g}}(\mathfrak{F})\mathcal{H}_{\vec{\lambda}}(\mathcal{B})) \subset \widehat{\mathfrak{g}}(\mathfrak{F})\mathcal{H}_{\vec{\lambda}}(\mathcal{B}).$$

Hence, $D(\vec{\ell})$ operates on $\mathcal{V}_{\vec{\lambda}}(\mathfrak{F})$. Moreover, it is a first order differential operator, if $\theta(\vec{\ell}) \neq 0$.

Proof. An element of $\widehat{\mathfrak{g}}(\mathfrak{F})\mathcal{H}_{\vec{\lambda}}(\mathcal{B})$ is a linear combination of elements of the form

$$F \otimes (\sum_{j=1}^{N} \rho_j(X \otimes t_j(h))|\Phi\rangle)$$

where

$$F \in \mathcal{O}_{\mathcal{B}}, \quad X \in \mathfrak{g}, \quad h \in \pi_* \mathcal{O}_{\mathcal{C}}(*S), \quad |\Phi\rangle \in \mathcal{H}_{\vec{\lambda}}$$

and $t_j(h)$ is the Laurent expansion of h at $S_j = s_j(\mathcal{B})$ with respect to the parameter ξ_j. First we shall show the following equality as operators on $\mathcal{H}_{\vec{\lambda}}(\mathcal{B})$.

(1.3-18)

$$[D(\vec{\ell}), \sum_{j=1}^{N} \rho_j(X \otimes t_j(h))] = \sum_{j=1}^{N} \rho_j\left(X \otimes \left\{\theta(\vec{\ell})(t_j(h)) + \underline{\ell}_j(t_j(h))\right\}\right)$$

where $\theta(\vec{\ell})$ operates on the coefficients of $t_j(h)$. By Proposition 1.3.8, 3) it is enough to show the equality (1.3-18) as operators on $\mathcal{H}_{\vec{\lambda}}$. For $|\Phi\rangle \in \mathcal{H}_{\vec{\lambda}}$, by (1.1-16) we have

$$\begin{aligned}
D(\vec{\ell})(\sum_{j=1}^{N} \rho_j(X \otimes t_j(h))|\Phi\rangle) &- \sum_{j=1}^{N} \rho_j(X \otimes t_j(h))(D(\vec{\ell})|\Phi\rangle) \\
&= \sum_{j=1}^{N} \{\rho_j(X \otimes \theta(\vec{\ell})(t_j(h)) - T[\underline{\ell}_j]\rho_j(X \otimes t_j(h))\}|\Phi\rangle \\
&\quad + \sum_{j=1}^{N} \rho_j(X \otimes t_j(h))T[\underline{\ell}_j]|\Phi\rangle \\
&= \sum_{j=1}^{N} \{\rho_j(X \otimes \theta(\vec{\ell})(t_j(h)) + \rho_j(X \otimes \underline{\ell}_j(t_j(h)))\}|\Phi\rangle.
\end{aligned}$$

But $\theta(\vec{\ell})(t_j(h)) + \underline{\ell}_j(t_j(h))$ is nothing but a Laurent expansion at $s_j(\mathcal{B})$ of a meromorphic function $\tau(h)$ where $\tau = p^{-1}(\vec{\ell}) \in \pi_*(\Theta'_{\mathcal{C}}(*S)_\pi$. Hence, we have the desired result. QED.

Similarly we can show that $D(\vec{\ell})$ acts on $\mathcal{V}^\dagger_{\vec{\lambda}}(\mathfrak{F})$.

Corollary 1.3.10. *The $\mathcal{O}_{\mathcal{B}}$-module $\mathcal{V}_{\vec{\lambda}}(\mathfrak{F})$ is locally free on $\mathcal{B} \setminus D$.*

Proof. By Lemma 1.3.5, for any point $x \in \mathcal{B} \setminus D$ we have an isomorphism

$$\mathbf{C}_x \otimes_{\mathcal{O}_{\mathcal{B}}} \mathcal{V}_{\vec{\lambda}}(\mathfrak{F}) \simeq \mathcal{V}_{\vec{\lambda}}(\mathfrak{F}_x)$$

where we put

$$\mathfrak{F}_x := (\, C_x = \pi^{-1}(x), s_1(x), \ldots, s_N(x); \eta_1|_{C_x}, \cdots, \eta_N|_{C_x}).$$

Let $v_1, \ldots, v_m$ be local holomorphic sections of $\mathcal{V}_{\vec{\lambda}}(\mathfrak{F})$ in a neighbourhood of x such that $\{v_1(x), \ldots, v_m(x)\}$ is a basis of $\mathcal{V}_{\vec{\lambda}}(\mathfrak{F}_x)$. Suppose that there is a non-trivial relation

$$a_1 v_1 + a_2 v_2 + \cdots + a_m v_m = 0 \tag{1.3-19}$$

where a_i's are holomorphic function in a neighbourhood of x. By our assumption

$$a_1(x) = a_2(x) = \cdots = a_m(x) = 0.$$

Changing the order of suffices if necessary , we may assume that there is a positive integer k such that

$$\begin{aligned} &a_1 \in \mathfrak{m}_x^k \setminus \mathfrak{m}_x^{k+1}, \\ &a_i \in \mathfrak{m}_x^l, \quad l \geq k, \quad i = 2, 3, \ldots, m. \end{aligned}$$

We choose a_i's in such a way that the positive number k is the smallest among the relations (1.3-19). Let τ be a nowhere vanishing local holomorphic vector field in a neighbourhood of x such that $\tau(a_1) \in \mathfrak{m}_x^{k-1}$. There exists $\vec{\ell} \in \mathcal{L}(\mathfrak{F})$ with $\theta(\vec{\ell}) = \tau$. Applying $\tau = \theta(\vec{\ell})$ to the equality (1.3-19) we obtain the equality

$$\sum_{i=1}^{m} (\tau(a_i) + \sum_{j=1}^{m} \alpha_{ji}) v_i = 0 \tag{1.3-20}$$

where

$$D(\vec{\ell})(v_i) = \sum_{j=1}^{m} \alpha_{ij} v_j.$$

Then, the relation (1.3-20) is non trivial and

$$\tau(a_1) + \sum_{j=1}^{m} \alpha_{j1} \in \mathfrak{m}_x^{k-1}.$$

This contradicts our assumption. Therefore, $v_1, \ldots, v_m$ are $\mathcal{O}_\mathcal{B}$-linearly independent. Hence, $\mathcal{V}_{\vec{\lambda}}(\mathfrak{F})$ is a locally free $\mathcal{O}_\mathcal{B}$-module at x. QED.

For a coherent $\mathcal{O}_\mathcal{B}$-module $\mathcal{G}$, the locus M consisting of points at which $\mathcal{G}$ is not locally free is a closed analytic subset of $\mathcal{B}$ of codimension at least 2. Therefore, we have the following corollary.

Corollary 1.3.11. *Let W be the maximal subset of $\mathcal{B}$ over which $\mathcal{V}_{\vec{\lambda}}(\mathfrak{F})$ is not locally free. Then, W is an analytic subset of $\mathcal{B}$ and*

$$W \subsetneqq D.$$

Since we defined

$$\mathcal{V}_{\vec{\lambda}}^\dagger(\mathfrak{F}) = \underline{\mathrm{Hom}}_{\mathcal{O}_\mathcal{B}}(\mathcal{V}_{\vec{\lambda}}(\mathfrak{F}), \mathcal{O}_\mathcal{B})$$

we have the following corollary.

Corollary 1.3.12. *$\mathcal{V}_{\vec{\lambda}}^\dagger(\mathfrak{F})|_{\mathcal{B}\setminus D}$ is a locally free $\mathcal{O}_\mathcal{B}$-module and for any subvariety Y of $\mathcal{B} \setminus D$ we have an $\mathcal{O}_Y$-module isomorphism*

$$\mathcal{O}_Y \otimes_{\mathcal{O}_\mathcal{B}} \mathcal{V}_{\vec{\lambda}}^\dagger(\mathfrak{F}) \simeq \mathcal{V}_{\vec{\lambda}}^\dagger(\mathfrak{F}|_Y)$$

These two corollaries play crucial role to prove local freeness in general. The above corollaries imply the following theorem.

Theorem 1.3.13. *If $\mathfrak{F}$ is a family of N-pointed smooth curves with formal neighbourhoods such that the induced family $\mathfrak{F}^{(0)}$ of N-pointed smooth curves is not necessarily versal, then $\mathcal{V}_{\vec{\lambda}}(\mathfrak{F})$ and $\mathcal{V}_{\vec{\lambda}}^\dagger(\mathfrak{F})$ are locally free $\mathcal{O}_\mathcal{B}$-modules and they are dual to each other.*

Another important consequence of the above discussions is the following.

Proposition 1.3.14. *For each element $\vec{\ell} \in \mathcal{L}(\mathfrak{F})$, $D(\vec{\ell})$ acts on $\mathcal{V}_{\vec{\lambda}}^\dagger(\mathfrak{F})$. Moreover, if $\theta(\vec{\ell}) \neq 0$, then $D(\vec{\ell})$ acts on $\mathcal{V}_{\vec{\lambda}}^\dagger(\mathfrak{F})$ as a first order differential operator.*

Note that for the natural bilinear pairing $\langle \quad | \quad \rangle : \mathcal{V}_{\vec{\lambda}}^\dagger(\mathfrak{F}) \times \mathcal{V}_{\vec{\lambda}}(\mathfrak{F}) \to \mathcal{O}_\mathcal{B}$, we have the equality

$$\{D(\vec{\ell})\langle\Psi|\}|\Phi\rangle + \langle\Psi|\{D(\vec{\ell})|\Phi\rangle\} = \theta(\vec{\ell})(\langle\Psi|\Phi\rangle). \tag{1.3-21}$$

§1.4 Projectively flat connection and flat sections

In this section, based on the arguments in §1.3, b) we shall define the projectively flat connection on the sheaf of conformal blocks. The connection is defined over the locus of smooth curves $\mathcal{B} \setminus D$ and it has regular singularities along the boundary D, the locus of the singular curves.

a) Projectively flat connection

Let $\mathfrak{F} = (\pi : \mathcal{C} \to \mathcal{B}; s_1, \ldots, s_N; \eta_1, \ldots, \eta_N)$ be a family of N-pointed curves with formal neighbourhoods such that $\mathfrak{F}^{(0)} := (\pi : \mathcal{C} \to \mathcal{B}; s_1, \ldots, s_N)$ is a versal family of N-pointed curves. We shall use freely the notations in the previous sections.

As was shown in Proposition 1.3.9 and Proposition 1.3.14 the sheaf $\mathcal{L}(\mathfrak{F})$ acts on $\mathcal{V}_{\vec{\lambda}}(\mathfrak{F})$ and $\mathcal{V}^{\dagger}_{\vec{\lambda}}(\mathfrak{F})$ from the left. Moreover, for $\vec{\ell} \in \mathcal{L}(\mathfrak{F})$, if $\theta(\vec{\ell}) \neq 0$, the action defines a first order differential operator. In this section first we shall study the action of $\vec{\ell}$ more closely when $\theta(\vec{\ell}) = 0$. Note that by (1.3-14) we have an exact sequence

$$(1.4\text{-}1) \qquad 0 \to \pi_*(\Theta_{\mathcal{C}/\mathcal{B}}(*S)) \xrightarrow{t} \mathcal{L}(\mathfrak{F}) \xrightarrow{\theta} \Theta_{\mathcal{B}}(-\log D) \to 0.$$

In the following for simplicity we shall discuss mainly a smooth family of N-pointed curves with formal neighbourhoods though we shall use general notation.

Lemma 1.4.1. *Assume that $\mathcal{B}$ is small enough so that we can find a symmetric bidifferential*

$$\omega \in H^0(\mathcal{C} \times_{\mathcal{B}} \mathcal{C}, \omega_{\mathcal{C} \times_{\mathcal{B}} \mathcal{C}/\mathcal{B}}(2\Delta))$$

satisfying $\mathrm{Res}^2(\omega) = 1$, *that is*

$$\omega = \left(\frac{1}{(z-w)^2} + \textit{holomorphic} \right) dz dw$$

at the diagonal Δ. Then there exists a unique $\mathcal{O}_{\mathcal{B}}$-module homomorphism

$$a : \pi_* \Theta_{\mathcal{C}/\mathcal{B}}(*S) \to \mathcal{O}_{\mathcal{B}}$$

independent of the choice of ω such that for any $\vec{\ell} \in \mathcal{L}(\mathfrak{F})$ which is the image of an element of $\pi_* \Theta_{\mathcal{C}/\mathcal{B}}(*S)$ by the exact sequence (1.4-1), we have

$$D(\vec{\ell}) = a(\vec{\ell}) \cdot \mathrm{id}$$

as a linear operator acting on $\mathcal{V}_{\vec{\lambda}}(\mathfrak{F})$ from the left and

$$D(\vec{\ell}) = -a(\vec{\ell}) \cdot \mathrm{id}$$

as a operator acting on $\mathcal{V}^{\dagger}_{\vec{\lambda}}(\mathfrak{F})$ from the left .

Proof. Since $\theta(\vec{\ell}) = 0$, by (1.3-15) for any $\langle\Psi| \in \mathcal{V}^{\dagger}_{\vec{\lambda}}(\mathfrak{F})$ and $|\Phi\rangle \in \mathcal{V}_{\vec{\lambda}}(\mathfrak{F})$ we have

$$\begin{aligned}\langle\Psi|\{D(\vec{\ell})|\Phi\rangle\} &= -\sum_{j=1}^{N}\langle\Psi|\rho_j(T[\underline{\ell}_j])|\Phi\rangle \\ &= -\sum_{j=1}^{N}\operatorname*{Res}_{\xi_j=0}(\ell_j(\xi_j)\langle\Psi|T(\xi_j)|\Phi\rangle d\xi_j)\end{aligned}$$

where $\underline{\ell}_j = \ell_j(\xi_j)\dfrac{d}{d\xi_j}$ and by Theorem 1.2.18 we have

$$\begin{aligned}&\langle\Psi|T(\xi_j)|\Phi\rangle(d\xi_j)^2 \\ &= \lim_{\zeta_j\to\xi_j}\left\{\frac{1}{2(\ell+g^*)}\sum_{a=1}^{\dim\mathfrak{g}}\langle\Psi|J^a(\zeta_j)J^a(\xi_j)|\Phi\rangle d\zeta_j d\xi_j\right. \\ &\qquad\left. -\frac{c_v}{2(\zeta_j-\xi_j)^2}\langle\Psi|\Phi\rangle d\zeta_j d\xi_j\right\}\end{aligned}$$

where

$$c_v = \frac{\ell\cdot\dim\mathfrak{g}}{g^*+\ell}.$$

Choose a symmetric bidifferential

$$\omega \in H^0(\mathcal{C}\times_{\mathcal{B}}\mathcal{C}, \omega_{\mathcal{C}\times_{\mathcal{B}}\mathcal{C}/\mathcal{B}}(2\Delta))$$

with $\mathrm{Res}^2(\omega) = 1$. The existence of such a bidifferential is well-known for smooth curves. Put

(1.4-2)

$$\begin{aligned}&\langle\Psi|\widetilde{T}(z)|\Phi\rangle(dz)^2 \\ &:= \lim_{w\to z}\left\{\frac{1}{2(\ell+g^*)}\sum_{a=1}^{\dim\mathfrak{g}}\langle\Psi|J^a(w)J^a(z)|\Phi\rangle dwdz\right. \\ &\qquad\left. -\frac{c_v}{2}\omega(w,z)\langle\Psi|\Phi\rangle dwdz\right\}.\end{aligned}$$

Also define $S_\omega(z)(dz)^2$ by

$$S_\omega(z)(dz)^2 = 6\lim_{w\to z}\left\{\omega(w,z)dwdz - \frac{dwdz}{(w-z)^2}\right\}.$$

The form $S_\omega(z)(dz)^2$ is called a projective connection. It depends not only on the choice of ω but also on the choice of local coordinates:

$$S_\omega(w)(dw)^2 = S_\omega(z)(dz)^2 + \{w; z\}(dz)^2.$$

Now we have

$$\langle\Psi|T(\xi_j)|\Phi\rangle(d\xi_j)^2 = \langle\Psi|\widetilde{T}(\xi_j)|\Phi\rangle(d\xi_j)^2 + \frac{c_v}{12}\langle\Psi|\Phi\rangle S_\omega(\xi_j)(d\xi_j)^2.$$

Thus we have

(1.4-3)

$$\langle\Psi|\{D(\vec{\ell})|\Phi\rangle\} = -\sum_{j=1}^{N} \operatorname*{Res}_{\xi_j=0}(\ell_j(\xi_j)\langle\Psi|\widetilde{T}(\xi_j)|\Phi\rangle\Phi d\xi_j)$$

$$- \frac{c_v}{12}\langle\Psi|\Phi\rangle \sum_{j=1}^{N} \operatorname*{Res}_{\xi_j=0}(\ell_j(\xi_j)S_\omega(\xi_j)d\xi_j).$$

Since $\ell_j(z)\langle\Psi|\widetilde{T}(z)|\Phi\rangle dz$ is a global meromorphic one-form which has poles only at $s_j(\mathcal{B})$, the first term of the right hand side of the above equality is zero. Therefore, if we put

$$a_\omega(\vec{\ell}) = -\frac{c_v}{12}\sum_{j=1}^{N} \operatorname*{Res}_{\xi_j=0}(\ell_j(\xi_j)S_\omega(\xi_j)d\xi_j) \tag{1.4-4}$$

then we have

$$\langle\Psi|\{D(\vec{\ell})|\Phi\rangle\} = a_\omega(\vec{\ell})\langle\Psi|\Phi\rangle.$$

Now $\mathcal{V}_{\vec{\lambda}}(\mathfrak{F})$ and $\mathcal{V}^\dagger_{\vec{\lambda}}(\mathfrak{F})$ are locally free and dual to each other, we conclude

$$D(\vec{\ell})|\Phi\rangle = a_\omega(\vec{\ell})|\Phi\rangle.$$

Let us show that $a_\omega(\vec{\ell})$ is independent of the choice of ω. If $\omega' \in H^0(\mathcal{C} \times_{\mathcal{B}} \mathcal{C}, \omega_{\mathcal{C}\times_{\mathcal{B}}\mathcal{C}/\mathcal{B}}(2\Delta))$ is another symmetric bidifferential with $\mathrm{Res}^2(\omega) = 1$, then $\omega - \omega'$ is a holomorphic bidifferential on $\mathcal{C} \times_{\mathcal{B}} \mathcal{C}$. Hence, $S_\omega - S_{\omega'}$ is also a holomorphic section of $\omega_{\mathcal{C}/\mathcal{B}}^{\otimes 2}$ on $\mathcal{C}$. Let τ be an element of $\pi_*\Theta_{\mathcal{C}/\mathcal{B}}(*S)$ with $t(\tau) = \vec{\ell}$. Then, $\tau(z)(S_\omega(z) - S_{\omega'}(z))dz$ is a meromorphic one form on $\mathcal{C}$. Hence, we have

$$\sum_{j=1}^{N} \operatorname*{Res}_{\xi_j=0}(\ell(\xi_j)(S_\omega(\xi_j) - S_{\omega'}(\xi_j))d\xi_j) = 0.$$

Thus, we conclude

$$a_\omega(\vec{\ell}) = a_{\omega'}(\vec{\ell}).$$

This shows the first part of the lemma. By (1.3-21) similarly we can show the second part. QED.

The above proof shows also the following corollary.

Corollary 1.4.2. *Under the same assumption as in Lemma 1.4.1 for each $\vec{\ell} \in \mathcal{L}(\mathfrak{F})$, if we define $a_\omega(\vec{\ell})$ by (1.4-4), then*

$$a_\omega : \mathcal{L}(\mathfrak{F}) \to \mathcal{O}_{\mathcal{B}}$$

is an $\mathcal{O}_{\mathcal{B}}$-module homomorphism.

Remark 1.4.3. In the above corollary, since $\vec{\ell}$ is not necessarily an image of a global meromorphic vector field, $a_\omega(\vec{\ell})$ does depend on the choice of ω.

Now we are ready to define the connections on $\mathcal{V}_{\vec{\lambda}}(\mathfrak{F})$ and $\mathcal{V}^\dagger_{\vec{\lambda}}(\mathfrak{F})$, if $\mathcal{B}$ is small enough. Let us fix a symmetric bidifferential $\omega \in H^0(\mathcal{C} \times_{\mathcal{B}} \mathcal{C}, \omega_{\mathcal{C}\times_{\mathcal{B}}\mathcal{C}/\mathcal{B}}(2\Delta))$ with $\operatorname{Res}^2(\omega) = 1$. For each element $X \in \Theta_{\mathcal{B}}(-\log D)$, there is an element $\vec{\ell} \in \mathcal{L}(\mathfrak{F})$ with $\theta(\vec{\ell}) = X$. Define an operator $\nabla^{(\omega)}_X$ acting on $\mathcal{V}_{\vec{\lambda}}(\mathfrak{F})$ from the left by

$$\nabla^{(\omega)}_X([|\Phi\rangle]) = D(\vec{\ell})([|\Phi\rangle]) - a_\omega(\vec{\ell})([|\Phi\rangle]) \tag{1.4-5}$$

where by $[|\Phi\rangle]$ we denote the element in $\mathcal{V}_{\vec{\lambda}}(\mathfrak{F})$ corresponding to $|\Phi\rangle$.

Proposition 1.4.4. *$\nabla^{(\omega)}_X$ is well defined and enjoys the following property.*

$$\nabla^{(\omega)}_X([f|\Phi\rangle]) = X(f)[|\Phi\rangle] + f\nabla^{(\omega)}_X([|\Phi\rangle]) \tag{1.4-6}$$

where $f \in \mathcal{O}_{\mathcal{B}}$. Hence, the correspondence $X \mapsto \nabla^{(\omega)}_X$ defines a connection on $\mathcal{V}_{\vec{\lambda}}(\mathfrak{F})$ with regular singularities along D.

Proof. Choose another $\vec{\ell}' \in \mathcal{L}(\mathfrak{F})$ with $\theta(\vec{\ell}') = X$. Then, we have

$$a(\vec{\ell} - \vec{\ell}') = a_\omega(\vec{\ell}) - a_\omega(\vec{\ell}').$$

By Lemma 1.4.1 we have

$$(D(\vec{\ell}) - D(\vec{\ell}'))[|\Phi\rangle] = -\sum_{j=1}^{N}[|\Phi\rangle] = a(\vec{\ell} - \vec{\ell}')[|\Phi\rangle].$$

Thus, as operators on $\mathcal{V}_{\vec{\lambda}}(\mathfrak{F})$ we have the equality

$$D(\vec{\ell}) - a_\omega(\vec{\ell}) \cdot \mathrm{id} = D(\vec{\ell}') - a_\omega(\vec{\ell}') \cdot \mathrm{id}.$$

Hence, $\nabla^{(\omega)}_X$ does not depend on the choice of $\vec{\ell}$. By Proposition 1.3.8, 3) the equality (1.4-6) holds. QED.

Theorem 1.4.5. *The connection $\nabla^{(\omega)}$ on $\mathcal{V}_{\vec{\lambda}}(\mathfrak{F})|_{B-D}$ is projectively flat.*

Remark 1.4.6. For $\mathbf{P}^1$, since the form

$$\omega = \frac{dwdz}{(w-z)^2}$$

is a global form, we have

$$a_\omega = 0.$$

Hence, in this case the connection is flat. For N-pointed $\mathbf{P}^1$, by (1.2-24) the connection is the inhomogeneous form of Knizhnik-Zamolodchikov equation ([KZ]).

b) Family of one-pointed stable curves of genus 1

In this subsection we shall show how to prove local freeness of the sheaf of conformal blocks over the locus of singular curves in case of the universal family of one-pointed stable curves of genus 1. The argument will also show how to construct multivalued flat sections from the data of a boundary. It is deeply related to monodromy representation of Teichmüller modular groups ([TK], [Ko], [W]) and the Verlinde conjecture ([Ve], [MS1]). For details we refer the reader to [U].

First let us construct a versal family of stable curves of genus one. Let z and w be coordinates of $\mathbf{P}^1$ with center 0 and ∞, respectively with

$$zw = 1.$$

Put

$$\begin{aligned} &D = \{q \in \mathbf{C} \mid |q| < 1\}, \quad X = \{R \in \mathbf{P}^1 \mid |z(R)| < 1\}, \\ &Y = \{P \in \mathbf{P}^1 \mid |w(R)| < 1\}. \end{aligned}$$

and

$$\begin{aligned} S &= \{(x,y,q) \in \mathbf{C}^3 \mid xy = q,\ |x| < 1, |y| < 1, |q| < 1\} \\ Z &= \{(R,q) \in \mathbf{P}^1 \times D \mid R \in \mathbf{P}^1 \setminus \{X \cup Y\} \quad \text{or} \quad R \in X \quad |z(R)| > |q|\,\} \\ W &= \{(P,q) \in \mathbf{P}^1 \times D \mid P \in \mathbf{P}^1 \setminus \{X \cup Y\} \quad \text{or} \quad P \in X \quad |w(P)| > |q|\,\} \end{aligned}$$

Let us introduce an equivalence relation on $S \cup Z \cup W$. The equivalence relation is generated by the following relations.

1) A point $(R,q) \in X \times D \cap Z$ and a point $(x,y,q') \in S$ are equivalent if

$$(x,y,q') = (z(R), \frac{q}{z(R)}, q).$$

2) A point $(P,q) \in Y \times D \cap W$ and a point $(x,y,q') \in S$ are equivalent if

$$(x,y,q') = (\frac{q}{w(P)}, w(P), q).$$

3) A point $(R,q) \in Z$ and a point $(P,q') \in W$ are equivalent if

$$q' = q, \quad z(R)w(P) = q.$$

Put

$$\mathcal{E} = (S \cup Z \cup W)/\sim.$$

Then $\mathcal{E}$ is a two-dimensional complex manifold. For points $(x,y,q) \in S$, $(R,q) \in Z$ and $(P,q) \in W$ we denote the corresponding points in $\mathcal{E}$ by $[(x,y,q)]$, $[(R,q)]$ and $[(P,q)]$, respectively. We have a natural holomorphic mapping

$$\pi : \mathcal{E} \to D.$$

For $q \in D$ put

$$\begin{aligned} S(q) &= \{(x,y) \in \mathbf{C}^2 \mid xy = q\}, \\ Z(q) &= \{\mathbf{P}^1 - X \cup Y\} \cup \{R \in X \mid |z(R)| > |q|\}, \\ W(q) &= \{\mathbf{P}^1 - X \cup Y\} \cup \{P \in Y : \mid |w(P)| > |q|\}. \end{aligned}$$

By the reason which will become clear later we use mainly the coordinate w so that

$$\mathbf{P}^1 = \{(w:1) \mid w \in \mathbf{C}\} \cup \{(1:0)\}.$$

Assume that $q \neq 0$. We can rewrite

$$\begin{aligned} \mathbf{P}^1 - X \cup Y &= \{(w:1) \mid |w| = 1\} \\ Z(q) &= \{(w:1) \mid 1 \leq |w| < 1/|q|\} \\ W(q) &= \{(w:1) \mid |q| < |w| \leq 1\} \end{aligned}$$

On $S(q) \cup Z(q) \cup W(q)$ the above equivalence relation induces the following one.

$(w:1) \in X \times q \cap Z(q) = \{w \mid 1 \leq |w| < 1/|q|\}$ and $(x,y,q) \in S(q)$ is equivalent if

$$(x,y,q) = (1/w, qw, q);$$

$(w':1) \in Y \times q \cap W(q) = \{w' \mid |q| < |w'| \leq 1\}$ $(x,y,q) \in S(q)$ is equivalent if

$$(x,y,q) = (q/w', w', q).$$

Note that if $q \neq 0$, a point $(x,y,q) \in S(q)$ is either equivalent to the point $(w:1) \in X \times q \cap Z(q)$ (when $|x| \leq 1$) or to the point $(w':1) \in Y \times q \cap W(q)$

(when $|y| \leq 1$). Therefore, the quotient manifold $\mathcal{E}(q) = (S(q) \cup Z(q) \cup W(q))/\sim$ is obtained by identifying $1 \leq |w| < 1/|q|$ and $|q| < |w'| \leq 1$ by $w' = qw$. In other word, $\mathcal{E}(q)$ is the quotient manifold $E_{\langle q \rangle} = \mathbf{C}^*/\langle q \rangle$ where $\langle q \rangle$ is the infinite cyclic group of analytic automorphism generated by

$$w \mapsto qw.$$

Hence, $\mathcal{E}(q)$ is the elliptic curve $E_{\langle q \rangle}$, if $q \neq 0$ and a rational curve with an ordinary double point for $q = 0$. Note that $\mathcal{E}(q)$ is the fibre $\pi^{-1}(q)$ of $\pi : \mathcal{E} \to D$ and that the point $[1]$ of $\mathcal{E}(q)$ corresponding to $w = 1$ is the origin of the elliptic curve. Moreover, if $q = 0$, the fibre $\pi^{-1}(0)$ of π is a rational curve with ordinary double point obtained by identifying the points $\infty = (0 : 1)$ and $0 = (1 : 0)$.

It is easy to show that $\dfrac{dw}{w}$ is a holomorphic one-form on $\mathcal{E}(q)$ for $q \neq 0$. and the family $\pi : \mathcal{E} \to D$ is versal at each point $q \in D$.

The mapping

$$\begin{aligned} o : D \to \quad & \mathcal{E} \\ q \mapsto & [(q, 1, q)], \quad \text{where } (q, 1, q) \in S \end{aligned}$$

defines a holomorphic section o of the family $\pi : \mathcal{E} \to D$.

Let us consider the family $\mathfrak{F} = (\mathcal{E} \to D; o; u)$ of one-pointed stable curves of genus one with coordinate, where $u = w - 1$. Let us consider the sheaf of conformal blocks $\mathcal{V}_0^\dagger(\mathfrak{F})$. First we shall construct formal sections of $\mathcal{V}_0^\dagger(\mathfrak{F})$ starting from the data on the fibre $\mathcal{E}(0)$ at the origin.

Let

$$\nu : \widetilde{\mathcal{E}}(0) = \mathbf{P}^1 \to \mathcal{E}(0)$$

be the normalization of $\mathcal{E}(0)$. Then, we have

$$\begin{aligned} &\nu(0) = \nu(\infty) = \text{the double point} \\ &\nu(1) = o(0). \end{aligned}$$

We know that $\mathcal{V}_0^\dagger((\mathcal{E}(0); o(0), u))$ is canonically isomorphic to

$$\bigoplus_{\lambda \in \Lambda} \mathcal{V}^\dagger_{\lambda^\dagger, \lambda, 0}((\mathbf{P}^1; \infty, 0, 1; \eta, \xi, \xi - 1)).$$

For simplicity, put

$$\mathcal{V}^\dagger_{\lambda^\dagger, \lambda, 0} = \mathcal{V}^\dagger_{\lambda^\dagger, \lambda, 0}((\mathbf{P}^1; \infty, 1, 0; \eta, \xi - 1, \xi)).$$

The vector space $\mathcal{V}^\dagger_{\lambda^\dagger, \lambda, 0}$ is one-dimensional and there is the element $\langle \Psi | \in \mathcal{V}^\dagger_{\lambda^\dagger, \lambda, 0}$ such that

$$\langle \Psi | v^\dagger \otimes v \otimes 0 \rangle = (v^\dagger | v)$$

for any $|v^\dagger\rangle \in \mathcal{H}_{\lambda^\dagger}$ and $|v\rangle \in \mathcal{H}_\lambda$ where $|0\rangle$ is the highest weight vector of $\mathcal{H}_0$ and $(\ |\ )$ is the bilinear pairing

$$\mathcal{H}_{\lambda^\dagger} \otimes \mathcal{H}_\lambda \to \mathbf{C}$$

defined in §2.2. Let us construct an element

$$\langle \Psi^{(\lambda)} | \in \mathcal{V}_0^\dagger(\mathfrak{F})$$

from $\langle \Psi |$.

For any non-negative integer d, by Lemma 1.1.11 there exist bases $\{v^k(d)\}$ and $\{v_k(d)\}$, $k = 1, 2, \ldots, m_d$ of $\mathcal{H}_{\lambda^\dagger}(d)$ and $\mathcal{H}_\lambda(d)$, respectively such that

$$(v^j(d)|v_k(d)) = \delta^j_k.$$

For each element $|\phi\rangle \in \mathcal{H}_0$ $\langle \Psi_d | \in \mathcal{H}_0^\dagger$ is defined by

$$\langle \Psi_d | \phi \rangle = \sum_{i=1}^{m_d} \langle \Psi | v^i(d) \otimes v_i(d) \otimes \phi \rangle.$$

Now $\langle \Psi^{(\lambda)} |$ is defined by

$$\langle \Psi^{(\lambda)} | \phi \rangle = \sum_{d=0}^{\infty} \langle \Psi_d | \phi \rangle q^d.$$

This is a formal power series in q and from our construction it is not clear that it converges. Later we shall show that $q^{\Delta_\lambda} \langle \Psi^{(\lambda)} | \phi \rangle$ is a formal solution of a differential equation of Fuchsian type so that it converges near the origin. If we put $|\phi\rangle = |0\rangle$, then

$$q^{\Delta_\lambda} \langle \Psi^{(\lambda)} | 0 \rangle = \sum_{d=0}^{\infty} \sum_{i=1}^{m_d} (v^i)d)|v_i(d))q^d = \sum_{d=0}^{\infty} \dim \mathcal{H}_\lambda(d) q^{\Delta_\lambda + d}. \tag{1.4-7}$$

This formal power series is the character of the integrable $\mathfrak{g}$-module $\mathcal{H}_\lambda$ and can be described by using theta functions ([Ka]). It is a multivalued holomorphic function on D^*.

At the moment we only know that $\langle \Psi^{(\lambda)} |$ is an element of $\mathcal{H}_0[[q]]$. We shall show that it satisfies the formal gauge condition so that it can be regarded as an element of the completion of $\mathcal{V}_0^\dagger(\mathfrak{F})$ at the origin.

To explain the formal gauge condition first we shall consider meromorphic functions on our elliptic curve $\mathcal{E}(q)$ having only poles at the origin [1]. Put

$$x(q, w) = -2 \sum_{n=1}^{\infty} \frac{q^n}{(1-q^n)^2} + \sum_{n=-\infty}^{+\infty} \frac{wq^n}{(1-wq^n)^2} \tag{1.4-8}$$

$$y(q, w) = \sum_{n=-\infty}^{+\infty} \frac{(1+wq^n)wq^n}{(1-wq^n)^3} \tag{1.4-9}$$

Then, if we put $x = \exp(2\pi\sqrt{-1}z)$, we have

$$x(q,w) = -\frac{1}{12} - \frac{1}{4\pi^2}\wp(z)$$
$$y(q,w) = \frac{\sqrt{-1}}{8\pi^3}\wp'(z)$$

where $\wp(z)$ is the Weierstrass $\wp$ function. Moreover, we have

$$x(q,qw) = x(q,w)$$
$$y(q,qw) = y(q,w)$$
$$x(q,w^{-1}) = x(q,w)$$
$$y(q,w^{-1}) = -y(q,w)$$

and they satisfy the following equation:

(1.4-10) $$y^2 - 4x^3 - x^2 + g_2(q)x + g_3(q) = 0,$$

where we put

$$x = x(q,w)$$
$$y = y(q,w)$$
$$g_2(q) = 20\sum_{n=1}^{\infty}\frac{n^3q^n}{1-q^n}$$
$$g_3(q) = \frac{1}{3}\sum_{n=1}^{\infty}\frac{(7n^5+5n^3)q^n}{1-q^n}.$$

Note that $x(q,w)$ has pole of order 2 at [1] and $y(q,w)$ has pole of order 3 at [1]. They are holomorphic except [1]. By the theory of elliptic functions we know that $H^0(\mathcal{E}(q), \mathcal{O}(*[1]))$ is spanned by monomials of $x(q,w)$ and monomials of $x(q,w)$ times $y(q,w)$. Note that $x(q,w)$ has an expansion

$$x(q,w) = \frac{w}{(1-w)^2} + \sum_{\ell=1}^{\infty}\left\{\sum_{m|\ell}(mw^m + mw^{-m} - 2)\right\}q^\ell.$$

This formula shows that each coefficient of q^ℓ can be regarded as a meromorphic function of $\mathbf{P}^1$ having only poles at the points ∞, 1 and 0. The same is true if we expand polynomials $P(x,y)$ of $x(q,w)$ and $y(q,w)$ in q. That is, if we have an expansion

(1.4-11)
$$P(x(q,w),y(q,w)) = Q_1(x(q,w)) + Q_2(x(q,w))y(q,w) = \sum_{k=0}^{\infty} f_k(w)q^k,$$

all coefficients $f_k(w)$ are in $H^0(\mathbf{P}^1, \mathcal{O}_{\mathbf{P}^1}(*(\infty+1+0)))$.

Now the gauge condition can be calculated formally as follows.

$$\langle\Psi^{(\lambda)}|X \otimes P(x(q,w), y(q,w))\phi\rangle = \sum_{d=0}^{\infty}\langle\Psi_d|\sum_{k=0}^{\infty}(X \otimes f_k(w))q^k|\phi\rangle q^d$$

$$(1.4\text{-}12) \qquad = \sum_{d,k=0}^{\infty}(\sum_{i=1}^{m_d}\langle\Psi|(X \otimes f_k(w))|v^i(d) \otimes v_i(d) \otimes \phi\rangle)q^{d+k}$$

By the gauge condition on $\langle\Psi| \in \mathcal{V}_{\lambda^\dagger,0,\lambda}$, we have

$$\langle\Psi|(X \otimes f_k(w))|v^i(d) \otimes v_i(d) \otimes \phi\rangle = 0.$$

This proves formally

$$\langle\Psi^{(\lambda)}|X \otimes P(x(q,w), y(q,w))$$

$$(1.4\text{-}13) \qquad = \sum_{d,k=0}^{\infty}(\sum_{i=1}^{m_d}\langle\Psi|(X \otimes f_k(w))|v^i(d) \otimes v_i(d) \otimes \phi\rangle)q^{d+k}.$$

Since we do not know the convergence of $\langle\Psi^{(\lambda)}|$, the above argument does not prove the fact

$$\langle\Psi^{(\lambda)}| \in \mathcal{V}_0^\dagger(\mathfrak{F}).$$

To go further we need some facts from algebraic geometry. First we take the completion $\widehat{\mathcal{V}}_0^\dagger(\mathfrak{F})_{/0}$ of our sheaf $\mathcal{V}_0^\dagger(\mathfrak{F})$ at the origin. Let $\mathfrak{m}$ be the defining ideal sheaf of the point $q = 0$ in D. Then, the completion is defined by

$$\widehat{\mathcal{V}}_0^\dagger(\mathfrak{F})_{/0} = \varprojlim_{n\to\infty} \mathcal{V}_0^\dagger(\mathfrak{F})/\mathfrak{m}^n\mathcal{V}_0^\dagger(\mathfrak{F}).$$

Let $\widehat{\mathcal{O}}_{D/0}$ be the completion of the structure sheaf $\mathcal{O}_D$ of D at the origin. Since $\widehat{\mathcal{O}}_{D/0}$ is faithfully flat over $\mathcal{O}_D$, we have

$$\widehat{\mathcal{V}}_0^\dagger(\mathfrak{F})_{/0} = \mathcal{V}_0^\dagger(\mathfrak{F}) \otimes_{\mathcal{O}_D} \widehat{\mathcal{O}}_{D/0}.$$

On the other hand, we can show that $\widehat{\mathcal{V}}_0^\dagger(\mathfrak{F})_{/0}$ is characterized as a subsheaf of $\mathcal{H}_0 \otimes_{\mathcal{O}_D} \widehat{\mathcal{O}}_{D/0}$ satisfying the formal gauge condition (1.4-13). Hence, we have

$$\langle\Psi^{(\lambda)}| \in \widehat{\mathcal{V}}_0^\dagger(\mathfrak{F})_{/0}.$$

It is also easy to show that $\langle\Psi^{(\lambda)}|$, $\lambda \in P_\ell$ are linearly independent over $\widehat{\mathcal{O}}_{D/0}$. This means that

$$(1.4\text{-}14) \qquad \dim_K \widehat{\mathcal{V}}_0^\dagger(\mathfrak{F})_{/0} \otimes K \geq \dim \mathcal{V}_0^\dagger((\mathcal{E}(0), o(0), u)),$$

where K is the field of fraction of $\widehat{\mathcal{O}}_{D/0}$. Note that we already proved that the sheaf $\mathcal{V}_0^\dagger(\mathfrak{F})$ is locally free over D^* and we can show that

$$\text{rank}\mathcal{V}_0^\dagger(\mathfrak{F})|_{D^*} = \dim_K \widehat{\mathcal{V}}_0^\dagger(\mathfrak{F})_{/0} \otimes K. \tag{1.4-15}$$

Now we need to consider the sheaf $\mathcal{V}_0(\mathfrak{F})$. In Lemma 1.3.3 we showed the isomorphism

$$\mathcal{V}_0(\mathfrak{F}) \otimes \mathbf{C}_0 \simeq \mathcal{V}_0((\mathcal{E}(0), o(0), u)). \tag{1.4-16}$$

The similar isomorphism is indeed true for $\mathcal{V}_0^\dagger(\mathfrak{F}) \otimes \mathbf{C}_0$, but this can be proved after we prove that $\mathcal{V}_0^\dagger(\mathfrak{F})$ is locally free at the origin. This is one of the main reasons we need to introduce the sheaf of covacua. ¿From (1.4-14), (1.4-15) and (1.4-16) we conclude

$$\begin{aligned}\text{rank}\mathcal{V}_0(\mathfrak{F})|_{D^*} &= \text{rank}\mathcal{V}_0^\dagger(\mathfrak{F})|_{D^*} \\ &\geq \dim \mathcal{V}_0^\dagger((\mathcal{E}(0), o(0), u)) = \dim \mathcal{V}_0((\mathcal{E}(0), o(0), u)) \\ &= \dim \mathcal{V}_0(\mathfrak{F}) \otimes \mathbf{C}_0.\end{aligned}$$

But, since our sheaf $\mathcal{V}_0(\mathfrak{F})$ is coherent, we have the opposite inequality:

$$\dim \mathcal{V}_0(\mathfrak{F}) \otimes \mathbf{C}_0 \geq \text{rank}\mathcal{V}_0(\mathfrak{F})|_{D^*}. \tag{1.4-17}$$

Therefore in (1.4.17) the equality holds. This means that our sheaf $\mathcal{V}_0(\mathfrak{F})$ is a locally free $\mathcal{O}_D$-module, hence so is $\mathcal{V}_0^\dagger(\mathfrak{F})$. This proves local freeness.

The essential ideas of proving the local freeness for general case are the same as above.

Next we write down explicitly the connection on the sheaf $\mathcal{V}_0^\dagger(\mathfrak{F})$.

Let us consider the exact sequence given in (1.4-1)

$$0 \to \pi_*\Theta_{\mathcal{E}/D}(*o(D)) \xrightarrow{t} \mathcal{L}(\mathfrak{F}) \xrightarrow{\theta} \Theta_D(-\log 0) \to 0.$$

The sheaf $\Theta_D(-\log 0)$ is free $\mathcal{O}_D$-module generated by $q\dfrac{d}{dq}$. We can show that we have

$$\theta(u^{-1}\frac{d}{du}) = -q\frac{d}{dq}.$$

Introducing the connection on $\mathcal{V}_0^\dagger(\mathfrak{F})$ with regular singularity at the origin is equivalent to giving the action of $\Theta_D(-\log 0)$ on $\mathcal{V}_0^\dagger(\mathfrak{F})$. For that purpose it is enough to show that the image of the mapping t acts trivially on $\mathcal{V}_0^\dagger(\mathfrak{F})$.

By (1.3-17) for any local section $\langle\Psi|$ of $\mathcal{V}_0^\dagger(\mathfrak{F})|_{D^*}$ and $\underline{\ell} \in \mathcal{L}(\mathfrak{F})$, we have

$$\{D(\underline{\ell})|\langle\Psi|\}\phi\rangle = -\langle\Psi|\{D(\underline{\ell})\phi\rangle\} + \theta(\underline{\ell})\langle\Psi|\phi\rangle, \tag{1.4-18}$$

for any local section $|\phi\rangle \in \mathcal{H}_0 \otimes \mathcal{O}_D$. By (1.3-15) we also have

$$D(\underline{\ell})|\phi\rangle = \theta(\underline{\ell})|\phi\rangle - T[\underline{\ell}]|\phi\rangle.$$

If $\underline{\ell}$ is in the image of t we have $\theta(\underline{\ell}) = 0$. Hence, it is enough to show

$$\langle\Psi|T[\underline{\ell}]|\phi\rangle = 0.$$

That is

$$\operatorname*{Res}_{u=0}\{\ell(u)\langle\Psi|T[u]|\phi\rangle du\} = 0 \tag{1.4-19}$$

where $\underline{\ell} = \ell(u)\dfrac{d}{du}$.

As was shown in §1.2, c) the form

$$\langle\Psi|T[z]|\phi\rangle dz^2$$

is a single valued meromorphic form. In our coordinate it can be expressed in the form

$$\langle\Psi|T[w]|\phi\rangle(dw/w)^2$$

where

$$w = \exp(2\pi\sqrt{-1}z).$$

Now in our case, since we have

$$\underline{\ell} = \theta(\tau)$$

for a relative vector field $\tau \in \pi_*\Theta_{\mathcal{E}/D}(*o(D))$ having pole only along $o(D)$, the form $\ell(u)\langle\Psi|T[u]|\phi\rangle du$ is a global meromorphic one form

$$\iota_\tau(\langle\Psi|T[w]|\phi\rangle(dw/w)^2)$$

which has pole only at $o(D)$, where ι is the interior product. Hence (1.4-19) holds. Thus we can define the connection on $\mathcal{V}_0^\dagger(\mathfrak{F})$. We can show that the connection is flat. If the genus of a curve is greater than one, the form $\langle\Psi|T[z]|\phi\rangle dz^2$ is no more global form on the curve. Therefore, in general we can only show that the image of t acts on $\mathcal{V}_{\vec{\lambda}}^\dagger(\mathfrak{F})$ as the multiplication of a certain holomorphic functions. This is the reason why we only have a projectively flat connection in general.

Finally we need to show that our formal solutions actually converge so that we have

$$\langle\Psi^{(\lambda)}| \in \mathcal{V}_0(\mathfrak{F}).$$

Put

$$\langle\widetilde{\Psi}^{(\lambda)}| = q^{\Delta_\lambda}\langle\Psi^{(\lambda)}|.$$

We can show that $\langle\widetilde{\Psi}^{(\lambda)}|$ satisfies the differential equation

$$q\frac{d}{dq}\langle\widetilde{\Psi}^{(\lambda)}|\phi\rangle - \langle\widetilde{\Psi}^{(\lambda)}|L_{-2}\phi\rangle = 0 \tag{1.4-20}$$

for any $|\phi\rangle \in V_0 \otimes \mathcal{O}_D$, where $V_0 = \mathcal{H}_0(0)$.

It is rather messy to prove this by direct calculation. A general method to prove this fact can be found in [U, §5.3]. Here we only prove the equation (1.4.20) when $|\phi\rangle = |0\rangle$. In this case, we have

$$\begin{aligned}\langle\widetilde{\Psi}^{(\lambda)}|0\rangle &= \sum_{d=0}^{\infty}\Big(\sum_{i=1}^{m_d}\langle\Psi|v^i(d)\otimes v_i(d)\otimes 0\rangle\Big)q^{\Delta_\lambda+d}\\ &= \sum_{d=0}^{\infty}(\dim\mathcal{H}_\lambda(d))q^{\Delta_\lambda+d}.\end{aligned}$$

First let us calculate

$$\langle\widetilde{\Psi}^{(\lambda)}|L_{-2}0\rangle.$$

For that purpose it is enough to calculate

$$\begin{aligned}&\langle\Psi|v^i(d)\otimes v_i(d)\otimes L_{-2}0\rangle\\ &= \frac{1}{2(g^*+\ell)}\sum_{a=1}^{\dim\mathfrak{g}}\sum_{m=-\infty}^{\infty}\langle\Psi|v^i(d)\otimes v_i(d)\otimes J^a(m)J^a(-2-m)0\rangle\end{aligned}$$

For $m \neq -1$, since m or $-2-m$ is non-negative, we have

$$J^a(m)J^a(-2-m)|0\rangle = J^a(-2-m)J^a(m)|0\rangle = 0.$$

For $m = -1$ the gauge condition implies

$$\begin{aligned}&\langle\Psi|v^i(d)\otimes v_i(d)\otimes J^a(-1)J^a(-1)0\rangle\\ &= \langle\Psi|v^i(d)\otimes v_i(d)\otimes J^a[\frac{1}{\xi-1}]J^a[\frac{1}{\xi-1}]0\rangle\\ &= \langle\Psi|J^a[\frac{\eta}{1-\eta}]J^a[\frac{\eta}{1-\eta}]v^i(d)\otimes v_i(d)\otimes 0\rangle\\ &\quad + 2\langle\Psi|J^a[\frac{\eta}{1-\eta}]v^i(d)\otimes J^a[\frac{1}{\xi-1}]v_i(d)\otimes 0\rangle\\ &\quad + \langle\Psi|v^i(d)\otimes J^a[\frac{1}{\xi-1}]J^a[\frac{1}{\xi-1}]v_i(d)\otimes 0\rangle\end{aligned}$$

We have

$$\begin{aligned}&\langle\Psi|J^a[\frac{\eta}{1-\eta}]J^a[\frac{\eta}{1-\eta}]v^i(d)\otimes v_i(d)\otimes 0\rangle\\ &= \sum_{m,n=1}^{\infty}\langle\Psi|J^a[\eta^m]J^a[\eta^n]v^i(d)\otimes v_i(d)\otimes 0\rangle\\ &= \sum_{m,n=1}^{\infty}(J^a(m)J^a(n)v^i(d)|v_i(d)) = 0,\end{aligned}$$

since $J^a(m)J^a(n)v^i(d) \notin \mathcal{H}_{\lambda^\dagger}(d)$. Similarly we have

$$\langle\Psi|v^i(d)\otimes J^a[\frac{1}{\xi-1}]J^a[\frac{1}{\xi-1}]v_i(d)\otimes 0\rangle = \sum_{m,n=0}^{\infty}(v^i(d)|J^a(m)J^a(n)v_i(d)) = (v^i(d)|J^aJ^av_i(d)).$$

Also we have

$$\begin{aligned}\langle\Psi|J^a[\frac{\eta}{1-\eta}]v^i(d)\otimes J^a[\frac{1}{\xi-1}]v_i(d)\otimes 0\rangle &= -\sum_{m,n=0}^{\infty}(J^a(m+1)v^i(d)|J^a(n)v_i(d))\\ &= -\sum_{n=1}^{\infty}(J^a(n)v^i(d)|J^a(n)v_i(d))\\ &= \sum_{n=1}^{\infty}(v^i(d)|J^a(-n)J^a(n)v_i(d))\end{aligned}$$

Thus we conclude

$$\langle\Psi|v^i(d)\otimes v_i(d)\otimes L_{-2}0\rangle = (v^i(d)|L_0v_i(d)) = (\Delta_\lambda+d)(v^i(d)|v_i(d)).$$

Thus we obtain

$$\langle\widetilde{\Psi}^{(\lambda)}|L_{-2}0\rangle = \sum_{d=0}^{\infty}(\Delta_\lambda+d)\dim\mathcal{H}_\lambda(d)q^{\Delta_\lambda+d}.$$

The right hand side of the formula is equal to

$$q\frac{q}{dq}\langle\widetilde{\Psi}^{(\lambda)}|0\rangle.$$

Thus we obtain the desired result. The argument can be generalized for general $|\phi\rangle\in\mathcal{H}_0\otimes\mathcal{O}_D$.

Remark1.4.6. In the physics literature, $\langle\widetilde{\Psi}^{(\lambda)}|\phi\rangle$ is written as

(1.4-21) $$\mathrm{tr}_{\mathcal{H}_\lambda}(q^{L_0}\langle\Psi|*\otimes*\phi\rangle).$$

Usually physicists add the term $q^{-\frac{c_v}{24}}$:

(1.4-22) $$\mathrm{tr}_{\mathcal{H}_\lambda}(q^{L_0-\frac{c_v}{24}}\langle\Psi|*\otimes*\phi\rangle).$$

This only changes the differential equation (1.4.15) to

$$q\frac{d}{dq}\langle\widetilde{\Psi}^{(\lambda)}|\phi\rangle-\langle\widetilde{\Psi}^{(\lambda)}|L_{-2}\phi\rangle-\frac{c_v}{24}\langle\widetilde{\Psi}^{(\lambda)}|\phi\rangle = 0.$$

The expressions (1.4-21) and (1.4-22) can be interpreted as follows. If we fix an element $|\phi\rangle \in \mathcal{H}_0$, then

$$(1.4\text{-}23) \qquad \langle \Psi | * \otimes * \phi\rangle : u \otimes \mapsto \langle \Psi | u^\dagger \otimes u \otimes \phi\rangle$$

defines an endomorphism of $\mathcal{H}_\lambda(d)$, where $u^\dagger$ is the dual of u with respect to the inner product defined in Lemma 1.1.11. Hence,

$$\sum_{i=1}^{m_d} \langle \Psi | v^i(d) \otimes v_i(d) \otimes \phi\rangle$$

can be interpreted as the trace on $\mathcal{H}_\lambda(d)$ of the mapping

$$\langle \Psi | * \otimes * \phi\rangle \in \mathrm{Hom}(\mathcal{H}_{\lambda,d}, \mathcal{H}_\lambda(d)).$$

Moreover, since L_0 acts on $\mathcal{H}_\lambda(d)$ by the multiplication of $\Delta_\lambda + d$, we may justify the notation

$$\mathrm{tr}_{\mathcal{H}_\lambda(d)}(q^{L_0}\langle \Psi | * \otimes * \phi\rangle) = q^{\Delta_\lambda + d} \sum_{i=1}^{m_d} \langle \Psi | v^i(d) \otimes v_i(d) \otimes \phi\rangle.$$

Hence, in (1.4-21) and (1.4-22) we may regard $\mathrm{tr}_{\mathcal{H}_\lambda}$ as

$$\sum_{d=0}^{\infty} \mathrm{tr}_{\mathcal{H}_\lambda(d)}.$$

Remark 1.4.7. A rational mapping

$$\mathcal{E} \to \mathbf{P}^2 \times D$$
$$[(q, (w : 1)] \mapsto ((1 : x(q, w) : y(q, w)), q)$$

gives a projective imbedding of $\mathcal{E}$ as a submanifold in $\mathbf{P}^2 \times D$ defined by the equation (1.4-10).

Chapter II Abelian Conformal Field Theory

In this chapter we shall briefly explain a generalization of abelian conformal field theory; this is joint work with A. Tsuchiya.

Abelian conformal field theory is usually discussed from the view point of the universal Grassmann manifold and Krichever maps ([KNTY]). Here, we consider it from the view point of non-abelian conformal field theory discussed in the previous chapter and generalize it. We take the Vertex operator algebra constructed from Heisenberg algebra as a gauge group. In the following we shall show that the main ideas of the non-abelian conformal field theory can be applied to our situation and the relationship between conformal blocks and theta functions of higher level will be clarified. The results in this section are joint work with A. Tsuchiya.

§2.1 Vertex operators

For a positive even integer M we let H_M be an algebra generated by operators $a(n)$, $n \in \mathbf{Z}$ with commutation relation

$$[a(n), a(m)] = Mn\delta_{n+m,0} \cdot id. \tag{1.1}$$

The algebra H_M is called the Heisenberg algebra. It is a universal enveloping algebra of an affine Lie algebra $\{a(n)\}$ associated with a one-dimensional abelian Lie algebra $\mathbf{C}$ with commutation relation (1.1). For each $p \in \mathbf{C}$, by $\mathcal{F}(p)$ we denote an irreducible highest weight module of H_M determined by

$$\begin{aligned} a(0)|p\rangle &= p|p\rangle \\ a(n)|p\rangle &= 0, \quad \text{if} \quad n \geq 1, \end{aligned}$$

where $|p\rangle$ is a highest weight vector. Let $t_0, t_1, t_2, \ldots$ be independent variables. Put

$$\begin{aligned} a(m) &= \frac{\partial}{\partial t_m}, \quad m = 0, 1, 2, \ldots \\ a(-n) &= nMt_n, \quad n = 1, 2, 3, \ldots \end{aligned}$$

Then, the Heisenberg algebra H_M and its irreducible module $\mathcal{F}(p)$ are realized as

$$\begin{aligned} H_M &= \mathbf{C}[t_1, t_2, \ldots t_n, \ldots, \frac{\partial}{\partial t_0}, \frac{\partial}{\partial t_1} \cdots \frac{\partial}{\partial t_m}, \ldots, \frac{\partial}{\partial t_m}, \ldots] \\ \mathcal{F}(p) &= \mathbf{C}[t_1, t_2, \ldots, t_n, \ldots]e^{pt_0}, \end{aligned}$$

where the highest weight vector $|p\rangle$ corresponds to e^{pt_0}. Using this realization, we introduce an operator $\widehat{q}$ by

$$\widehat{q} = Mt_0.$$

Put

$$\begin{aligned} \phi(z) &= \widehat{q} + a(0)\log z - \sum_{n \neq 0} \frac{a(n)}{n} z^{-n} \\ a(z) &= \sum_{n \in \mathbf{Z}} a(n) z^{-n-1} \end{aligned}$$

Then we have

$$d\phi(z) = a(z)dz$$

For each integer k, the Vertex operator $V_k(z)$ is defined as

$$V_k(z) = {}^{\circ}_{\circ} e^{k\phi(z)} {}^{\circ}_{\circ}$$

where ${}^{\circ}_{\circ}\ \ {}^{\circ}_{\circ}$ is a normal ordering defined by putting $a(n)$, $n \geq 0$ the right hand side, and $\widehat{q}$ and $a(-n)$,$n \geqslant 1$ the left hand side. Hence, we have

$$V_k(z) = \exp\left\{k\sum_{n=1}^{\infty}\frac{a(-n)}{n}z^n\right\}e^{k\widehat{q}}e^{ka(0)\log z}\exp\left\{-k\sum_{n=1}^{\infty}\frac{a(n)}{n}z^{-n}\right\}$$

The Vertex operator $V_k(z)$ is an intertwiner between the representations $\mathcal{F}(p)$ and $\mathcal{F}(kM+p)$. The energy-momentum tensor $T(z)$ is defined as

$$T(z) = \frac{1}{2M}{}^{\circ}_{\circ} a(z)a(z) {}^{\circ}_{\circ}$$

There is a formal expansion

$$T(z) = \sum_{n\in\mathbf{Z}} L_n z^{-n-2},$$

and $\{L_n\}$ is a Virasoro algebra. These operators have the following operator product expansions.

$$\text{(2.1-1)}\qquad \begin{cases} a(z)a(w) & \sim \dfrac{M}{(z-w)^2} \\ T(z)a(w) & \sim \dfrac{1}{(z-w)^2}a(w) + \dfrac{1}{z-w}\partial_w a(w) \\ T(z)V_k(w) & \sim \dfrac{\frac{k^2}{2}M}{(z-w)^2}V_k(w) + \dfrac{1}{z-w}\partial_w V_k(w) \\ a(z)V_k(w) & \sim \dfrac{k}{z-w}V_k(w) \end{cases}$$

These formulas show that $a(z)$ behaves as a one-form and $V_k(z)$ behaves as a $\frac{k^2}{2}M$-form. Also we have

$$\text{(2.1-2)}\qquad V_{k_1}(z)V_{k_2}(w) = (z-w)^{Mk_1k_2}\, {}^{\circ}_{\circ} V_{k_1}(z)V_{k_2}(w) {}^{\circ}_{\circ}.$$

In the following we only consider irreducible highest weight representations of H_M with highest weight vectors $|p\rangle$ where p's are *integers*.

Let $\Lambda = \{\overline{0}, \overline{1}, \dots, \overline{M-1}\}$ be representatives of the module $\mathbf{Z}/M\mathbf{Z}$. For each $\overline{p} \in \{\overline{0}, \overline{1}, \dots, \overline{M-1}\}$, put

$$\mathcal{H}(\overline{p}) := \bigoplus_{\overline{p}=p \mod M} \mathcal{F}(p).$$

Let $\mathfrak{X} = (C; Q_1, \dots, Q_N; \xi_1, \dots, \xi_N)$ be an N-pointed stable curve of genus g with formal neighbourhoods. To each point Q_j we associate an element $\overline{p_j} \in \Lambda$ and put

$$\vec{\overline{p}} = (\overline{p}_1, \overline{p}_2, \dots, \overline{p}_N),$$
$$\mathcal{H}(\vec{\overline{p}}) = \mathcal{H}(\overline{p}_1) \otimes \mathcal{H}(\overline{p}_2) \otimes \cdots \otimes \mathcal{H}(\overline{p}_N)$$

Put also

$$\mathcal{H}^\dagger(\vec{\overline{p}}) = \mathrm{Hom}_{\mathbf{C}}(\mathcal{H}(\vec{\overline{p}}), \mathbf{C}).$$

We have a natural pairing

$$\begin{array}{ccc} \mathcal{H}^\dagger(\vec{\overline{p}}) \times \mathcal{H}(\vec{\overline{p}}) & \to & \mathbf{C} \\ (\langle\psi|, |\phi\rangle) & \mapsto & \langle\psi|\phi\rangle \end{array}$$

where $\langle\psi|\phi\rangle$ means $\psi(|\phi\rangle)$.

§2.2 Conformal blocks and theta functions

Definition 2.2.1. Conformal block (or space of vacua) $\mathcal{V}^\dagger_{\vec{\overline{p}}}(\mathfrak{X})$ attached to the N-pointed stable curve $\mathfrak{X} = (C; Q_1, \dots, Q_N; \xi_1, \dots, \xi_N)$ with formal neighbourhoods $\mathfrak{X}$ is a subspace of $\mathcal{H}^\dagger(\vec{\overline{p}})$ consisting of vectors $\langle\psi|$ satisfying the following conditions.

(1) For each $|\phi\rangle \in \mathcal{H}(\vec{\overline{p}})$, the data

$$\langle\psi|\rho_j(a(\xi_j))|\phi\rangle d\xi_j, \quad j = 1, 2, \dots, N$$

are the Laurent expansions of an element

$$\omega \in H^0(C, \omega_C(*\sum Q_j))$$

at Q_j's with respect to the formal coordinates ξ_j's,

(2) For each $|\phi\rangle \in \mathcal{H}(\vec{\overline{p}})$, the data

$$\langle\psi|\rho_j(V_{\pm 1}(\xi_j)|\phi\rangle(d\xi_j)^{\frac{M}{2}}, \quad j = 1, 2, \dots, N$$

are the Laurent expansions of an element $\tau \in H^0(C, \omega_C^{\otimes\frac{M}{2}}(*\sum Q_j))$ at Q_j's with respect to the formal coordinates ξ_j.

Theorem 2.2.2. *We have*

$$\dim_{\mathbf{C}} \mathcal{V}^{\dagger}_{\vec{p}}(\mathfrak{X}) = \begin{cases} M^g, & \text{if } \overline{p}_1 + \cdots + \overline{p}_N = \overline{0} \\ 0, & \text{otherwise} \end{cases}$$

where g is the genus of the stable curve C.

For a proof of the above theorem we first rewrite the conditions (1), (2) in Definition 2.2.1. Analogue of Lemma 1.2.5 shows that the condition (1) is equivalent to the condition

$$(1^*) \qquad \sum_{j=1}^{n} \operatorname*{Res}_{\xi_j=0}(\langle\psi|\rho_j(a(\xi_j))|\phi\rangle g(\xi_j)d\xi_j) = 0$$

for every $g \in H^0(C, \mathcal{O}_C(*\sum Q_j))$, where $g(\xi_j)$ is the Laurent expansion of g at Q_j. The condition (2) is equivalent to the condition

$$(2^*) \qquad \sum_{j=1}^{N} \operatorname*{Res}_{\xi_j=0}(\langle\psi|\rho_j(V_{\pm 1}(\xi_j))|\phi\rangle h(\xi_j)d\xi_j) = 0$$

for every $h \in H^0(C, \omega_C^{\otimes(1-\frac{M}{2})}(*\sum Q_j))$, where $h(\xi_j)(d\xi_j)^{\frac{M}{2}}$ is the Laurent expansion of h at Q_j. Thus, the conditions (1*) and (2*) are analogue of the gauge condition (1.2-8).

In the following we choose integers p_j such that $\overline{p}_j = p_j \mod M$, and put

$$|p_1, p_2, \cdots, p_N\rangle = |p_1\rangle \otimes |p_2\rangle \otimes \cdots \otimes |p_N\rangle.$$

Since we have

$$\operatorname*{Res}_{\xi_j=0}\{(a(\xi_j)|p_j\rangle d\xi_j\} = a(0)|p_j\rangle = p_j|p_j\rangle,$$

applying the condition (1*) to $\langle\psi| \in \mathcal{V}^{\dagger}_{\vec{p}}(\mathfrak{X})$ and $1 \in H^0(C, \mathcal{O}_C(*\sum Q_j))$, we have

$$(\sum_{j=1}^{N} p_j)\langle\psi|p_1, p_2, \cdots, p_N\rangle = 0.$$

Hence, if $\langle\psi|p_1, p_2, \cdots, p_N\rangle \neq 0$, then $\sum_{j=1}^{N} p_j = 0$.

Let us consider an N-pointed projective line $(\mathbf{P}^1; z_1, z_2, \ldots, z_N)$ with $z_1 = 0, z_2 = 1, z_N = \infty$. Let z (resp. w) be a coordinate of an affine line in $\mathbf{P}^1$ containing 0 (resp.∞) with $z \cdot w = 1$. Put

$$(2.2\text{-}1) \qquad \xi_j = \begin{cases} z - z_j, & j = 1, 2, \cdots, N-1 \\ w, & j = N, \end{cases}$$

and

$$\mathfrak{X} = (\mathbf{P}^1; z_1, z_2, \ldots, z_N; \xi_1, \xi_2, \ldots, \xi_N).$$

First we shall prove the following proposition.

Proposition 2.2.3.

$$\dim_{\mathbf{C}} \mathcal{V}^{\dagger}_{\vec{\overline{p}}}(\mathfrak{X}) = \begin{cases} 1, & \text{if } \overline{p}_1 + \overline{p}_2 + \cdots + \overline{p}_N = 0 \\ 0, & \text{otherwise} \end{cases}$$

Let $F_0\mathcal{H}(\overline{p}_j)$ be a subspace of $\mathcal{H}(\overline{p}_j)$ spanned by the highest weight vectors $|lM + p_j\rangle$, $l \in \mathbf{Z}$ over $\mathbf{C}$. Put

$$F_0\mathcal{H}(\vec{\overline{p}}) = F_0\mathcal{H}(\overline{p}_1) \otimes F_0\mathcal{H}(\overline{p}_2) \otimes \cdots \otimes F_0\mathcal{H}(\overline{p}_N).$$

To prove the above proposition we need the following lemma.

Lemma 2.2.4. *Under a natural mapping*

$$j : \mathrm{Hom}_{\mathbf{C}}(\mathcal{H}(\vec{\overline{p}}), \mathbf{C}) \longrightarrow \mathrm{Hom}_{\mathbf{C}}(F_0\mathcal{H}(\vec{\overline{p}}), \mathbf{C}),$$

the conformal block $\mathcal{V}^{\dagger}_{\vec{\overline{p}}}(\mathfrak{X})$ of the N-pointed projective line with coordinates (2.2-1) is mapped injectively.

The lemma and the above consideration imply

$$\mathcal{V}^{\dagger}_{\vec{\overline{p}}}(\mathfrak{X}) = 0$$

if $\overline{p}_1 + \overline{p}_2 + \ldots + \overline{p}_N \neq 0$. Therefore, assume $\overline{p}_1 + \overline{p}_2 + \ldots + \overline{p}_N = 0$. Choose p_j's in such a way that

$$p_1 + p_2 + \ldots + p_N = 0.$$

For an element $\langle\psi| \in \mathcal{V}^{\dagger}_{\vec{\overline{p}}}(\mathfrak{X})$, put

$$\psi_{l_1, l_2, \ldots, l_N} = \langle\psi|(|l_1 M + p_1\rangle \otimes |l_2 M + p_2\rangle \otimes \cdots \otimes |l_N M + p_N\rangle).$$

If $\psi_{l_1, l_2, \ldots, l_N} \neq 0$, then $l_1 + l_2 + \ldots l_N = 0$. The condition (1*) implies that $\psi_{l_1, l_2, \ldots, l_N}$ determines uniquely the values

$$\langle\psi|(a(-n^{(1)}_1) \ldots a(-n^{(1)}_{k_1})|l_1 M + p_1\rangle \otimes a(-n^{(2)}_1) \ldots a(-n^{(2)}_{k_2})|l_2 M + p_2\rangle \otimes \\ \cdots \otimes a(-n^{(N)}_1) \ldots a(-n^{(N)}_{k_N})|l_N M + p_M\rangle),$$

for any positive integers $n^{(i)}_j$. Also,the condition (2*) implies that $\psi_{l_1, l_2, \ldots, l_N}$ can be uniquely determined by the value $\psi_{0,0,\ldots,0}$. Thus, we conclude that

$$\dim_{\mathbf{C}} \mathcal{V}^{\dagger}_{\vec{\overline{p}}}(\mathfrak{X}) = 1$$

This proves Proposition 2.2.3.

In the following we assume

$$\overline{p}_1 + \overline{p}_2 + \ldots + \overline{p}_N = 0$$

Let us define a subspace $\mathcal{V}^\dagger_{\vec{\overline{p}}}(n)$ of $\mathcal{H}^\dagger(\vec{\overline{p}})$. An element $\langle\psi|$ is in $\mathcal{V}^\dagger_{\vec{\overline{p}}}(n)$, if $\langle\psi|$ satisfies the following two conditions (1^{**}_n) and (2^{**}_n).

$$(1^{**}_n) \qquad \sum_{j=1}^{N} \operatorname*{Res}_{\xi_j=0}(\langle\psi|a(\xi_j)|\phi\rangle g_j(\xi_j)d\xi_j) = 0$$

for all

$$(g_j(\xi_j)d\xi_j) \in \mathbf{C} \bigoplus \oplus_{j=1}^{N}(\mathbf{C}[\xi_j]\xi_j^{-n}), \quad \text{and} \quad |\phi\rangle \in \mathcal{H}(\vec{\overline{p}}),$$

where an element $c \in \mathbf{C}$ in the right hand side can be considered as $(c, c, \ldots, c)$.

$$(2^{**}_n) \qquad \sum_{j=1}^{N} \operatorname*{Res}_{\xi_j=0}(\langle\psi|V_{\pm1}(\xi_j)|\phi\rangle g_j(\xi_j)d\xi_j) = 0$$

for all

$$(h_j(\xi_j)(d\xi_j)^{\frac{M}{2}}) \in \bigoplus_{j=1}^{N}(\mathbf{C}[\xi_j]\xi_j^{-n})(d\xi_j)^{(1-\frac{M}{2})}, \quad \text{and} \quad |\phi\rangle \in \mathcal{H}(\vec{\overline{p}}).$$

Key Lemma. *Under the above notation we have*

$$\dim \mathcal{V}^\dagger_{\vec{\overline{p}}}(n) < \infty.$$

To prove the Key Lemma we first prove Theorem 2.2.2 for non-singular curve. In this case we can prove a stronger theorem.

Theorem 2.2.5. *Let $\mathfrak{X}$ be an N-pointed* smooth *curve of genus g with formal neighbourhoods. If we have*

$$\overline{p}_1 + \overline{p}_2 + \cdots + \overline{p}_N = 0,$$

then there is a canonical isomorphism

$$\mathcal{V}^\dagger_{\vec{\overline{p}}}(\mathfrak{X}) \simeq H^0(J(C), \mathcal{O}(M\Theta)).$$

where Θ is the theta divisor.

Since the vector space $H^0(J(C), \mathcal{O}(M\Theta))$ of the theta functions of level M is of dimension M^g, this theorem implies Theorem 2.2.2 for non-singular underlying curve C.

To prove Theorem 2.2.5 we need to consider the correlation functions of Vertex operators. For $\vec{k} = (k_1, k_2, \dots, k_m)$, $\vec{p} = (p_1, p_2, \dots, p_N)$ with $\overline{p}_j = p_j \mod M$ and $\langle\Psi| \in \mathcal{V}^{\dagger}_{\vec{\overline{p}}}(\mathfrak{X})$ put

(2.2-1)

$$\Psi_{\vec{k}\vec{p}}(z)(dz_1)^{\frac{k_1^2}{2}M} \cdots (dz_N)^{\frac{k_N^2}{2}M}$$
$$= \langle\Psi|V_{k_1}(z_1)V_{k_2}(z_2)\cdots V_{k_m}(z_m)|\vec{p}\rangle(dz_1)^{\frac{k_1^2}{2}M} \cdots (dz_N)^{\frac{k_N^2}{2}M}.$$

Then, by the operator product expansions (2.1-1) we can show the following Lemma.

Lemma 2.2.6. 1) *If $\Psi_{\vec{k}\vec{p}}(z) \not\equiv 0$, then we have*

$$\sum_{i=1}^{m} p_i + M\sum_{a=1}^{N} k_a = 0.$$

2) *Near $z_a = z_b$ we have the expansion*

$$\Psi_{\vec{k}\vec{p}}(z) = (z_a - z_b)^{Mk_a k_b}\Psi_{\vec{k}'\vec{p}}(z') + \quad \textit{higher order terms}$$

where

$$\vec{k}' = (k_1, \cdots, \overset{\vee}{k_a}, \cdots, k_m)$$
$$z' = (z_1, \cdots, \overset{\vee}{z_a}, \cdots, z_m).$$

3) *Near $z_a = Q_j$ we have the expansion*

$$\Psi_{\vec{k}\vec{p}}(z) = (z_a - z_b)^{k_a p_j}\Psi_{\vec{k}_{\hat{a}}\vec{p}_{\hat{a}}}(z) + \quad \textit{higher order terms}$$

where

$$\vec{k}_{\hat{a}} = (k_1, \cdots, \overset{\vee}{k_a}, \cdots, k_m)$$
$$\vec{p}_{\hat{a}} = (p_1, \cdots, p_a + k_a, \cdots, p_N).$$

Conversely, for $\langle\Psi| \in \mathrm{Hom}_{\mathbf{C}}(\mathcal{H}(\vec{\overline{p}}), \mathbf{C})$ if we define

$$\Psi_{\vec{k}\vec{p}}(z)(dz_1)^{\frac{k_1^2}{2}M} \cdots (dz_N)^{\frac{k_N^2}{2}M}$$

by (2.2-1) and if it has the above properties 1), 2), 3) for all $\vec{k}$ and $\vec{p}$, then

$$\langle\Psi| \in \mathcal{V}_{\vec{\overline{p}}}(\mathfrak{X}).$$

Furthermore, $\langle\Psi|$ *is uniquely determined by*

$$\Psi_{\vec{k}\vec{p}}(z)(dz_1)^{\frac{k_1^2}{2}M}\cdots(dz_N)^{\frac{k_N^2}{2}M}$$

for all $\vec{k}$ *and* $\vec{p}$.

Let

$$E(z,w)(dz)^{-1/2}(dw)^{-1/2}$$

be the prime form of the non-singular curve C. Put

$$\psi_{\vec{k}\vec{p}}(z)=\prod_{1\le a<b\le m}E(z_a,z_b)^{Mk_ak_b}\prod_{a=1}^{m}\prod_{j=1}^{N}E(z_a,Q_j)^{k_ap_j}.$$

Then, by Lemma 2.2.6 it is easy to show that

$$\phi_{\vec{k}\vec{p}}=\frac{\Psi_{\vec{k}\vec{p}}(z)}{\psi_{\vec{k}\vec{p}}(z)}$$

is a multivalued holomorphic functions on C^m. Multivaluedness comes from that of the prime form. Let $\{\alpha_1,\cdots,\alpha_g,\beta_1,\cdots,\beta_g\}$ and $\{\omega_1,\cdots,\omega_g\}$ are a basis of the first homology group of the curve C and a basis of holomorphic one-forms of C with

$$\int_{\alpha_i}\omega_j=\delta_{ij}\quad\int_{\beta_i}\omega_j=\tau_{ij},$$

respectively. Let

$$\begin{array}{rcl} j:C & \rightarrow & J(C)\\ P & \mapsto & [(\int_{P_0}^{P}\omega_1,\cdots,\int_{P_0}^{P}\omega_g)]\end{array}$$

be a natural mapping of the curve C into its Jacobian. Then, quasi periodicity of the prime form is given by

$$E(P+\alpha_i,Q)=E(P,Q)$$
$$E(P+\beta_i,Q)=\exp(-2\pi\sqrt{-1}(\frac{\tau_{ii}}{2}+\int_Q^P\omega_i))E(P,Q)$$

where $P+\alpha_i$ means the analytic continuation along the cycle α_i. Then, $\phi_{\vec{k}\vec{p}}(z)$ has the following quasi periodicity.

$$\phi_{\vec{k}\vec{p}}(z_1,\cdots,z_a+\alpha_i,\cdots z_m)=\phi_{\vec{k}\vec{p}}(z_1,\cdots,z_a,\cdots,z_m)$$
$$\phi_{\vec{k}\vec{p}}(z_1,\cdots,z_a+\beta_i,\cdots z_m)=\exp(\frac{k_a^2}{2}M\tau_{ii}$$
$$+k_a\sum_{b+1}^{m}Mk_b\int_{P_0}^{z_b}\omega_i+k_a\sum_{j=1}^{N}p_j\int_{P_0}^{Q_j}\omega_i)\phi_{\vec{k}\vec{p}}(z_1,\cdots,z_a,\cdots,z_m)$$

Let us define a holomorphic mapping of C^m to the Jacobian $J(C)$ by

$$\begin{array}{rccc} \varpi_n : & C^m & \rightarrow & J(C) \\ & (P_1, \cdots, P_n) & \mapsto & [(\sum_{b=1}^{m} k_b \int_{P_0}^{P_b} \vec{\omega} + (\sum_{j=1}^{N} \frac{p_j}{M} \int_{P_0}^{Q_j} \vec{\omega})]. \end{array}$$

It is well-known that if m is sufficiently large, then ϖ is surjective. Moreover, we can show the following lemma.

Lemma 2.2.7. *The multivalued holomorphic function $\phi_{\vec{k}\vec{p}}$ is a holomorphic section of the line bundle $\varpi_m^*[\Theta]$, where Θ is the theta divisor. Moreover, if m is sufficiently large, we have*

$$H^0(C^m, \mathcal{O}(\varpi_m^*[M\Theta])) = \varpi_m^* H^0(J(C), \mathcal{O}(M\Theta)).$$

This implies Theorem 2.2.5.

To prove Key Lemma we also need the following lemma.

Lemma 2.2.8. *For positive integers n and N there exist a smooth curve D of genus g and points $Q_1, \cdots, Q_N$ on D with local coordinates $\xi_1, \xi_2, \cdots, \xi_N$ such that*

$$Gr_\bullet^F H^0(D, \mathcal{O}_D(*\sum Q_j)) \subset \mathbf{C} \bigoplus \oplus_{j=1}^N \mathbf{C}[\xi_j^{-1}]\xi_j^{-n}$$

$$Gr_\bullet^F H^0(D, \omega_D^{\otimes(1-\frac{M}{2})}(*\sum Q_j)) \subset \mathbf{C} \bigoplus \oplus_{j=1}^N \mathbf{C}[\xi_j^{-1}]\xi_j^{-n}(d\xi_j)^{1-\frac{M}{2}}$$

where the filtration F can be defined by the order of poles at Q_j.

The first inclusion can be proved, if the divisor $n(Q_1 + Q_2 + \cdots + Q_N)$ is non-special on a curve D. The second inclusion is trivially true, if we have $(2g-2)(1-\frac{M}{2}) > nN$.

Now introducing the filtration on $\mathcal{H}(\vec{p})$ and $\mathcal{H}^\dagger(\vec{p})$ compatible with the filtration in Lemma 2.2.8, we can show Key Lemma which also implies *finite dimensionality* of $\mathcal{V}_{\vec{p}}(\mathfrak{X})$ for all N-pointed stable curve with formal neighbourhoods.

Now let us consider a semi-stable curve C. For a double point $P \in C$ we let $\pi : \tilde{C} \rightarrow C$ be the normalization at the point P. Then, the inverse image $\pi^{-1}(P)$ of the point P consists of two points P_+, P_-. Let η_+, η_- be formal coordinates of P_+ and P_-respectively such that C is defined formally in a neighbourhood of the origin of $\mathbf{C}^2$ by an equation $\eta_+ \cdot \eta_- = 0$. Let $\mathfrak{X} = (C; Q_1, \ldots, Q_N; \xi_1, \ldots, \xi_N)$ be an N-pointed stable curve with formal neighbourhoods whose underling curve is the semi-stable curve C. Put

$$\tilde{\mathfrak{X}} = (\tilde{C}; Q_1, \ldots, Q_N, P_+, P_-; \xi_1, \ldots, \xi_N, \eta_+, \eta_-).$$

Then,we have the following theorem.

Theorem 2.2.9. *Under the above notation and assumptions, we have a canonical isomorphism.*

$$\bigoplus_{\overline{q}\in\mathbf{Z}/M\mathbf{Z}} \mathcal{V}^{\dagger}_{\overline{q},-\overline{q},\overline{\overline{p}}}(\tilde{\mathfrak{X}}) \simeq \mathcal{V}^{+}_{\overline{\overline{p}}}(\mathfrak{X}).$$

From this theorem and Proposition 2.2.3 we infer the following lemma.

Lemma 2.2.10. *Let $\mathfrak{X} = (C; Q_1, \dots, Q_N; \xi_1, \dots, \xi_N)$ be an N-pointed stable curve with formal neighbourhoods. Assume that all the irreducible component of the semi-stable curve C are $\mathbf{P}^1$ and the genus of C is g. Then, we have*

$$\dim_{\mathbf{C}} \mathcal{V}^{\dagger}_{\overline{\overline{p}}}(\mathfrak{X}) = \begin{cases} M^g, & \text{if } \ \overline{p}_1 + \cdots + \overline{p}_N = 0 \\ 0, & \text{otherwise}. \end{cases}$$

Now we need to show that $\dim_{\mathbf{C}} \mathcal{V}^{+}_{\overline{\overline{p}}}(\mathfrak{X})$ depends only on the genus of the underlying curve C. For that purpose we need to consider the sheaf of conformal blocks attached to a family of N-pointed stable curves with formal neighbourhoods. Similar arguments as in the previous chapter give a proof of Theorem 2.2.2.

The family $\mathcal{V}_{\overline{\overline{p}},N} = \bigcup_{\mathfrak{X}} \mathcal{V}^{\dagger}_{\overline{\overline{p}}}(\mathfrak{X})$ over the moduli space $\overline{\mathcal{M}}^{(\infty)}_{g,N}$ of N-pointed curves of genus g with formal neighbourhoods. By a similar method as the one in [TUY], we can show that $\overline{\mathcal{V}}_{\overline{\overline{p}},N}$ comes from a sheaf $\mathcal{V}^{(1)}_{\overline{\overline{p}},N}$ on $\mathcal{M}^{(1)}_{g,N}$, the moduli space of N-pointed curves of genus g with first order neighbourhoods. Then, by Key Lemma we can show that $\mathcal{V}^{(1)}_{\overline{\overline{p}},N}$ is a coherent $\mathcal{O}_{\overline{\mathcal{M}}^{(1)}_{g,N}}$ -module and it carries a logarithmic projectively flat connection. From these facts we infer that $\mathcal{V}^{(1)}_{\overline{\overline{p}},N}$ is locally free on the open part of $\overline{\mathcal{M}}^{(1)}_{g,N}$ corresponding to non-singular curves.

Again, using similar arguments as in Chapter I §1.3 and §1.4, we can show that $\mathcal{V}^{(1)}_{\overline{\overline{p}},N}$ is locally free. By Lemma 2.2.11 this implies our Theorem 2.2.2.

REFERENCES

[BF] D. Bernard and G. Felder, *Fock representations and BRST cohomology in SL(2) current algebra*, Commun. Math. Phys. **127** (1990), 145 – 168.

[BL] A. Beauville and Y. Laszalo, *Conformal blocks and generalized theta functions*, preprint (1993).

[BPZ] A. A. Belavin, A. M. Polyakov and A. B. Zamolodchikov, *Infinite conformal symmetry in two-dimensional quantum field theory*, Nucl. Phys. **B241** (1984), 333 – 380.

[DV] R. Dijkgraaf and E. Verlinde, *Modular invariance and the fusion algebra*, Nucl. Phys. B(Proc. Suppl.) **5B** (1988), 87 – 97.

[EO] T. Eguchi and H. Ooguri, *Conformal and current algebras on a general Riemann surface*, Nucl. Phys. **B282** (1987), 308 –328.

[Fa] J. D. Fay, *Theta Functions on Riemann Surfaces*, Lecture Notes in Math. **352**, Springer-Verlag, 1973.

[Fe] G. Felder, *BRST approach to minimal models*, Nucl. Phys. **B317** (1989), 215 - 236.

[FS] D. Friedan and S. Shenker, *The analytic geometry of two dimensional conformal field theory*, Nucl. Phys. **B281** (1987), 509 - 545.

[Ga] O. Gabber, *The integrability of the characteristic variety*, Amer. J. Math. **103** (1981), 445.

[H] N. J. Hitchin, *Flat connections and geometric quantization*, Commun. Math. Phys. **131** (1990), 347 - 380.

[Ka] V. Kac,, *Infinite Dimensional Lie Algebras*, Cambridge University Press., Third edition, 1990.

[KSU1] T. Katsura, Y. Shimizu and K. Ueno, *New bosonization and conformal field theory over* **Z**, Commun. Math. Phys. **121** (1988), 603 - 627.

[KSU2] T. Katsura, Y. Shimizu and K. Ueno, *Formal groups and conformal field theory over* **Z**, Adv. Stud. Pure Math. **19** (1989), 347 - 366.

[KSU3] T. Katsura, Y. Shimizu and K. Ueno, *Complex cobordism ring and conformal field theory over* **Z**, Math. Ann. **291** (1991), 551 - 571.

[KNTY] N. Kawamoto, Y. Namikawa, A. Tsuchiya and Y. Yamada, *Geometric realization of conformal field theory on Riemann surfaces*, Commun. Math. Phys. **116** (1988), 247 - 308.

[KZ] V. G. Knizhnik and A. B. Zamolodchikov, *Current algebra and Wess-Zumino model in two dimensions*, Nucl. Phys. **B247** (1984), 83 - 103.

[Ko] T. Kohno, *Three-manifold invariants derived from conformal field theory and projective representations of modular groups*, Intern. J. of Modern Phys. **6** (1992), 1795 –1805.

[KNR] S. Kumar, M. S. Narasimhan and A. Ramanathan, *Infinite Grassmannian and moduli space of G-bundles*, preprint (1993).

[Ku] G. Kuroki, *Fock space representations of affine Lie algebras and integral representations in the Wess-Zumino-Witten models*, Commun. Math. Phys. **142** (1991), 511 - 542.

[MS1] G. Moore and N. Seiberg, *Polynomial equations for rational conformal field theories*, Phys. Lett. **B212** (1988), 451 - 460.

[MS2] G. Moore and N. Seiberg, *Classical and quantum conformal field theory*, Commun. Math. Phys. **123** (1989), 177 - 254.

[S] G. Segal, *The definition of conformal field theory*, preprint.

[TK] A. Tsuchiya and Y. Kanie, *Vertex operators in conformal field theory on* $\mathbf{P}^1$ *and monodromy representations of braid group*, Advanced Studies in Pure Math. **16** (1988), 297 – 372 Erratum, *ibid.* **19**(1989), 675 – 682.

[TUY] A.Tsuchiya, K. Ueno & Y. Yamada, *Conformal field theory on universal family of stable curves with gauge symmetries*, Adv. Stud. in Pure Math. **19** (1989), 459–566.

[U] K. Ueno, *Lectures on conformal field theory with gauge symmetries*, to appear.

[Ve] E. Verlinde, *Fusion rules and modular transformations in 2-D conformal field theory*, Nucl. Phys. **B300** **[FS22]** (1988), 360 – 376.

[W] E. Witten, *Quantum field theory and the Jones polynomial*, Commun. Math. Phys. **121** (1989), 351 – 399.

Printed in the United States
By Bookmasters